INNOVATIVE APPROACHES IN DIAGNOSIS AND MANAGEMENT OF CROP DISEASES

Volume 1
The Mollicutes

Innovative Approaches in Diagnosis and Management of Crop Diseases
Volume 1: The Mollicutes / hardback ISBN: 978-1-77463-024-2

Innovative Approaches in Diagnosis and Management of Crop Diseases
Volume 2: Field and Horticultural Crops / hardback ISBN: 978-1-77463-025-9

Innovative Approaches in Diagnosis and Management of Crop Diseases
Volume 3: Nanomolecules and Biocontrol Agents / hardback ISBN: 978-1-77463-026-6

Innovative Approaches in Diagnosis and Management of Crop Diseases, 3-volume set
hardback ISBN: 978-1-77463-027-3

INNOVATIVE APPROACHES IN DIAGNOSIS AND MANAGEMENT OF CROP DISEASES

Volume 1
The Mollicutes

Edited by
R. K. Singh, PhD
Gopala, PhD

AAP | APPLE ACADEMIC PRESS

First edition published 2022

Apple Academic Press Inc.
1265 Goldenrod Circle, NE,
Palm Bay, FL 32905 USA

4164 Lakeshore Road, Burlington,
ON, L7L 1A4 Canada

CRC Press
6000 Broken Sound Parkway NW,
Suite 300, Boca Raton, FL 33487-2742 USA

2 Park Square, Milton Park,
Abingdon, Oxon, OX14 4RN UK

© 2022 Apple Academic Press, Inc.

Apple Academic Press exclusively co-publishes with CRC Press, an imprint of Taylor & Francis Group, LLC

Library and Archives Canada Cataloguing in Publication

Title: Innovative approaches in diagnosis and management of crop diseases, Volume 1: The Mollicutes / edited by Rakesh Kumar Singh, PhD, Gopala, PhD.

Names: Singh, Rakesh Kumar (Plant pathologist), editor. | Gopala (Plant pathologist), editor.

Description: First edition. | Includes bibliographical references and index. | Contents: Volume 1. The mollicutes

Identifiers: Canadiana (print) 20210171766 | Canadiana (ebook) 20210172266 | ISBN 9781774630242 (v. 1 ; hardcover) | ISBN 9781774639252 (v. 1 ; softcover) | ISBN 9781774630273 (set) | ISBN 9781003187608 (v. 1 ; ebook)

Subjects: LCSH: Phytopathogenic microorganisms—Control. | LCSH: Plant diseases. | LCSH: Mycoplasmatales.

Classification: LCC SB731 .I56 2022 | DDC 632/.3—dc23

Library of Congress Cataloging-in-Publication Data

CIP data on file with US Library of Congress

ISBN: 978-1-77463-024-2 (hbk)
ISBN: 978-1-77463-925-2 (pbk)
ISBN: 978-1-00318-760-8 (ebk)

About the Editors

R. K. Singh, PhD

Head, Plant Pathology, Rajmata Vijayaraje Scindia Krishi Vishwa Vidyalaya, College of Agriculture, Indore, M.P., India

R. K. Singh, PhD, is presently working as a Head, Plant Pathology Section at RVSKVV, College of Agriculture, Indore, Madhya Pradesh, India. He is an esteemed member of national and international research societies and a fellow of the Indian Society of Pulses Research and Development. He has been honored with four awards from national societies. Dr. Singh has guided 25 MSc (Ag) and 3 PhD students on modern integrated areas in biology and has published over 65 research papers along with 6 book chapters, 3 instructional manuals, 1 text book and popular articles. He has experience in teaching, research, and extension work in agriculture in various capacities for over 20 years. He was PI of a project funded by the Japan International Cooperation Agency (JICA), ICAR, and evaluated several newly evolved agrochemicals. He is actively working in detection of plant pathogens and identification of sources of Genetic resistance in chickpea and mungbean. Dr. Singh has set up a molecular plant pathology laboratory, wilt sick plot for chickpea and mushroom cultivation and a value addition laboratory at college.

Gopala, PhD

Rajmata Vijayaraje Scindia Krishi Vishwa Vidyalaya (RVSKVV), College of Agriculture, Indore, M.P., India

Gopala, PhD, is affiliated with Rajmata Vijayaraje Scindia Krishi Vishwa Vidyalaya (RVSKVV), College of Agriculture, Indore, M.P., India. He earned his MSc and PhD degrees in Plant Pathology from IARI, New Delhi, India. He has cleared his exams, such as ICAR PGS-JRF, and had an IARI merit scholarship for his PhD. He also cleared ICAR NET in 2014. He has received three awards from national societies. During his MSc work, he has developed a new screening technique and rating scale for stalk rot of maize caused by *Macrophomina phaseolina*. During his PhD work, he has reported new phytoplasma diseases in *Cucurbita pepo*, bougainvillea, dianthus, petunia, and ornamental kale, along with its associated vectors. He has submitted more than 50 sequences to the National Center for Biotechnology Information and received accession numbers. Dr. Gopala has published more than 15 research papers in journals of national and international repute and 1 textbook.

Contents

Contributors

Dilshad Ahmad
ICAR-Indian Agricultural Research Institute, New Delhi–110012, India

Iqbal Ahmed
ICAR-Indian Agricultural Research Institute, New Delhi–110012, India

Alan C. Antony
College of Horticulture, Kerala Agricultural University, Thrissur, Kerala–680656, India

Rahul L. Chavhan
Assistant Professor, Department of Plant Biotechnology, Vilasrao Deshmukh College of Agricultural Biotechnology, Latur; Vasantrao Naik Marathwada Krishi Vidyapeeth (VNMKV), Parbhani–431402, Maharashtra, India, E-mail: rlchavhan@gmail.com

Devendra K. Choudhary
ICAR-Indian Agricultural Research Institute, New Delhi–110012, India

Nandlal Choudhary
Amity Institute of Virology and Immunology, Amity University Uttar Pradesh, Noida–201313, Uttar Pradesh, India, E-mail: nchoudhary@amity.edu

K. C. Darsana Dilip
College of Horticulture, Kerala Agricultural University, Thrissur, Kerala–680656, India

Gopala
Department of Plant Pathology, College of Agriculture, Indore–452001, Madhya Pradesh, India, E-mail: gplpatho33@gmail.com

Nitika Gupta
ICAR-Directorate of Floricultural Research, Pune–411005, Maharashtra, India, E-mail: nitika.iari@gmail.com

Vidya R. Hinge
Assistant Professor, Department of Plant Biotechnology, Vilasrao Deshmukh College of Agricultural Biotechnology, Latur; Vasantrao Naik Marathwada Krishi Vidyapeeth (VNMKV), Parbhani–431402, Maharashtra, India

Somnath K. Holkar
ICAR-Indian Institute of Sugarcane Research, Biological Control Center, Pravaranagar, Ahmednagar, Maharashtra–413712, India, E-mail: somnathbhu@gmail.com

Pritam Jadhav
ICAR-Directorate of Floricultural Research, Pune–411005, Maharashtra, India

R. K. Jain
Division of Plant Pathology, ICAR-Indian Agricultural Research Institute, Pusa Campus, New Delhi–110012, India

Sandeep. P. Kale
Senior Research Fellow, USAID-BIRAC International Project, School of Life Sciences, S.R.T.M.U., Nanded, Maharashtra, India

Shoaib N. Kirmani
ICAR-Central Institute of Temperate Horticulture, Srinagar–191132, Jammu and Kashmir, India

Atul Kumar
ICAR-Indian Institute of Sugarcane Research, Biological Control Center, Pravaranagar, Ahmednagar, Maharashtra–413712, India; Amity Institute of Biotechnology, Amity University, Lucknow Campus, Lucknow–226028, Uttar Pradesh, India

Sunil Kumar
Department of Plant Pathology, SKN College of Agriculture, Jobner, Jaipur–303329, Rajasthan, India, E-mail: Khaliasunil1987@gmail.com

Vimi Louis
Banana Research Station, Kerala Agricultural University, Kannara, Thrissur, Kerala–680652, India, E-mail: vimilouis@gmail.com

Shivam Maurya
Department of Plant Pathology, SKN College of Agriculture, Jobner, Jaipur–303329, Rajasthan, India

Sajad Un Nabi
ICAR-Central Institute of Temperate Horticulture, Srinagar–191132, Jammu and Kashmir, India, E-mail: sajad_patho@rediffmail.com

Kishor P. Panzade
Indian Agriculture Research Institute, New Delhi–110012, India

Richa Rai
Division of Plant Pathology, ICAR-Indian Agricultural Research Institute, New Delhi–110012, India

Parveez Sheikh
Sher-e-Kashmir University of Agricultural Science and Technology, Srinagar, Jammu and Kashmir, India

R. K. Singh
Department of Plant Pathology, College of Agriculture, Indore–452001, Madhya Pradesh, India

Pankhuri Singhal
Division of Plant Pathology, ICAR-Indian Agricultural Research Institute, New Delhi–110012, India, E-mail: pankhuri.agri47@gmail.com

Manoj K. Yadav
ICAR-National Rice Research Institute, Cuttack–753006, Odisha, India

Nida Yousuf
ICAR-Central Institute of Temperate Horticulture, Srinagar–191132, Jammu and Kashmir, India

Abbreviations

2D-PAGE	two-dimensional denaturing PAGE	BVY	blackberry virus Y
AB	applied biosystems	BYD	barley yellow dwarf
ABA	abscisic acid	BYMV	barley yellow mosaic virus
ACFSVd	apple chlorotic fruit spot viroid	BYV	beet yellow virus
		CaMV	cauliflower mosaic virus
ACMV	African cassava mosaic virus	CarNFV	carnation necrotic fleck virus
ADDT	agar double diffusion test	CBCVd	citrus bark cracking viroid
AFMoV	Angelonia flower mottle virus	CCCVd	coconut cadang-cadang viroid
		CCR	central conserved region
AGO	argonaute	CCYV	cucurbit chlorotic yellows virus
AI	artificial intelligence		
AMV	alfalfa mosaic virus	CDVd	citrus dwarfing viroid
ApMV	apple mosaic virus	CERV	carnation etched ring virus
ARC	apaf-1/R protein/CED 4	CEVd	citrus exocortis viroid
ArMV	Arabis mosaic virus	CF	co-factors
ASBVd	avocado sun blotch viroid	CFMMV	cucumber fruit mottle mosaic virus
ASSVd	apple scar skin viroid		
AST	aerated steam therapy	CGMMV	cucumber green mottle mosaic virus
AuNRs	gold nanorods		
AuNSs	gold nanospheres	ChaYMV	chayote yellow mosaic virus
Avr	aviarulence	CHS	chalcone synthase gene
AWMV	Algerian watermelon mosaic virus	CLBV	citrus leaf smudge infection
		CLVd	columnea latent viroid
AY	aster yellow	ClYVV	clover yellow vein virus
BaMV	bamboo mosaic potyvirus	CMV	cucumber mosaic virus
BBrMD	banana bract mosaic disease	Co-PCR	cooperative PCR
BBTV	banana bunchy top virus	CP	capsid protein
BBWV	broad bean wilt	CRISPR	clustered regularly inter-spaced short palindromic repeats
BCMV	bean common mosaic virus		
BCTV	beet curly top virus		
BeYDV	bean yellow dwarf virus	CRSV	carnation ringspot virus
BGMV	bean golden mosaic virus	CSNV	chrysanthemum stem necrosis virus
BIP	reverse internal preliminary		
BMV	brome mosaic virus	CSVd	chrysanthemum stunt viroid
BPYV	beet pseudo-yellows virus	CTV	citrus tristeza clostero virus
BRs	brassinosteroids	CVB	chrysanthemum virus B
BSCTV	beet severe curly top virus	CVBV	cucurbit vein-banding virus
BSGFV	banana streak GF virus	CVMoV	carnation vein mottle virus
BSMyV	banana streak Mysore virus	CVYV	cucumber vein yellowing virus
BSOLV	banana streak OL virus	CYD	cucurbit yellows disease

CymMV	cymbidium mosaic virus	HRP	horseradish peroxidase
CyRSV	cymbidium ring spot virus	HSVd	hop stunt viroid
CYSDV	cucurbit yellow stunting	HTS	high throughput sequencing
	disorder virus	ICAN	isothermal and chimeric
DCL	dicer-like		primer-initiated amplification
DIBA	dot immuno binding assay		of nucleic acids
DIG	digoxigenin	ICMV	Indian cassava mosaic virus
DNA	deoxyribonucleic acid	IC-PCR	immunocapture-polymerase
DRB	double-stranded RNA		chain reaction
	binding	IC-RT-PCR	immunocapture RT-PCR
DSB	double stranded break	ICTV	international committee on
dsRNA	double-stranded RNA		taxonomy of viruses
DTBIA	direct tissue blotting	IDM	integrated disease
	immunoassay		management
ECL	electro-chemiluminescence	IE	immunoelectrophoresis
ELISA	enzyme linked immunosor-	IF	immunofluorescence
	bant assays	IFMV	iris fulva mosaic virus
EM	electron microscopy	Ig	immunoglobulin
EPRV	endogenous pararetro	IgG	immunoglobulin G
	infections	IGL	immuno-gold marking
ER	extreme resistance	IMMV	iris mild mosaic virus
ESCRT	endosomal sorting complexes	INSV	impatiens necrotic spot virus
	required for transport	INT	infectivity balance tests
ETI	effector triggered immunity	IO	immuno-osmophoresis
FIB	forward inward preliminary	IR	intergenic region
FISH	fluorescence in-situ	ISEM	immunosorbent electron
	hybridization		microscopy
FOPPR	fiber optic particle plasmon	ISH	*in-situ* hybridization
	resonance	ISR	intergenic spacer regions
FrMV	frangipani mosaic virus	ITP	immuno-tissue printing
GA	genome analyzer	JA	jasmonic acid
GBNV	groundnut bud necrosis virus	KGMMV	Kyuri green mottle mosaic
GC-MS	gas chromatography-mass		virus
	spectrometry	LAI	leaf area index
GFP	green fluorescent protein	LAMP	loop-mediated isothermal
GLRaV-3	grapevine leafroll associated		amplification
	virus 3	LC/MS	liquid chromatography/mass
GLVd	grapevine latent viroid		spectrometry
GYSVd-1	grapevine yellow speckle	LFA	lateral flow assay
	viroid 1	LFD	lateral flow device
GYSVd-2	grapevine yellow speckle	LIYV	lettuce infectious yellows
	viroid 2		virus
HcPro	helper component proteases	LMoV	lily mottle virus
HDR	homology directed repair	LMV	lettuce mosaic virus
HEN1	hue enhancer1	LRR	leucine-rich-repeat
HNB	hydroxyl napthol blue	LSV	lily symptomless virus
HR	hypersensitive response	LT	latex test
HRM	high-resolution melting	MAbs	monoclonal antibodies

MAPK	mitogen-activated protein kinase	OY	onion yellows
MBD	maltose binding domain	PABP	poly(A)-binding protein
MCLuCV	melon chlorotic leaf curl virus	PAGE	polyacrylamide gel electrophoresis
MCMV	maize chlorotic mottle virus	PAM	protospacer-adjacent motif
MDMV	maize dwarf mosaic virus	PAMP/MAMP	pathogen or microbe-associated molecular patterns
MeMV	merremia mosaic virus		
MeSMV	melon severe mosaic virus		
miRNA	microRNA	PBCVd	pear blister canker viroid
MN	microneedle	PCFVd	pepper chat fruit viroid
MP	movement protein	PCR	polymerase chain reaction
M-PCR	multiplex PCR	PCR-MPH	PCR-microplate hybridization
MPSS	massively parallel signature sequencing	PD	plasmodesmata
MPT	microprecipitation test	PDR	pathogen derived resistance
mRT-PCR	multiplex RT-PCR	PDV	prune dwarf virus
MSN	mesoporous silica nanoparticles	PEMV-1	pea enations mosaic virus-1
		PepMoV	pepper mottle infection
MSV	maize streak virus	PFBV	pelargonium flower break virus
MVBMV	melon vein-banding mosaic virus		
		PFOR	progressive filtering of overlapping small RNAs
MYMIV	mung bean yellow mosaic India virus		
		PGPR	plant growth-promoting rhizobacteria
MYSV	melon yellow spot virus		
NAIMA	nucleic acid implemented microarray analysis	PIO	Pacific Indian Oceans
		PLRV	potato leaf roll virus
NASBA	nucleic acid sequence-based amplification	PluMV	plumeria mosaic virus
		PMMoV	pepper mild mottle virus
NASH	nucleic acid spot hybridization	PNPP	p-nitrophenyl phosphate
		PNRSV	prunus necrotic ringspot virus
NB	nucleotide-binding	PPV	plum pox virus
NBH	northern-blot hybridization	PRRs	pattern recognition receptors
NCBI	National Center for Biotechnology Information	PRSV	papaya ring spot virus
		PRSV-W	papaya ringspot virus-W
NCM	nitrocellulose film	PSTVd	potato spindle tuber viroid
NCS-TCP	National certification system for tissue culture raised plants	PTA	plate-trapped antigen
		PTA-ELISA	plate trapped antibody ELISA
NGS	next generation sequencing	PTGS	post-transcriptional gene silencing
NHEJ	non-homologous end joining		
NLV	narcissus latent virus	PTI	pathogen triggered immunity
NO	nitric oxide	PVY	potato virus Y
NP	nucleocapsid	qPCR	quantitative-PCR
nPCR	nested PCR	RBP	RNA binding proteins
ODDT	ouchterlony twofold dispersion test	RCA	rolling circle amplification
		RDBH	reverse dot blot hybridization
ORMV	oilseed rape mosaic virus		
ORSV	odontoglossum ring spot virus	RDR	RNA dependent RNA
		Rep	replication protein

REPs	repetitive palindromes	SLRV	strawberry latent ringspot
RFLP	restriction fragment length		virus
	polymorphism	SMRT	single molecule real-time
RHPs	host reticulon homology	SMV	soybean mosaic
	proteins	SNPs	single nucleotide
RIA	radioimmuno measure		polymorphisms
RIE	rocket	S-PAGE	sequential-PAGE
	immunoelectrophoresis	SPMMV	sweet potato mild mottle virus
RISC	RNA induced silencing	SSB	single stranded DNA binding
	complex	SSEM	serologically explicit
RLKs	receptor-like kinase		electron microscopy
RLPs	receptor-like proteins	STOL	stolbur
RNAi	RNA interference	TALENs	transcription activator-like
RoCV1	rose cryptic virus-1		effector nuclease
ROS	reactive oxygen species	TAS	trans-acting siRNA
RPA	recombinase polymerase	TASVd	tomato apical stunt viroid
	amplification	TAV	tomato aspermy virus
RPA	ribonuclease protection assay	TBIA	tissue smudge immunoassay
R-PAGE	return-PAGE	TBSV	tomato bushy stunt virus
RRV	rose rosette virus	TCDVd	tomato chlorotic dwarf viroid
RT	reverse transcriptase	TEM	transmission electron
RTBV	rice tungro bacilliform virus		microscopy
RTD	rice tungro disease	TeMV	telfairia mosaic virus
RT-LAMP	reverse transcriptase-loop	TEV	tobacco etch virus
	mediated isothermal	TGGE	temperature gradient gel
	amplification		electrophoresis
RT-PCR	real-time PCR	TGS	transcriptional gene silencing
RT-PCR	reverse transcription-	TIBA	tissue immuno blot assay
	polymerase chain reaction	TMB	tetramethylbenzidine
RT-RPA	reverse transcriptase-	TMV	tobacco mosaic virus
	recombinase polymerase	TNS	total nucleic acids
	amplification	TNV	tobacco necrosis virus
RTSV	resistant for tungro spherical	TobBRV	tobacco black ring
	virus	TobNV	tobacco necrosis
SA	salicylic acid	ToMV	tomato mosaic virus
SBS	sequencing by synthesis	ToRSV	tomato ringspot virus
SCMV	sugarcane mosaic virus	TPMVd	tomato planta macho viroid
SC-RT-PCR	squash catch	TPT	tube precipitation test
SDS	sodium dodecyl suphate	TRBP	trans-acting RNA-binding
SEA	South East Asian		proteins
SGSD	sugarcane grassy shoot	TRSV	tomato ring spot
	disease	TSV	tobacco streak virus
SHMV	Sunn-hemp mosaic virus	TSWV	tomato spotted wilt virus
siRNA	small interfering RNA	TuMV	turnip Mosaic Virus
SLCCV	squash leaf curl china virus	TVBMV	tobacco vein banding mosaic
SLCMV	squash leaf curl mild virus		infection
SLCV	squash leaf curl virus	TYLCV	tomato yellow leaf curl virus
SLCYNV	squash leaf curl Yunnan virus	UAVs	unmanned aerial vehicles

Ub	ubiquitin	WMV-2	watermelon mosaic virus 2
UHT	uninvolved hemagglutina-tion test	WMV-MO	watermelon mosaic virus morocco strain
UPS	ubiquitin-proteasome pathway	WSMoV	watermelon silver mottle virus
VLPs	virus-like particles	WSMV	wheat streak mosaic virus
VOC	volatile organic compound	WYMV	wheat yellow mosaic virus
VRCs	viral replication complexes	YMD	yellow mosaic disease
vRNP	viral ribonucleoprotein	ZFNs	zinc finger nucleases
VSRs	viral suppressors of RNA silencing	ZGMMV	zucchini green mottle mosaic virus
WB	western smudging	ZLCV	zucchini lethal chlorosis virus
WBNV	watermelon bud necrosis virus	ZSSV	zucchini shoestring virus
		ZYFV	zucchini yellow fleck virus
WLMV	watermelon leaf mottle virus	ZYMV	zucchini yellow mosaic virus

Foreword

Over the years, the science of plant diseases has grown by leaps and bounds, and the contributions made by scientists all over the world have immensely enhanced our understanding of the subject and helped humanity in facing the challenges posed by plant pathogens to various crops. But further, we need to bring in the advanced technologies such as host-induced gene silencing against many pathogens through RNAi, gene silencing for virus disease management, the science of omics in pathogenicity, and several such other molecular approaches for our enhanced understanding of the subject. Pathogenomics may show the way for the management of many plant diseases. The exploitation of safe secondary metabolites from beneficial microbes and integration of the same in disease management is another area to explore. With climate change a reality now, as well as the host-plant agrochemical resistance, food safety issues are more concerning than ever before.

On one side, we are proud of the advances made in molecular taxonomy, state-of-the art diagnostic techniques, identification of newer fungicide molecules, nanotechnology, understanding the mechanism of host-plant resistance, specifically effector genes, whole-genome sequencing of important pathogens, deciphering the hitherto unknown functions of many genes in plant pathogens, developing new formulations of biocontrol agents including plant growth promoters, etc. On the other side, we have the challenges to be addressed *viz.*, invasion and emergence of new diseases, biosecurity, inappropriate use of fungicides or antibiotics and development of fungicides resistance, lack of fool-proof management practices for virus and phytoplasma diseases, especially the viruses transmitted by sucking pests cum vectors.

Further, the principles of quarantine need to be practically implemented in letter and spirit. In spite of our enhanced understanding of the subject, some of the emerging plant diseases. I am aware that the editors Dr. R. K. Singh and Dr. Gopala are editing a book titled *Innovative Approaches in Diagnosis and Management of Crop Diseases*, published by Apple Academic

Press, Inc. I congratulate both editors and hope that the book would provide knowledge on plant diseases for the benefit of faculty, students, farmers, and society at large.

—Prof. S. K. Rao
Vice-Chancellor
Rajmata Vijayraje Scindia Krishi Vishwa Vidyalaya
Raja Pancham Singh Marg, Gwalior (M.P.)–474002, India

Preface 1

There are many kinds of plant pathogens ranging from ultramicroscopic entities to well-defined multicellular organisms with wide variations in pathogenic potential. They are known to be the principal causes of destructive diseases of many economically important crops cultivated all over the world. Many of them are widely distributed and survive in varied habitats. It is well recognized that the development of effective crop disease management depends on the rapid detection and precise identification of the pathogens(s) causing the disease in question. In this context, knowledge of the different methods needs to be available for the successful management of diseases(s) affecting the various crops in any location. The idea for this edited book arose from what we perceived as the need for an up-to-date guide to the detection, diagnosis, and management of plant viruses, phytoplasma, and viroid. We were encouraged to proceed after receiving nine positive responses from colleagues worldwide, in which, one chapter on phytoplasma and viroids out of its nine chapters. The remaining seven chapters discuss plant viruses.

This volume provides vital information on currently applied methods of detection, diagnosis, and management of plant viruses. The volume covers *viz.,* (i) recent insight into the detection and management of phytoplasma diseases, (ii) techniques for detection of viroids, (iii) CRISPR Cas9 genome editing in plant for virus resistance: opportunities and challenges, (iv) recent insights in detection and diagnosis of plant viruses using next generation sequencing technologies, (v) innovative diagnostic tools for plant pathogenic virus, (vi) development of recombinant coat protein for detection of banana viruses, (vii) management of plant viruses through host RNAi defense mechanism, (viii) diagnosis and management of viral diseases infecting ornamental plants, and (ix) global status on diagnosis, geographical distribution and integrated disease management strategies for major viruses infecting cucurbitaceous crops.

We have tried to cover all relevant branches of plant virology; including phytoplasma, viroids, and next-generation sequencing (NGS) have been described in an easily understandable manner to enable researchers in both the laboratory and the field. We are very happy to express our sincere thanks to all contributors and publishers.

—**R. K. Singh, PhD**
Gopala, PhD

Preface 2

Diagnosis and management of biotic stresses play a pivotal role in the foundation of agriculture production. The changing scenario of agriculture from localized technology to modern innovative scenario has focused our attention on ways to manage plant diseases, reduce crop losses, avoid wide fluctuations in production, and sustain the higher levels of productivity. During the last few decades, the science of crop production has developed rapidly, resulting in many diverse techniques that evolved to manage a variety of menaces incited by pathogenic microbes and nematodes. Currently, the prospects of chemical control do not appear good because of accidental and incidental trauma encountered in the field of chemical pesticides. Therefore, vigorous research efforts for exploring newer but pragmatic tools for the control of biotic stresses have become imperative under the climate change scenario. The scenario of climate change drastically fluctuates the pathogenic nature of the microbes. Climate change may activate sleeper pathogens to become more aggressive fungus/bacteria/virus/nematode and attract the attention of scientists due to considerable losses incited by them. However, it is disheartening to observe that the crop grown with the hard labor of the farmers is damaged by various biotic and abiotic stresses. The crop losses estimate range from 15 to 100% due to climate calamities coupled with damage by harmful pests and diseases. Sustaining food securities in developing countries under the enormous threat of emerging new strains of virulent pathogens is an uphill task, due to more recombination in pathogen and depleting trend of available nutrients in the soil. The global change in climatic and frequent emergence of new races/strains of pathogens may lead to influence the infection process, pathogenic behavior, perpetuation habit, and severity, which may directly affect crop production. Reliable and authentic information about the diagnosis and detection of the pathogen is the backbone of management strategies. The early and accurate diagnosis of plant physiological abnormalities is a crucial component of any crop-management strategy preparedness. Plant health abnormalities can be managed most effectively if control measures are introduced at an initial stage of disease development. Visualization of sign or symptoms on the plant parts have only appeared after getting colonization of infection, before that defense mechanism of the plant has been defeated by the virulent pathogen.

Reliance on symptoms is often not adequate in this regard, since the disease may be well underway when symptoms first appear, and symptom expression can be highly variable. General pathological biological techniques for disease diagnosis and pathogen detection are usually highly accurate but too slow and not amenable to large-scale application. Acceptable accuracy and precision of naked-eye visual disease assessments have often been achieved using traditional disease scales during the past 80 years. The assessment of visual symptoms is essential for the diagnosis of plant diseases.

However, these methods are too subjective. New technologies offer the opportunity to assess disease with greater reliability, precision, and accuracy. Visible light photography and digital image analysis have been increasingly used over the last 30 years, as the software has become more sophisticated and user-friendly for tentative disease diagnosis and suggesting probable management options to farmers. The inception of polymerase chain reaction (PCR) by Nobel laureate Kary Mullis had a profound impact on nucleic acid-based detection and biodiversity analysis of all types of fungi, bacteria, viruses, viroids, nematode, fastidious bacteria, phytoplasma, and algae. While nucleic acid technology is the only choice for detecting pathogens that have not been cultured or biographic in nature, DNA-based methods and serological techniques have not yet completely replaced classical microbiology and visual inspection. The complementary information generated by those techniques may be utilized for confirmation of individual pathogens by way of conducting pathogenicity. Current and future methodology for detection of plant disease include immunological and DNA-based methods, proximate detection approaches based on the analysis of volatile compounds and genes as biomarkers of disease, sensors based on phage display biophotonics, and remote sensing (RS) technologies in combination with spectroscopy-based methodologies.

This book, *Innovative Approaches in Diagnosis and Management of Crop Diseases,* comprises critical updated reviews and research articles on important diseases of different crops and their recent innovation in the detection of plant pathogens and most appropriated management strategies with recent technologies. This edited book consists of selected 40 chapters, contributed by the renowned and expert scientists working on the different aspects of plant pathology in India. The information on various topics is at advanced as well as comprehensive levels, covering important diseases of crops, incited by bacteria, fungi, viruses, viroids, phytoplasma, and nematode of commercial field and horticultural crops and their management. Chapters cover recent advances in diagnosis and detection of diseases of rice, wheat, pulses, guava, aonla, cruciferous, cucurbits, ginger, sesame, cotton, pigeonpea, field pea, small millets, and maize, in addition of that book thoroughly accommodated the multidimensional ways of extraordinary recent technologies generated and their future prospects are comprising in the chapter of individual disease.

This book aims to be a standard reference work, offering available basic facts, re-evaluating, and reviewing the past research, and providing the new and current discoveries on the subject and up-to-date information on nano-technologies, green nanotechnologies, soil-borne diseases, host resistance, bio-intensive management options, biodiversity, the ecology of the seed-borne pathogen, bio-fumigation, versatile biocontrol agent, plant quarantine, plant immunization, and climate change.

We are grateful to all the authors for contributing their valuable original article to this book. The editorial help of the staff at Apple Academic Press sincerely appreciated.

It is our hope that this book will serve the science fraternity as well, and equally hope that the book will stimulate young scholars to work on biological control, nanotechnology, host RNAi defense, recombinant coat protein, NGS, CRISPR Cas9 genome editing, plant quarantine, biodiversity, plant immunization, and recent detection and diagnosis of plant pathogens.

—**R. K. Singh, PhD**
Gopala, PhD

CHAPTER 1

Recent Insight into the Detection and Management of Phytoplasma Diseases

GOPALA,[1] R. K. SINGH,[1] and KISHOR P. PANZADE[2]

[1]*Department of Plant Pathology, College of Agriculture, Indore–452001, Madhya Pradesh, India, E-mail: gplpatho33@gmail.com (Gopala)*

[2]*Indian Agriculture Research Institute, New Delhi–110012, India*

ABSTRACT

Phytoplasmas are phytopathogenic prokaryotes without a cell wall and inhabit phloem sieve elements in infected plants belongs to class mollicutes. Phytoplasma diseases are increasingly important worldwide, with a high economic impact on crop production and quality, causing losses of millions of dollars in several economic fields and horticultural crops that affect their market values. The plants infected with phytoplasmas are of major concern as they are easily vectored by insects and very difficult to manage once established. The remedial measures advocated so far focused mostly on vector management and prophylactic measures rather than curative measures. Prevention of the infection through control of insect vectors is the most probable solution available. Once the plant is infected, it is difficult to target the pathogen and control it as it spreads through vegetative propagation. The successful elimination of phytoplasma from infected plants by the use of tetracycline derivatives in various crops. The use of plant growth regulators, especially auxins in different combinations, have shown a reduction in phytoplasma induced symptoms on plants. A revival of symptoms in infected plants by use of plant immunity activators like benzothiadiazole, acibenzolar methyl, etc., has been observed in fruit trees from abroad.

1.1   INTRODUCTION

In 1967, Japanese scientists discovered that plant pathogens known as phytoplasmas (previously termed mycoplasma-like organisms), were the probable causes of plant yellows diseases (Doi et al., 1967). Numerous yellows-type diseases of plants earlier thought to be caused by viruses in view of their symptomatology and the nature of transmission by insects (Lee et al., 1992), are now associated with phytoplasma pathogens in several plant species (IRPCM, 2004). Phytoplasmas are obligate parasites that reside mainly in plant phloem tissues and in insect vectors. They are probably evolved and diverged from Gram-positive bacteria and have been given the status of the genus '*Candidatus* Phytoplasma' (IRPCM, 2004). They are pleomorphic in shape (Figure 1.1), with diameters less than 1 micrometer, and have the smallest genomes (530–1350 kb) in the bacterial kingdom. The sequence analysis of 16S rDNA and other housekeeping genes suggests that they are most closely related to the *Acholeplasma* spp. than to Spiroplasmas (Gasparich et al., 2004; Lim et al., 1992). Advances in molecular-based diagnostics have made it convenient to develop systems for their accurate identification and taxonomic classification. Molecular-based probes, antibodies against membrane proteins, and cloned phytoplasma DNA fragments developed in the 1980s were authentically used to detect diverse phytoplasmas associated with plants and insects and to know their genetic inter-relationships (Chen et al., 1992). A comprehensive classification scheme was constructed based on restriction fragment length polymorphism (RFLP) patterns of PCR-amplified 16S rDNA sequence (Lee et al., 1993, 1994). This approach provided a reliable tool for broad differentiation among phytoplasmas strains and classified phytoplasmas into 34 groups and 120 subgroups (Lee et al., 2004a, b, 2006; Arocha et al., 2005b; Al-Saady et al., 2008).

1.2   ECONOMIC IMPORTANCE OF PHYTOPLASMA DISEASES

Phytoplasmas are associated with plant diseases in seven hundred plant species, including cereals, fruit, vegetable, ornamentals, oil crops, timber, and fodder (Bertaccini et al., 2018a). Typical symptoms include leaf yellowing, phyllody, proliferation of axillary buds (witches' broom), and sterility of flowers, flat stem, sterile flower buds, and stunting (Figure 1.2 (A–L)). The list of diseases caused by phytoplasma has been increasing

(Bertaccini et al., 2018a). Regularly some newly emerging diseases of uncertain etiology have been identified to be associated with phytoplasma, for example, citrus Huanglongbing disease in China (16SrI) (Teixeira et al., 2009) and pigeon pea witches' broom disease in Brazil (Chen et al., 2008). Phytoplasmas severely affect many plants all over the world (Bertaccini et al., 2007). The major economic losses in vegetable crops and ornamental plants are caused by the aster yellows (AYs) group of phytoplasmas in North America and Europe (Bertaccini et al., 2011). In the 1990s, peach, citrus, and cherry orchards have many losses due to affected by peach yellows and X-disease in the United States and the Middle East and also in Iran and Oman, the lime witches' broom has completely eliminated from its traditional cultivation of lime.

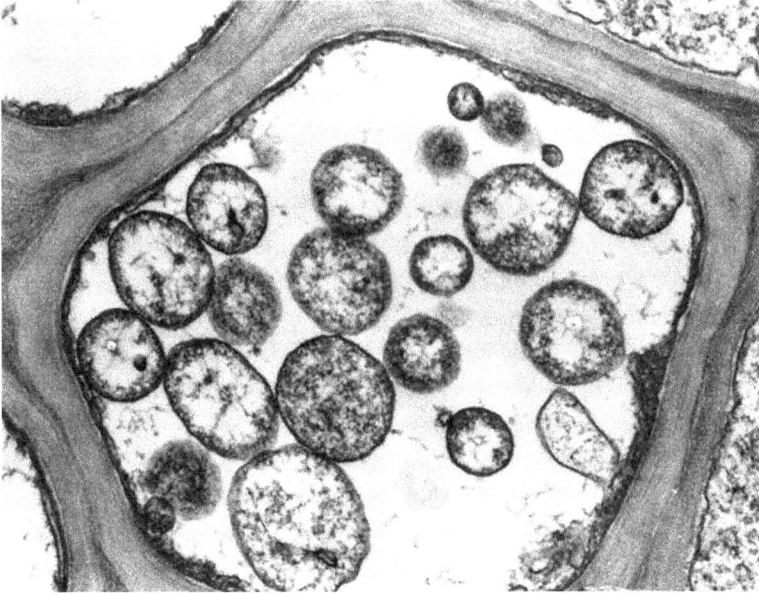

FIGURE 1.1 Transmission electron micrograph showing phytoplasma bodies present in phloem sieve plates of plant.

Source: Courtesy: https://alchetron.com/Phytoplasma.

Potato witches' broom, Rice yellow dwarf, maize bushy stunt, grapevine yellows, and Sweet potato witches' broom severely affects crops in several regions of southwestern Asia, central and south America, Asia, Europe, and Australia (Bertaccini et al., 2011; Li et al., 2015; Yu et al., 1998). Legume diseases such as peanut witches' broom, sesame, and soybean phyllody cause considerable losses of these crops in Asia. In 1995, severe symptoms appeared on, *Bellisperenni, Callistephus chinensis, Gypsophila paniculata, Helichrysum bracteatum, Limonium sinuatum, Tagetes patula*, and some wild plants are grown at different locations in Poland. The symptoms include chlorosis, leaf discoloration, dieback, production of secondary shoots, and flower malformation (Kamińska et al., 1996). The severe economic losses were reported in poinsettias (*Euphorbia pulcherrima*) and *Silene nicaeensis* affected with phytoplasmas (Bertaccini et al., 1996a; Cozza et al., 2008). Significant economic losses were also reported in *Anemone, Alstroemeria, Calendula* sp., *Digitalis lanata,* and *Asclepias physocarpa* showing virescence in flowers, plant malformation, and stunting caused by different strains of phytoplasmas (Bertaccini et al., 1996b, 2006; Bellardi et al., 2007; Tolu et al., 2006; Pavlovic et al., 2014).

1.3   PHYTOPLASMA CLASSIFICATION

Phytoplasmas cannot be differentiated and classified using the classical biophysical and biochemical phenotypic criteria utilized routinely for cultivable microorganisms. In the 1980s and early 1990s, improved protein and nucleic acid-based assay techniques provided new insights into the diversity and genetic interrelationships of phytoplasmas (Bertaccini et al., 2014). Later several phytoplasma groups could be distinguished and characterized by their conserved (16S rDNA) ribosomal phytoplasma region by utilization of molecular tools viz., PCR/RFLP, and nested-PCR (Lee et al., 1998; Seemüller et al., 1998). The different classified group/clade of phytoplasma based on RFLP and phylogenetically by utilization of full-length sequences of *16S rRNA* gene sequence. Phylogeny-based on 16S rRNA classified into different clades *viz.*, Subclade I (AYs, 16SrI, and stolbur, 16SrXII), subclade II (apple proliferation (16SrX), subclade III Western X (WX, 16SrIII), palm lethal yellowing (LY, 16SrIV) and elm yellows (EY, 16SrV) (Hogenhout et al., 2008). The utilization of multilocus genes like *LeuS, rp, tuf, dnaB, polc, secA,* and *secB* for finer differentiation and identification of phytoplasmas. (Smart et al., 1996; Davis et al., 2003; Dickinson et al., 2010, 2013; Rao et al., 2017e).

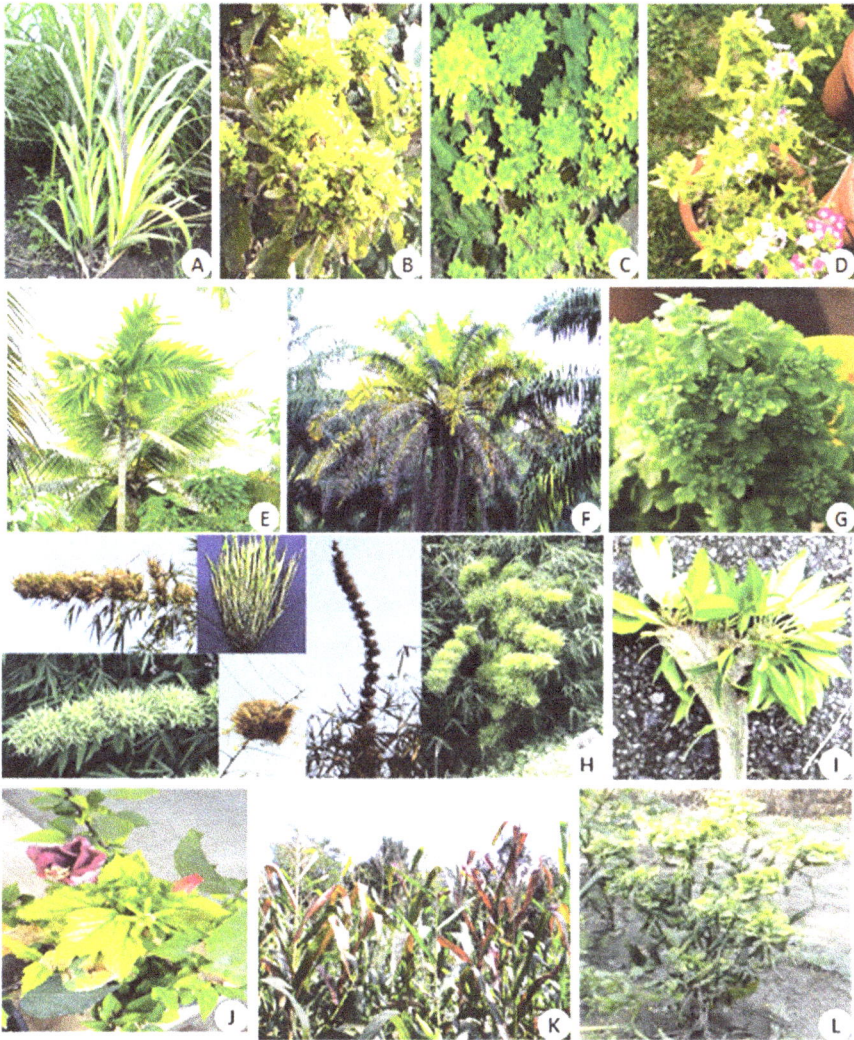

FIGURE 1.2 Phytoplasma disease symptoms: (A) sugarcane affected with grassy shoot disease; (B) little leaf disease of brinjal; (C) little leaf in *Catharanthus roseus*; (D) *Phlox* yellowing and little leaf; (E-F) areca nut leaf yellowing; (G) *Chrysanthemum* phyllody; (H) witches' broom on bamboo plants; (I) flat stem and witches' broom in sapota; (J) phyllody in *Hibiscus rosa-Chinensis*; (K) maize leaf redness disease; (L) witches' broom in chili.

Source: Reprinted with permission from the author. Rao et al. (2017a).

The idea of using illustrative RFLP patterns to differentiate and classify phytoplasmas was conceived in the early 1990s (Lee et al., 19938; Schneider et al., 1993). Based on RFLP analyzes of PCR amplified 16S rDNA, a comprehensive classification scheme has been constructed for phytoplasmas (Lee et al., 1993). The scheme has been periodically updated (Lee et al., 1998, 2000; Seemüller et al., 1998; Wei et al., 2007; Fránová et al., 2011). By 1998, based on similarity coefficients derived from RFLP analyzes, a total of 34 representative phytoplasma strains were differentiated into 14 groups and 32 sub-groups, and later it's extended to 34 groups and 120 subgroups (Lee et al., 1998; Bertaccini et al., 2011; Dickinson et al., 2013; Davis et al., 2016; Naderali et al., 2017). According to ICSB, previously described *Candidatus* species and new '*Candidatus*' species have different similarity sequences reason that when they: (a) are transmitted by different insect vectors; (b) have different alternate natural plant host; and (c) there is evidence for molecular diversity between two phytoplasmas (IRPCM, 2004; Firrao et al., 2005). An interactive phytoplasma research tool, *iPhyClassifier*, has been launched on the internet (http://www.plantpathology.ba.ars.usda.gov/cgibin/resource/iphyclassifier.cgi). Based on calculated virtual RFLP pattern, similarity coefficients, and generally sequence similarity scores, *iPhyClassifier* makes instant suggestions on tentative phytoplasma 16Sr group/subgroup clas-sification status and '*Candidatus* Phytoplasma' species. RFLP patterns that have a similarity coefficient of 0.99 or 0.98 with the standard pattern type of the designated representative or reference member in a given subgroup are considered as variants of the standard pattern type (Zhao et al., 2009, 2010).

1.4 PHYTOPLASMA GENOME

The genomic properties of uncultured phytoplasmas have been determined using partially purified phytoplasmas or phytoplasma-enriched preparations from the infected plants or insect vectors. The size of phytoplasma genomes varies considerably, ranging from 350 to 1350 kb using chromosomal DNA linearized by gamma irradiation and pulsed-field gel electrophoresis for DNA separation (Neimark et al., 1993; Marcone et al., 1999). The Bermuda grass white leaf phytoplasma represents the smallest genome size (530 kb) found in phytoplasmas to date (Razin et al., 1998). The genome sizes of phytoplasmas are similar to those of members of the genus *Mycoplasma* (580–1300 kb), order *Mycoplasmatales,* but are lesser than their closest relatives, members of the genus *Acholeplasma* (~1600 kb), order *Acholeplasmatales* (Neimark et al., 1993; Razin et al., 1998).

Phytoplasmas are differentiated from mycoplasmas because they have a spacer region (about 300 base pairs) between the inter-spacer region of 16S and 23S ribosomal regions, which codes for isoleucine tRNA and part of the sequences for alanine tRNA Ala. The G+C contents estimated from buoyant density centrifugation show phytoplasma chromosomal DNA ranges from 23 and 29 mol% (Razin et al., 1998; Kollar et al., 1989).

Phytoplasma genomes contain large numbers of insertion sequences and transposon genes that are unique to these mollicutes. The unique transposon genes are responsible for variability in diverse environments of plants and insects that make phytoplasmas to survive. These sequences were named as variable mosaics (SVM) (Jomantiene et al., 2006, 2007) and as potential mobile units (PMUs) (Bai et al., 2006). They also contain a unique family of repetitive palindromes (REPs) called PhREPS, which suggested to may play a role in transcription termination or genome stability (Jomantiene et al., 2006, 2007). These have been described as being at the roots of phytoplasma evolution (Wei et al., 2008b). Phytoplasma contains plasmids ranges from 1.7–7.4 kb found in all members of the AYs group, X-disease, clover proliferation, and stolbur groups. Plasmids play an important role in the virulence and pathogenicity of plant pathogenic bacteria (Kuboyama et al., 1998; Rekab et al., 1999; Nishigawa et al., 2000a, 2002b).

Phytoplasma chromosome size differences are due to the occurrence of gene duplication and redundancy. For example, '*Ca*. P. cynodontis' and a tomato strain of the stolbur phytoplasma that have a chromosomal size of 530 and 1,350 kb, respectively, and belongs to different 16Sr group; similarly, rape virescence and the hydrangea phyllody from the same subgroup (16SrI-B) but differ in their respective genome sizes: 1,130 kb vs. 660 kb (Marcone et al., 1999). Genome of the onion yellows (OY) phytoplasma (Oshima et al., 2004), 18% of the total genes were multiple redundant copies of only five genes: *uvrD, hflB, tmk, dam,* and *ssb*. All of these are commonly single copies in the other Mollicutes whose genomes have been sequenced (Miyata et al., 2003).

1.5 MULTIPLE GENE-BASED SYSTEMS FOR CLASSIFICATION OF PHYTOPLASMAS

Molecular-based characterization introduced during the last two decades has proven to be more accurate and reliable than biological criteria long in earlier days for phytoplasma identification (Lee et al., 2000). PCR-based assays developed in the 1990s further facilitate the classification and detection of phytoplasmas by provides a much more accurate tool for phytoplasma detection (Lee et al., 2000). The *16S rRNA* gene is the most widely used marker in the phytoplasma research community and proves to be very useful in the preliminary classification of phytoplasmas. A total of 34 distinct groups, termed *16S rRNA* groups (16Sr groups), based on actual RFLP analysis of PCR-amplified 16S rDNA sequences and *in silico.*

Several molecular markers have been employed for differentiation and classification of phytoplasmas viz. *sec*A, *sec*Y, *tuf, groEL, gyr*A, *gyrB, rp,* etc., from less well-conserved gene sequences, have been identified for finer differentiation of closely related strains that cannot be distinguished by *16S rRNA* gene sequence alone (Schneider et al., 1993; Hodgetts et al., 2010; Bertaccini et al., 2015). Multilocus sequence analyzes provided additional information to the *16S rRNA* gene-based phylogenetic backbone and enhanced the resolving power in delineating distinct phytoplasma lineages and closely related strains (Hodgetts et al., 2010; Malembic et al., 2011; Davis et al., 2013).

The *16S rRNA* gene is the most widely used marker in the phytoplasma research community and proves to be very useful in the preliminary classification of phytoplasmas. Several universal or generic oligonucleotide primer pairs based on the *16S rRNA* gene, the *16S-23S* intergenic spacer region, and partial *23S rRNA* gene sequences have been designed, gives amplification of >1200 base pair to near full-length *16S rRNA* gene sequences of all phytoplasmas related with various plants and insect vectors (Lee et al., 1993; Schneider et al., 1993; Smart et al., 1996; Namba et al., 1993; Gundersen et al., 1996). RFLP analysis of PCR-amplified *16S rRNA* gene sequences using selected restriction enzymes was employed by Schneider et al. (1993) and Lee et al. (1993) for the classification of phytoplasmas. The highly conserved nature of *the 16S rRNA*-based system for classification is of many merits because the universal oligonucleotide primers are relatively easily designed, and in the wealth of sequences available in the GenBank database, which makes it plausible to conduct comprehensive phylogenetic studies. However, because of its highly conserved nature, the *16S rRNA* gene is inadequate for finer differentiation of closely related but distinct phytoplasmas strains. It was evident that some subgroups contained more than one biologically significant strain.

The intergenic spacer regions (ISR) serve as a tool for the differentiation of phytoplasmas subgroups and groups. The conserved gene, *16S-23S rRNA,* contains a portion that codes for the highly conserved tRNAIIe (Smart et al., 1996). Because of limited informative characters available in its relatively short sequences, ISR cannot be used to differentiate all the 16Sr subgroups. On the other hand, combined analysis of the entire *16S rRNA* gene plus ISR sequence proved to be useful in several cases for differentiating distinct strain types within a given 16Sr subgroups (Griffiths et al., 1999; Marcone et al., 2000; Padovan et al., 2000; Andersen et al., 2006).

Ribosomal protein (*rp*) genes are more variable than *16S rRNA* genes and have more phylogenetically informative characters, which substantially enhance the resolving power in delineating distinct phytoplasma strains. In recent times, (Martini et al., 2007) making a phylogenetic tree based on the analysis of two ribosomal protein genes, *rplV* and *rpsC*, from 46 phytoplasma strains representing 12–16Sr groups. This *rp* gene-based phylogenetic tree, more distinct phytoplasma subclades, and distinct lineages than those resolved by the *16S rRNA* gene-based tree (Lee et al., 2007).

The *tuf* gene, encoding the elongation factor, EF-Tu is another highly conserved gene that has been frequently used for differentiation and classification of phytoplasmas. Designed primer pairs that can be used for amplification of *tuf* gene sequences from most phytoplasma groups. It was found that the *tuf* gene, like the *16S rRNA* gene, represents a potential marker for the classification of phytoplasma groups (Schneider et al., 1997). The nucleotide sequence similarities among the AYs, peach X-disease, and stolbur (STOL) phytoplasma groups and subgroups can be differentiated based on RFLP analyzes using several restriction enzymes. The resolving efficacy for separation of distinct lineages among phytoplasmas is slightly lower than that of the *16S rRNA* gene (Marcone et al., 2000; Schneider et al., 1997). However, in some cases, the tuf gene was found to be useful in the differentiation of various ecological strains or strains variants within *16S rRNA* subgroups (Langer et al., 2004). For example, several strain variants were recognized within 16XII-A and 16XII-B, based on analysis of tuf gene sequences (Firrao et al., 2005; Andersen et al., 2006; Langer et al., 2004; Pacifico et al., 2007; Riolo et al., 2007; Iriti et al., 2008).

The *secY* gene, encoding for a protein translocase submit, is another molecular marker that is useful for finer differentiation of phytoplasma strains. The *secY* gene sequence variability is similar to that of *rp* genes. The average *secY* gene sequence similarities between two given 16Sr phytoplasma groups ranges from 57.4 to 76.0%. *SecY* subgroups delineated based on RFLP analyzes of *secY* gene sequences from groups 16SrI and 16SrV phytoplasma generally coincided with those delineated with *rp* gene sequences (Lee et al., 2004a, b, 2006; Martini et al., 2007). However, due to more informative characters, the resolving power of *secY* is slightly better than *rp* gene sequences. Complete characterization of the majority of phytoplasma groups and their representative strains is in progress. The *secY* gene, like the *rp* gene, could represent a good candidate marker for the classification of phytoplasma strains.

1.6   MANAGEMENT OF PHYTOPLASMA DISEASES

Phytoplasma ecology is complex. It is affected by the host range and geographic distribution of both the phytoplasma and vector insects transmitting them, and affected much by weather conditions. Epidemiology of phytoplasma-associated diseases is closely tied to their insect vector. Identification of a vector is crucial in determining management strategies, as their control can prevent spread of disease agents. There is no authorized curative treatment for phytoplasma diseases and, thus, these diseases are not easy to control (Jarausch et al., 1999). Control of phytoplasma-associated diseases relies on prevention rather than cure (Constable et al., 2010; Rao et al., 2011). Although phytoplasma disease symptoms may be delayed or alleviated by treatment with certain classes of antibiotics, this approach is not usually practical. Plants are not really cured, and symptoms reappear when treatments cease. Management of phytoplasma-infected plants has mainly focused on controlling the insect vectors and rouging out infected plants and weeds (Rao et al., 2011). It is possible to contain phytoplasma diseases with an integrated approach that includes control measures based on clean propagating material, vector control, weed management, etc. A vector monitoring system has to be developed for population dynamics of the phytoplasma vector species, monitor their ability to spread. Besides, coordinate research into these and other factors governing phytoplasma spread. Identification of alternative control strategies against these diseases, such as usage of biocontrol organisms or mild strains of phytoplasma that could provide effective tools for reducing phytoplasma effect or spread in an environmentally sustainable approach. The most promising strategy for avoiding phytoplasma disease is the identification or development of resistant plant varieties. Phytoplasmas are normally controlled by the breeding and planting of disease-resistant varieties of crops (believed to the most economically viable option) and by the control of the insect vector (Lee et al., 2000). A novel approach based on cryotherapy was developed in for efficient elimination of phytoplasmas and production of pathogen-free plant stocks (Wang et al., 2009). Some of the virus-infected plants are *Prunus*, banana, grapevine, and potato produced through cryotherapy. Thus, cryotherapy has many advantages, such as produce virus-free plants and for long-term storage of plant germplasm, a simultaneously alternate tool for elimination of obligate parasites such as phytoplasma from plants (Brison et al., 1997; Wang et al., 2003, 2006). The production of genetically engineered plants by introducing disease-resistance genes into cultivated crops together with the use of resistance inducer microorganisms to reduce the intensity of disease symptoms represent effective tools to control phytoplasma (Musetti et al., 2008). The application of insecticides for the control of leafhoppers is the most effective management option available to farmers to stop the spread of AYs phytoplasma (Mahr et al., 1993). To give the most efficient control, the timing of insecticide applications is correlated with both their being virulence and leafhopper numbers. Timely and correct detection of plant diseases is a critical factor of the disease management system. At the primary stage of disease inception, plant diseases can be successfully managed if preventive measures are applied. Outbreaks of phytoplasma disease epidemics can be controlled either by controlling the vectors or by eliminating the pathogens from infected plants by meristem tip culture, by antibiotics, or by other chemicals (Bertaccini et al., 2007). It is difficult to achieve results when wild reservoir plants are sources for acquiring phytoplasma contamination for polyphagous leafhoppers (feed on various different types of host) such as stolbur/*Hyalesthes obsoletus*/bindweed/nettle, or when reservoirs and/or vectors are unknown. The use of antibiotics for management of phytoplasma diseases is not practicable because prohibited in many countries, stimulating the production of antibodies, and too costly so, it do not always provide long-time control. Besides the production of transgenic plants producing antibodies or resistance to these pathogens is still far away (Lee et al., 1998; Chen et al., 1998). Tetracyclines are bacteriostatic to phytoplasmas, inhibit their growth. However, without continuous use of the antibiotic, disease symptoms will reappear. Thus, tetracycline is not a viable control agent in agriculture, but it is used to protect ornamental and coconut trees to some extent (Chung et al., 2007). The four antibiotics tested until now (tetracycline, oxytetracycline, streptomycin, erythromycin A) were all capable to induce a delay in symptom appearance and phytoplasma multiplication although not active in blocking phytoplasma infection. Heat or hot water treatment combined with or without tissue culture technique was also reported to eliminate phytoplasmas from different hosts like sugarcane, vines, and chrysanthemums (Perica et al., 2008). Shoot tips, either alone or combined with other strategies, is an efficient method for obtaining phytoplasma-free plants (Tiwari et al., 2011). Production of virus-free healthy clones or phytoplasma free healthy clones from infected plants through tissue culture. The probability of gaining healthy plants in this manner can be enhanced by the use of cryotherapy, cryopreservation, before using them for tissue culture (Wang et al., 2009). Work has also been carried out investigating the effectiveness of plantibodies targeted against phytoplasmas (Chen et al., 1998). Since no single effective control measures are available for the phytoplasma-associated diseases, an effective integrated management schedule has to be adapted. Limited attempts have been made to manage the phytoplasma disease in India except in citrus, sandal, and potato. Amelioration of phytoplasma symptoms by treatment with a tetracycline antibiotic in *Portulaca grandiflora* plants infected by *Candidatus* phytoplasma sp. by foliar spray with tetracycline hydrochloride or penicillin at 250, 500, and 1000 µg/mL concentration at weekly intervals (Ajayakumar et al., 2007). Production of phytoplasma-free plants from yellow leaf diseased *Catharanthus roseus* by employing *in vitro* chemotherapy with kinetin, 6-benzyl aminopurine, and indole-3-butyric acid at the rate of 2.0 mg/L, 0.75 mg/L, and 0.1 mg/L, respectively (Singh et al., 2007). Use of MS medium for maintaining of phytoplasma-infected shoot-tip explants then treated with different concentrations of antibiotics such as oxytetracycline (25–100 mg/L) for two weeks and after that transferred to oxytetracycline free medium for 6 succeeding serial transfers. Meristem cultures are found to be the best way to remove pathogens from the plants. Sugarcane streak mosaic virus, sugarcane mosaic virus (SCMV), sugarcane bacilliform virus, and sugarcane yellow leaf virus have been eliminated from infected sugarcane plants by meristem tip culture (Fitch et al., 2001; Parmessur et al., 2002). Reports are also available for the elimination of sugarcane phytoplasma through in vitro culture techniques (Parmessur et al., 2002). Other techniques like chemotherapy, cryotherapy, and heat therapy are also useful but often fail to eliminate pathogens.

When used alone while their combinations with the meristem culture technique have given satisfactory results (Balamuralikrishnan et al., 2002; Wang et al., 2008; Ramgreeb et al., 2010). Aerated steam therapy (AST) treatment of seed cane for a period of 1 h at 50°C (Vishwanathan et al., 2001) with hot water at 50°C for 2 h (Singh et al., 1968), or by moist hot air at 54°C for 2 h (Singh et al., 1968) was used to eliminate SGSD. Use of apical meristem culture technique for elimination of sugarcane grassy shoot disease (SGSD) for producing clean planting material of sugarcane. The meristems length of 2 and 3 mm were free from the sugarcane grassy shoot disease phytoplasma at higher than larger meristem length or 4 mm. However, the survival of explants during initiation of shoot cultures was lesser in smaller meristems (40%) in comparison to larger ones (60%). The meristems of length of 2 and 3 mm were freed from the SCGS pathogen. Heat or hot water treatment combined with or without tissue culture technique was also applied in the attempts to eliminate phytoplasmas from different hosts like sugarcane, vines, and chrysanthemum (Perica et al., 2008; Rao et al., 2012). Addition of antibiotic oxytetracycline to the medium has been suggested to be more effective for elimination of phytoplasma from infected tissues (Singh et al., 2007). Growing of pear micropropagules on oxytetracycline medium for four weeks, shown free from a pear decline phytoplasma (Davis et al., 1998). Elimination of sugarcane white leaf phytoplasma by *in vitro* culture of plantlets on antibiotic (oxytetracycline) amended media through five to eight subcultures (Wongkaew et al., 2004). The use of antibiotic treatment is valuable for the control of yellows diseases especially in the areas where the pathogen is endemic and causes extreme crop losses (Montasser et al., 2012). Tetracycline had a bacteriostatic effect on phytoplasmas (Kaminska et al., 2003; Andersen et al., 2006; Wongkaew et al., 2004), but symptoms in treated plants mostly reappeared after the transfer of plants to antibiotic-free medium. The bactericidal effect of tetracyclines on the phytoplasma in infected hosts *Euphorbia pulcherrima* and *Catharanthus roseus*, respectively (Bradel et al., 2000). Successful elimination of phytoplasmas by spraying oxytetracycline (0, 100, 200, 400, 800, and 1200 mg/L) in *Dendranthema grandiflora* at three days interval for four months. However, the higher concentrations of oxytetracycline @ 800 and 1200 mg/L proved to be lethal for the treated plants (Chung et al., 2002). Shoots showing no phytoplasma symptoms after phytoplasma-infected shoot-tip explants were subjected to different concentrations of oxytetracycline (20–100 mg/L) for 2 weeks and then transferred to antibiotic-free medium for six subsequent serial transfers, where 80 mg/L was found to be the most effective for elimination of phytoplasmas (Madhupriya, 2016). Fifty percent of regenerated plants as phytoplasma free plants as confirmed by polymerase chain reaction (PCR), which remained healthy for more than 3 years (Singh et al., 2007). Exogenously supplemented auxins on healthy and phytoplasma-infected catharanthus plants are free from phytoplasma diseases. Different concentration of auxins are needed for efficient rooting of healthy as opposed to phytoplasma-infected cathatanthus that phytoplasmas may block the transport of auxin and affect endogenous auxin levels in infected periwinkles. The electron micrograph revealed that increased endogenous concentration of indole-3-acetic acid in phytoplasma-infected plants, as well as on decreased number of phytoplasma cells present in phloem tissue of infected catharanthus plants treated with high concentrations of exogenous auxins (Pertot et al., 1998). The actual mechanism by which high concentration of exogenously supplemented auxins affect the phytoplasma tested is not clear, but they revealed some kind of interdependence.

1.7 CONCLUSION(S)

The phytoplasmas are an important group of pathogens that drastically damage growth and marketing parameters of different plants and affect their commercial value. Knowledge of the importance of phytoplasmas as plant disease agents has advanced rapidly over the last few decades with significant economic losses. Phytoplasmas have been associated with more than 100 plant species in India including vegetables, spices, fruits, medicinal plants, oil crops, ornamentals, cash crops, palm species, fruits, and weed species. Timely and correct identification of plant diseases and vector is a critical factor of disease management system. At initial stages, plant diseases can be successfully managed if preventive measures are applied. Since no effective control measures are available for the phytoplasma-associated diseases. Initially identification of potential vector which spread the phytoplasma disease and prevents it from application of insecticides. An alternate method of phytoplasma disease management is the use of resistance plants. Last but not least, understand their epidemiology and interaction with host and vector.

ACKNOWLEDGMENTS

The authors express sincere thanks to the Dean Faculty of Agriculture, Rajmata Vijayaraje Scindia Krishi Vishwavidyalaya (RVSKVV), Gwalior, Dr. Ashok Krishna, Dean, College of Agriculture, Indore, and Dr. G. P. Rao, Principal Scientist, IARI, New Delhi for a kind help.

CONFLICT OF INTEREST STATEMENT

The authors declare that they have no conflict of interest.

KEYWORDS

- **aerated steam therapy**
- **growth regulators**
- **intergenic spacer regions**
- **phytoplasma**
- **restriction fragment length polymorphism**
- **tetracycline**

REFERENCES

Ajayakumar, P. V., Samad, A., Shasany, A. K., Gupta, M. K., Alam, M., & Rastogi, S., (2007). First record of a '*Candidatus phytoplasma*' associated with little leaf disease of *Portulaca grandiflora*. *Aus. Plant Dis. Notes, 2,* 67–69.

Al-Saady, N. A., Khan, A. J., Calari, A., Al-Subhi, A. M., & Bertaccini, A., (2008). '*Candidatus phytoplasma* omanense,' associated with witches' broom of *Cassia italica* (Mill.) spreng. in Oman. *Int. J. Syst. Evol. Microbiol., 58*(2), 461–466. doi: https://doi.org/10.1099/ijs.0.65425-0.

Andersen, M. T., Newcomb, R. D., Liefting, L. W., & Beever, R. E., (2006). Phylogenetic analysis of '*Candidatus phytoplasma* australiense' reveals distinct populations in New Zealand. *Phytopathology, 96,* 838–845. doi: 10.1094/phyto-96-0838.

Arocha, Y., Lopez, M., Pinol, B., Fernandez, M., Picornell, B., Almeida, R., Palenzuela, I., Wilson, M. R., & Jones, P., (2005b). '*Candidatus phytoplasma* graminis' and '*Candidatus phytoplasma* caricae,' two novel phytoplasmas associated with diseases of sugarcane, weeds and papaya in Cuba. *Int. J. Syst. Evol. Microbiol., 55*(6), 2451–2463. doi: 10.1099/ijs.0.63797-0.

Bai, X., Zhang, J., Ewing, A., Miller, S. A., Radek, A. J., Shevchenko, D. V., Tsukerman, K., et al., (2006). Living with genome instability: The adaptation of phytoplasmas to diverse environments of their insect and plant hosts. *J. Bacteriol., 188*(10), 3682–3696. doi: 10.1128/JB.188.10.3682-3696.2006.

Balamuralikrishnan, M., Dorisamy, S., Ganapathy, T., & Viswanathan, R., (2002). Combined effect of chemotherapy and meristem culture on sugarcane mosaic virus elimination in sugarcane. *Sugar Technol., 4,* 19–25.

Bellardi, M. G., Benni, A., Paltrinieri, S., & Bertaccini, A., (2007). A severe disease induced by '*Candidatus phytoplasma* Asteris' in *Digitalis lanata* Ehrh. *Bull. Insectol., 60,* 275–276. doi: http://www.bulletinofinsectology.org/pdfarticles/vol60-2007-275-276bellardi.pdf.

Bellardi, M. G., Vicchi, V., & Bertaccini, A., (1985). Mycoplasma disease of *Gladiolus*. *Inf. Fitopatol., 35,* 35–39. doi: https://doi.org/10.1007/978-3-319-39670-5-47.

Bertaccini, A., & Duduk, (2011). Phytoplasma and phytoplasma diseases: A review of recent research. *Phytopathol. Mediterr., 48,* 355–378. doi: http://dx.doi.org/10.14601/Phytopathol_Mediterr-3300.

Bertaccini, A., & Lee, I. M., (2018a). Phytoplasmas: An update. In: Rao, G. P., Bertaccini, A., Fiore, N., & Liefting, L. W., (eds.), *Characterization and Epidemiology of Phytoplasma-Associated Diseases (Phytoplasmas: Plant Pathogenic Bacteria-I)* (pp. 1–30, 345). Springer Nature Singapore Pte Ltd., Singapore. ISBN: 978-981-13-0118-6. doi: 10.1007/978-981-13-0119-3.

Bertaccini, A., (1990a). Cyclamen: A new host of mycoplasma-like organisms. *Phytopathol. Mediterr., 29*(3), 213–214. doi: 10.1556/APhyt.45.2010.1.3.

Bertaccini, A., (2007). Phytoplasmas: Diversity, taxonomy, and epidemiology. *Frontieres in Bioscience, 1,* 673–689. doi: 10.2741/2092.

Bertaccini, A., (2015). Phytoplasma research between past and future: What directions? *Phytopathogenic Mollicutes, 5*(1), S1–S4. doi: 10.5958/2249-4677.2015.00001.8.

Bertaccini, A., Bellardi, M. G., Botti, S., Paltrinieri, S., & Restuccia, P., (2006). Phytoplasma infection in *Asclepias physocarpa*. *Acta Hortic., 722,* 229–234. doi: 10.17660/ActaHortic.2006.722.44.

Bertaccini, A., Davis, R. E., & Lee, I. M., (1990b). Distinctions among mycoplasma like organisms (MLOs) in *Gladiolus, Ranunculus, Brassica,* and *Hydrangea* through detection with nonradioactive cloned DNA probes. *Phytopathol. Mediterr.,* 107–113. doi: https://www.cabdirect.org/cabdirect/abstract/19912305722.

Bertaccini, A., Duduk, B., Paltrinieri, S., & Contaldo, N., (2014). Phytoplasmas and phytoplasma diseases: A severe threat to agriculture. *Am J Plant Sci., 5*(12), 46–62. doi: 10.4236/ajps.2014.512191.

Bertaccini, A., Marani, F., & Rapetti, F., (1988). Phyllody and virescence in *Ranunculus* hybrids. *Acta Hortic., 234,* 123–128.

Bertaccini, A., Vibio, M., & Bellardi, M. G., (1996a). Virus diseases of ornamental shrubs. X. *Euphorbia pulcherrima* Willd. Infected by viruses and phytoplasmas. *Phytopathol Mediterr., 35*(2), 129–132. doi: https://www.cabdirect.org/cabdirect/abstract/19961005995.

Bertaccini, A., Vibio, M., Bellardi, M., & Danielli, A., (1996b). Identification of phytoplasmas in *Alstroemeria. Acta Hortic., 432,* 312–319. doi: 10.17660/ActaHortic.1996.432.38.

Bradel, B. G., Preil, W., & Jeske, H., (2000). Remission of the free-branching pattern of Euphorbia pulcherrima by tetracycline treatment. *Journal of Phytopathology, 148,* 587–590.

Brison, M. D., Boucaud, M. T., Pierronnet, A., & Dosba, F., (1997). Effect of cryopreservation on the sanitary state of a cv *Prunus* rootstock experimentally contaminated with plum pox potyvirus. *Plant Science, 123,* 189–196.

Chen, J., Chang, C. J., & Jarret, R. L., (1992). DNA probes as molecular markers to monitor the seasonal occurrence of walnut witches'-broom mycoplasma like organisms. *Plant Dis., 76,* 1116–1119. doi: 10.1094/PD-76-1116.

Chen, J., Pu, X., Deng, X., Liu, S., Li, H., & Civerolo, E., (2008). A phytoplasma closely related to the pigeon pea witches' broom phytoplasma (16SrIX) is associated with citrus huanglongbing symptoms in the state of São Paulo, Brazil. *Phytopathology, 98,* 977–984. doi: 10.1094/phyto-99-3-0236.

Chen, Y. D., & Chen, T. A., (1998). Expression of engineered antibodies in plants: A possible tool for spiro plasma and phytoplasma disease control. *Phytopathology, 88,* 1367–1371.

Chung, B. N., & Choi, G. S., (2002). Elimination of aster yellows phytoplasma from Dendranthema grandiflorum by application of oxytetracycline as a foliar spray. *Plant Pathology J., 18,* 93–97.

Chung, B. N., (2007). Phytoplasma diseases in chrysanthemum and Lilly. In: Harrison, N. A., Rao, G. P., & Marcone, C., (eds.), *Characterization, Diagnosis and Management of Phytoplasma* (p. 175). Studium Press LLC, Texas, USA.

Constable, F. E., (2010). Phytoplasma epidemiology. Grapevines as a model. In: Weintraub, P. G., & Jones, P., (eds.), *Phytoplasmas Genomes, Plant Hosts and Vectors* (pp. 188–212). CAB International, UK.

Cozza, R., Bernardo, L., Calari, A., Silvestro, G., Duduk, B., & Bertaccini, A., (2008). Molecular identification of 'Candidatus phytoplasma asteris' inducing histological anomalies in *Silene nicaeensis. Phytoparasitica, 36,* 290–293. doi: 10.1007/BF02980775.

Davis, R. E., & Sinclair, W. A., (1998). Phytoplasma identity and disease etiology. *Phytopathology, 88,* 1372–1276.

Davis, R. E., Harrison, N. A., Zhao, Y., Wei, W., & Dally, E. L., (2016). 'Candidatus phytoplasma hispanicum,' a novel taxon associated with Mexican periwinkle virescence disease of *Catharanthus roseus. Int. J. Syst. Evol. Microbiol., 66*(9), 3463–3467. doi: 10.1099/ijsem.0.001218.

Davis, R. E., Jomantiene, R., Zhao, Y., & Dally, E. L., (2003). Folate biosynthesis pseudogenes, ψfolP and ψfolK, and an O-sialoglycoprotein endopeptidase gene homolog in the phytoplasma genome. *DNA Cell Biology, 22,* 697–706. doi: doi/abs/10.1089/104454903770946674.

Davis, R. E., Zhao, Y., Dally, E. L., Lee, M., Jomantiene, R., & Douglas, S. M., (2013). '*Candidatus phytoplasma* pruni,' a novel taxon associated with X-disease of stone fruits, *Prunus* spp.: multilocus characterization based on *16S rRNA*, *secY*, and ribosomal protein genes. *Inter. J. Syst. Evol. Microbiol., 63*(2), 766–776. doi: 10.1099/ijs.0.041202-0.

Dickinson, M., & Hodgetts, J., (2013). PCR analysis of phytoplasmas based on the *secA* gene. In: Dickinson, M., & Hodgetts, J., (eds.), *Phytoplasma. Methods in Molecular Biology (Methods and Protocols)* (p. 938). Humana Press, Totowa, NJ. doi: 10.1007/978-1-62703-089-2_17.

Dickinson, M., (2010). Mobile units of DNA in phytoplasma genomes. *Mol. Microbiol., 77*(6), 1351–1353. doi: https://doi.org/10.1111/j.1365-2958.2010.07308.x.

Doi, Y., Teranaka, M., Yora, K., & Asuyama, H., (1967). Mycoplasma- or PLT group-like microorganisms found in the phloem elements of plants infected with mulberry dwarf, potato witches' broom, aster yellows, or paulownia witches' broom. *Ann. Phytopathol. Soc. Japan, 33,* 259–266. doi: https://doi.org/10.3186/jjphytopath.33.259.

Firrao, G., Gibb, K., & Streten, C., (2005). Short taxonomic guide to the genus '*Candidatus phytoplasma.*' *J. Plant Pathol., 87,* 249–263. doi: http://dx.doi.org/10.4454/jpp.v87i4.926.

Fitch, M. M. M., Lehrer, A. T., Komor, E., & Moore, P. H., (2001). Elimination of *Sugarcane yellow leaf virus* from infected sugarcane plants by meristem tip culture visualized by tissue blot immunoassay. *Plant Pathology., 50,* 676–680. doi: https://doi.org/10.1046/j.1365-3059.2001.00639.x.

Fránová, R., (2011). Difficulties with conventional phytoplasma diagnostic using PCR/RFLP analyses. *Bull. Insectol., 64,* 287–288.

Gasparich, G. E., Whitcomb, R. F., Dodge, D., French, F. E., Glass, J., & Williamson, D. L., (2004). The genus spiro plasma and its non-helical descendants: Phylogenetic classification, correlation with phenotype and roots of the mycoplasma mycoides clade. *Int. J. Syst. Evol. Microbiol., 54*(3), 893–918. doi: 10.1099/ijs.0.02688-0.

Griffiths, H. M., Sinclair, W. A., Smart, C. D., & Davis, R. E., (1999). The phytoplasma associated with ash yellows and lilac witches'-broom: '*Candidatus phytoplasma* fraxini.' *Int. J. Syst. Evol. Microbiol., 49*(4), 1605–1614. doi:10.1099/00207713-49-4-1605.

Gundersen, D. E., & Lee, I. M., (1996). Ultrasensitive detection of phytoplasmas by nested-PCR assays using two universal primer pairs. *Phytopathol. Mediter., 1,* 144–151. doi: https://www.jstor.org/stable/42685262?seq=1#page_scan_tab_contents.

Hiruki, C., Romg, X. D., & Deng, S. J., (1994). *Hydrangea* virescence, a disease associated with mycoplasma like organisms in Canada. *Acta Hortic., 377,* 325–333. doi: 10.17660/ActaHortic.1994.377.35.

Hodgetts, J., & Dickinson, M., (2010). Phytoplasma phylogeny and detection based on genes other than *16S rRNA*. In: Weintraub, P. G., & Jones, P., (eds.), *Phytoplasmas Genomes, Plant Hosts and Vectors* (pp. 93–113). CAB International, UK. doi: https://www.researchgate.net/publication/292521186_Phytoplasmas_Genomes_plant_hosts_and_vectors.

Hogenhout, S. A., Oshima, K., Ammar, E. D., Kakizawa, S., Kingdom, H. N., & Namba, S., (2008). Phytoplasmas: Bacteria that manipulate plants and insects. *Mol. Plant Pathol., 9,* 403–423. doi: https://doi.org/10.1111/j.1364-3703.2008.00472.x.

Iriti, M., Quaglino, F., Maffi, D., Casati, P., Bianco, P. A., & Faoro, F., (2008). *Solanum malacoxylon*: A new natural host of stolbur phytoplasma. *J. Phytopathol., 156*(1), 8–14.

IRPCM, (2004). Phytoplasma/spiro plasma working team-phytoplasma taxonomy group. '*Candidatus phytoplasma,*' a taxon for the wall-less, nonhelical prokaryotes that colonize plant phloem and insects. *Int. J. Syst. Evol. Microbiol., 54,* 1243–1255. doi: 10.1099/ijs.0.02854-0.

Jarausch, W., Eyquard, J., Mazy, K., Lansac, M., & Dosba, F., (1999). High level of resistance of sweet cherry (*Prunus avium* L.) towards European stone fruit yellows phytoplasmas. *Advances in Horticultural Science, 13,* 108–112.

Jomantiene, R., & Davis, R. E., (2006). Clusters of diverse genes existing as multiple, sequence-variable mosaics in a phytoplasma genome. *FEMS Microbiology Letters., 255*(1), 59–65. doi: 10.1111/j.1574-6968.2005.00057.x.

Jomantiene, R., Zhao, Y., & Davis, R. E., (2007). Sequence-variable mosaics: Composites of recurrent transposition characterizing the genomes of phylogenetically diverse phytoplasmas. *DNA Cell Biology, 26,* 557–56410. doi: 10.1089/dna.2007.0610.

Kaminska, M., & Sliwa, H., (2004). First report of phytoplasma belonging to apple proliferation group in roses in Poland. *Plant Dis., 88,* 1283. doi: http://dx.doi.org/10.1094/PDIS.2004.88.11.1283A.

Kamińska, M., Malinowski, T., Komorowska, B., & Rudzińska-Langwald, A., (1996). Etiology of yellows and witches' broom symptoms in some ornamental plants. *Acta Hort., 432,* 96–106. doi: 10.17660/ActaHortic.1996.432.11.

Kaminska, M., Sliwa, H., Malinowski, T., & Skrzypczak, C. Z., (2003). The association of aster yellows phytoplasma with rose dieback disease in Poland. *J. Phytopathol., 151,* 469–476. doi: 10.1046/j.1439-0434.2003.00756.x.

Kollar, A., & Seemüller, E., (1989). Base composition of the DNA of mycoplasma-like organisms associated with various plant diseases. *J. Phytopathol., 127*(3), 177–186. doi: 10.1111/j.1439-0434.1989.tb01127.x.

Kuboyama, T., Huang, C. C., Lu, X., Sawayanagi, T., Kanazawa, T., Kagami, T., Matsuda, I., et al., (1998). A plasmid isolated from phytopathogenic onion yellows phytoplasma and its heterogeneity in the pathogenic phytoplasma mutant. *Mol. Plant Microbe. Inteactr., 11*(11), 1031–1037. doi: 10.1094/MPMI.1998.11.11.1031.

Langer, M., & Maixner, M., (2004). Molecular characterization of grapevine yellows associated phytoplasmas of the stolbur group based on RFLP-analysis of non-ribosomal DNA. *Vitis, 43,* 191–200. doi: https://doi.org/10.1094/phyto-07-17-0253-R.

Lee, I. M., Davis, R. E., & Gundersen, R. D. E., (2000). Phytoplasma: Phytopathogenetic mollicutes. *Ann. Rev. Microbiol., 54,* 221–255. doi: https://doi.org/10.1146/annurev.micro.54.1.221.

Lee, I. M., Davis, R. E., Chen, T. A., Chiykowski, L. N., Fletcher, J., Hiruki, C., & Schaff, D. A., (1992). A genotype-based system for identification and classification of mycoplasma like organisms (MLOs) in the aster yellows MLO strain cluster. *Phytopathology, 82,* 977–986. doi: 10.1094/Phyto-82-977.

Lee, I. M., Gundersen, D. E., Hammond, R. W., & Davis, R. E., (1994). Use of mycoplasma like organism (MLO) group-specific oligonucleotide primers for nested-PCR assays to detect mixed-MLO infections in a single host plant. *Phytopathology, 84,* 559–566. doi: 10.1094/Phyto-84-559.

Lee, I. M., Gundersen-Rindal, D. E., Davis, R. E., & Bartoszyk, I. M., (1998). Revised classification scheme of phytoplasmas based on RFLP analyses of *16S rRNA* and ribosomal protein gene sequences. *Int. J. Syst. Evol. Microbiol., 48*(4), 1153–1169. doi:10.1099/00207713-48-4-1153.

Lee, I. M., Gundersen-Rindal, D. E., Davis, R. E., Bottner, K. D., Marcone, C., & Seemüller, E., (2004a). '*Candidatus phytoplasma* asteris,' a novel phytoplasma taxon associated with aster yellows and related diseases. *Int. J. Syst. Evol. Microbiol.*, *54*(4), 1037–1048. doi: 10.1099/ijs.0.02843-0.

Lee, I. M., Hammond, R. W., Davis, R. E., & Gundersen, D. E., (1993). Universal amplification and analysis of pathogen 16S rDNA for classification and identification of mycoplasma-like organisms. *Phytopathology, 83*, 834–842. doi: 10.1094/Phyto-83-834.

Lee, I. M., Martini, M., Marcone, C., & Zhu, S. F., (2004b). Classification of phytoplasma strains in the elm yellows group (16SrV) and proposal of '*Candidatus phytoplasma* ulmi' for the phytoplasma associated with elm yellows. *Int. J. Syst. Evol. Microbiol.*, *54*, 337–347. doi: 10.1099/ijs.0.02697-0.

Lee, I. M., Zhao, Y., & Bottner, K. D., (2006). *sec*Y gene sequence analysis for finer differentiation of diverse strains in the aster yellows phytoplasma group. *Mol. Cell Probes.*, *20*, 87–91. doi: 10.1016/j.mcp.2005.10.001.

Lee, I. M., Zhao, Y., Davis, R. E., Wei, W., & Martini, M., (2007). Prospects of DNA-based systems for differentiation and classification of phytoplasmas. *Bull. Insectol., 60*(2), 239–244. doi: http://www.bulletinofinsectology.org/pdfarticles/vol60-2007-239-244lee.pdf.

Li, S., Hao, W., Lu, G., Huang, J., Liu, C., & Zhou, G., (2015). Occurrence and identification of a new vector of rice orange leaf phytoplasma in South China. *Plant Dis., 99*(11), 1483–1487. doi: https://doi.org/10.1094/PDIS-12-14-1243-RE.

Lim, P. O., & Sears, B. B., (1992). Evolutionary relationships of a plant-pathogenic mycoplasma like organism and *Acholeplasma laidlawii* deduced from two ribosomal protein gene sequences. *J. Bacteriol., 174*(8), 2606–2611. doi: 10.1128/jb.174.8.2606-2611.1992.

Madhupriya, (2016). *Molecular Characterization of Phytoplasmas Associated with Important Ornamental Plant Species in Northern India [Thesis]*. Amity University, Haryana, India.

Mahr, S. E. R., Wyman, J. A., & Chapman, R. K., (1993). Variability in aster yellows infectivity of local populations of the aster leafhopper (Homoptera: Cicadellidae) in Wisconsin. *Journal of Economic Entomology, 86*, 1522–1526.

Malembic-Maher, S., Salar, P., Filippin, L., Carle, P., Angelini, E., & Foissac, X., (2011). Genetic diversity of European phytoplasmas of the 16SrV taxonomic group and proposal of '*Candidatus phytoplasma* Rubi.' *Int. J. Syst. Evol. Microbiol., 61*(9), 2129–2134. doi: 10.1099/ijs.0.025411-0.

Marcone, C., Lee, I. M., Davis, R. E., Ragozzino, A., & Seemüller, E., (2000). Classification of aster yellows-group phytoplasmas based on combined analyses of ribosomal RNA and *tuf* gene sequences. *Int. J. Syst. Evol. Microbiol., 50*, 1703–1713. doi: 10.1099/00207713-50-5-1703.

Marcone, C., Neimark, H., Ragozzino, A., Lauer, U., & Seemüller, E., (1999). Chromosome sizes of phytoplasmas composing major phylogenetic groups and subgroups. *Phytopathology, 89*(9), 805–810. doi: 10.1094/phyto.1999.89.9.805.

Martini, M., Lee, I. M., Bottner, K. D., Zhao, Y., Botti, S., Bertaccini, A., Harrison, N. A., et al., (2007). Ribosomal protein gene-based phylogeny for finer differentiation and classification of phytoplasmas. *Int. J. Syst. Evol. Microbiol., 57*(9), 2037–2051. doi: 10.1099/ijs.0.65013-0.

Miyata, S. I., Oshima, K., Kakizawa, S., Nishigawa, H., Jung, H. Y., Kuboyama, T., Ugaki, M., & Namba, S., (2003). Two different thymidylate kinase gene homologues, including one that has catalytic activity, are encoded in the onion yellows phytoplasma genome. *Microbiol., 149*(8), 2243–2250. doi: 10.1099/mic.0.25834-0.

Montasser, M. S., Hanif, A. M., Al-Awadhi, H. A., & Suleman, P., (2012). Tetracycline therapy against phytoplasma causing yellowing disease of date palms. *The Federation of American Societies for Experimental Biology Journal, 26*(800.1). doi: https://www.fasebj.org/doi/abs/10.1096/fasebj.26.1_supplement.800.1.

Musetti, R., (2008). Management and ecology of phytoplasma diseases of grapevine and fruit crops: Integrated management of diseases caused by fungi, phytoplasma and bacteria. *Integrated Management of Plant Pests and Diseases, 3,* 43–60.

Naderali, N., Nejat, N., Vadamalai, G., Davis, R. E., Wei, W., Harrison, N. A., & Zhao, Y., (2017). 'Candidatus phytoplasma wodyetiae,' a new taxon associated with yellow decline disease of foxtail palm (*Wodyetia bifurcata*) in Malaysia. *Int. J. Syst. Evol. Microbiol., 67*(10), 3765–3772. doi: 10.1099/ijsem.0.002187.

Namba, S., Kato, S., Iwanami, S., Oyaizu, H., Shiozawa, H., & Tsuchizaki, T., (1993). Phylogenetic diversity of phytopathogenic organisms. *Int. J. Syst. Bacteriol., 43,* 461–467. doi: 10.1099/00207713-43-3-461.

Neimark, H., & Kirkpatrick, B. C., (1993). Isolation and characterization of full-length chromosomes from non-culturable plant-pathogenic mycoplasma-like organisms. *Mol. Microbiol., 7*(1), 21–28. doi: 10.1111/j.1365-2958.1993.tb01093.x.

Nishigawa, H., Oshima, K., Kakizawa, S., Jung, H. Y., Kuboyama, T., Miyata, S., Ugaki, M., & Namba, S., (2002a). A plasmid from a non-insect-transmissible line of a phytoplasma lacks two open reading frames that exist in the plasmid from the wild-type line. *Gene., 298,* 195–201. doi: 10.1006/viro.2001.0938.

Nishigawa, H., Oshima, K., Kakizawa, S., Jung, H. Y., Kuboyama, T., Miyata, S., Ugaki, M., & Namba, (2002b). Evidence of intermolecular recombination between extra chromosomal S. DNAs in phytoplasma: A trigger for the biological diversity of phytoplasma. *Microbiology, 148,* 1389–1396. doi: 10.1099/00221287-148-5-1389.

Oshima, K., Kakizawa, S., Nishigawa, H., Jhung, H. Y., Wei, W., Suzuki, S., Arashida, R., et al., (2004). Reductive evolution suggested from the complete genome sequence of a plant. *Nat. Gene., 36,* 27–29. doi: 10.1038/ng1277.

Pacifico, D., Foissac, X., Veratti, F., & Marzachi, C., (2007). Genetic diversity of Italian and French "bois noir" phytoplasma isolates. *Bull. Insectol., 60*(2), 345. doi: http://www.bulletinofinsectology.org/pdfarticles/vol60-2007-345-346pacifico.pdf.

Padovan, A., Gibb, K., & Persley, D., (2000). Association of 'Candidatus phytoplasma australiense' with green petal and lethal yellows diseases in strawberry. *Plant Pathol., 49,* 362–369. doi: https://doi.org/10.1046/j.1365-3059.2000.00461.x.

Parmessur, Y., Aljanabi, S., Saumtally, S., & Dookun-Saumtally, A., (2002). Sugarcane yellow leaf virus and sugarcane yellows phytoplasma: Elimination by tissue culture. *Plant Pathology, 51*(5), 561–566.

Parrella, G., Paltrinieri, S., Botti, S., & Bertaccini, A., (2008). Molecular identification of phytoplasmas from virescent *Ranunculus* plants and from leafhoppers in Southern Italian crops. *J. Plant Pathol., 90,* 537–543. doi: http://dx.doi.org/10.4454/jpp.v90i3.698.

Pavlovic, S., Starovic, M., Stojanovic, S., Aleksic, G., Kojic, S., Zdravkovic, M., & Josic, D., (2014). The first report of stolbur phytoplasma associated with phyllody of *Calendula officinalis* in Serbia. *Plant Dis., 98*(8), 1152–1152. doi: http://dx.doi.org/10.1094/PDIS-01-14-0085-PDN.

Perica, M. C., (2008). Auxin-treatment induces recovery of phytoplasma infected periwinkle. *Journal of Applied Microbiology, 105,* 1826–1834.

Pertot, I., Musetti, R., Pressacco, L., & Osler, R., (1998). Changes in indole-3-acetic acid level in micro propagated tissues of *Catharanthus roseus* L. infected by the agent of the clover phyllody and effect of exogenous auxins on phytoplasma morphology. *Cytobios., 95,* 13–23.

Ramgreeb, S., Snyman, S. J., Antwerpen, T., & Rutherford, R. S., (2010). Elimination of virus and rapid propagation of disease-free sugarcane (*Saccharum* spp.) using apical meristem culture. *Plant Cell Tiss. Organ Cult., 100,* 175–181.

Rao, G. P., Madhupriya, Thorat, V., Manimekalai, R., Tiwari, A. K., & Yadav, A., (2017a). A century progress of research on phytoplasma diseases in India. *Phytopathogenic Mollicutes., 7*(1), 1–38.

Rao, G. P., Mall, S., Raj, S. K., & Snehi, S. K., (2011). Phytoplasma diseases affecting various plant species in India. *Acta Phytopathologica et Entomologica Hungarica., 46,* 59–99.

Rao, G. P., Prakasha, T. L., Priya, M., Thorat, V., Kumar, M., Baranwal, V. K., & Yadav, A., (2017e). First report of association of *Candidatus phytoplasma* cynodontis (16SrXI-B group) with streak, yellowing, and stunting disease in durum and bread wheat genotypes from Central India. *Plant Dis., 101*(7), 1314–1314. doi: 10.1094/PDIS-02-17-0163-PDN.

Rao, G. P., Tiwari, A. K., Singh, M., Chaturvedi, Y., & Madhupriya, (2012). Molecular characterization and phylogenetic relationships of an isolate of *Candidatus phytoplasma asteris* affecting *Zinnia elegans* in India. *Phytopathogenic Molecutes, 2,* 33–36.

Razin, S., Yogev, D., & Naot, Y., (1998). Molecular biology and pathology of myco-plasmas. *Microbiol Mol. Biol. Rev., 62,* 1094–1156. doi: https://www.ncbi.nlm.nih.gov/pubmed/9841667.

Rekab, D., Carraro, L., Schneider, B., Seemüller, E., Chen, J., Chang, C. J., Locci, R., & Firrao, G., (1999). Gemini virus-related extra chromosomal DNAs of the X-clade phytoplasmas share high sequence similarity. *Microbiology, 145*(6), 1453–1459. doi: 10.1099/13500872-145-6-1453.

Riolo, P., Landi, L., Nardi, S., & Isidoro, N., (2007). Relationships among *Hyalesthes obsoletus*, its herbaceous host plants and "bois noir" phytoplasma strains in vineyard ecosystems in the Marche region (central-eastern Italy). *Bull. Insectol., 60*(2), 353.

Schneider, B., Ahrens, U., Kirkpatrick, B. C., & Seemüller, E., (1993). Classification of plant-pathogenic mycoplasma-like organisms using restriction-site analysis of PCR-amplified 16S rDNA. *Microbiol., 139*(3), 519–527. doi: 10.1099/00221287-139-3-519.

Schneider, B., Gibb, K. S., & Seeümller, E., (1997). Sequence and RFLP analysis of the elongation factor *Tuf* gene used in differentiation and classification of phytoplasmas. *Microbiol., 143,* 3381–3389. doi: 10.1099/00221287-143-10-3381.

Seemüller, E., Marcone, C., Lauer, U., Ragozzino, A., & Göschl, M., (1998). Current status of molecular classification of the phytoplasmas. *J. Plant Pathol., 1,* 3–26. doi: http://dx.doi.org/10.4454/jpp.v80i1.789.

Singh, K., (1968). Grassy shoot disease of sugarcane: II. Hot air therapy. *Curr. Sci., 37,* 592–594.

Singh, S. K., Aminuddin, Srivastava, P., Singh, B. R., & Khan, J. A., (2007). Production of phytoplasma-free plants from yellow leaf diseased *Catharanthus roseus* (L.) G. Don. *Journal of Plant Disease Protection, 114,* 2–5.

Smart, C. D., Schneider, B., Blomquist, C. L., Guerra, L. J., Harrison, N. A., Ahrens, U., Lorenz, K. H., Seemüller, E., & Kirkpatrick, B. C., (1996). Phytoplasma-specific PCR primers based on sequences of the *16S-23S rRNA* spacer region. *Appl. Env. Microbiol., 62,* 2988–2993. doi: https://aem.asm.org/content/aem/62/8/2988.full.pdf.

Smookeler, M., & Dabush, S., (1974). Mycoplasma like bodies in tulips. *Plant Dis. Repot., 58,* 1142. doi: 10.17660/ActaHortic.1994.377.35.

Teixeira, D. C., Wulff, N. A., Martins, E. C., Kitajima, E. W., Bassanezi, R., Ayres, A. J., Eveillard, S., et al., (2009). A phytoplasma related to '*Candidatus phytoplasma* asteri' detected in citrus showing huanglongbing (yellow shoot disease) symptoms in Guangdong, P. R. China. *Phytopathology, 99,* 236–242. doi: 10.1094/PHYTO-98-9-0977.

Tiwari, A. K., Tripathi, S., Lal, M., Sharma, M. L., & Chiemsombat, P., (2011). Elimination of sugarcane grassy shoot disease through apical meristem culture. *Archives of Phytopathology and Plant Protection, 44,* 1942–1948.

Tolu, G., Botti, S., Garau, R., Prota, V. A., Sechi, A., Prota, U., & Bertaccini, A., (2006). Identification of 16SrIIE phytoplasmas in *Calendula arvensis* L., *Solanum nigrum* L. and *Chenopodium* spp. *Plant Dis., 90,* 325–330. doi: https://doi.org/10.1094/PD-90-0325.

Vishwanathan, R., (2001). Different aerated steam therapy (AST) regimes on the development of grassy shoot disease symptoms in sugarcane. *Sugar Technol., 3*(3), 83–91.

Wang, Q. C., & Valkonen, J. P. T., (2008). Efficient elimination of sweet potato little leaf phytoplasma from sweet potato by cryotherapy of shoot tips. *Plant Pathol., 5,* 338–347.

Wang, Q. C., Munir, M., Li, P., Gafny, R., Sela, I., & Tanne, E., (2003). Elimination of grapevine virus A (GVA) by cryopreservation of *in vitro*-grown shoot tips of *Vitis vinifera* L. *Plant Science, 165,* 321–327.

Wang, Q. C., Panis, B., Engelmann, F., Lambardi, M., & Valkonen, J. P. T., (2009). Cryotherapy of shoot tips: A technique for pathogen eradication to produce healthy planting materials and prepare healthy plant genetic resources for cryopreservation. *Annals of Applied Biology, 154,* 351–363.

Wang, Q., Liu, Y., Xie, Y., & You, M., (2006). Cryotherapy of potato shoot tips for efficient elimination of potato leaf roll virus (PLRV) and potato virus Y (PVY). *Potato Research, 49,* 119–129.

Wei, W., Davis, R. E., Lee, M., & Zhao, Y., (2007). Computer-simulated RFLP analysis of *16S rRNA* genes: Identification of ten new phytoplasma groups. *Int. J. Syst. Evol. Microbiol., 57*(8), 1855–1867. doi: 10.1099/ijs.0.65000-0.

Wei, W., Lee, M., Davis, R. E., Suo, X., & Zhao, Y., (2008b). Automated RFLP pattern comparison and similarity coefficient calculation for rapid delineation of new and distinct phytoplasma 16Sr subgroup lineages. *Int. J. Syst. Evol. Microbiol., 58*(10), 2368–2377. doi: 10.1099/ijs.0.65868-0.

Wongkaew, P., & Fletcher, J., (2004). Sugarcane white leaf phytoplasma in tissue culture: Long-term maintenance, transmission, and oxytetracycline remission. *Plant Cell Rep., 23,* 426–434. doi: 10.1007/s00299-004-0847-2.

Yu, Y., Yeh, K., & Lin, C., (1998). An antigenic protein gene of a phytoplasma associated with sweet potato witches' broom. *Microbiology, 144,* 1257–1262. doi: 10.1099/00221287-144-5-1257.

Zhao, Y., Sun, Q., Wei, W., Davis, R. E., Wu, W., & Liu, Q., (2009). '*Candidatus phytoplasma tamaricis,*' a novel taxon discovered in witches'-broom-diseased salt cedar (*Tamarix chinensis* Lour.). *Int. J. Syst. Evol. Microbiol., 59*(10), 2496–2504. doi: 10.1099/ijs.0.010413-0.

Zhao, Y., Wei, W., Davis, R. E., & Lee, I. M., (2010). Recent advances in *16S rRNA* gene-based phytoplasma differentiation, classification and taxonomy. In: Weintraub, P. G., & Jones, P., (eds.), *Phytoplasmas Genomes, Plant Hosts and Vectors* (pp. 64–92). CAB International, UK. doi: 10.1079/9781845935306.0064.

CHAPTER 2

Techniques for Detection of Viroids

PANKHURI SINGHAL,[1] MANOJ K. YADAV,[2] and SAJAD UN NABI[3]

[1]Division of Plant Pathology, ICAR-Indian Agricultural Research Institute, New Delhi–110012, India, E-mail: pankhuri.agri47@gmail.com

[2]ICAR-National Rice Research Institute, Cuttack–753006, Odisha, India

[3]ICAR-Central Institute of Temperate Horticulture, Srinagar–191132, Jammu and Kashmir, India

ABSTRACT

Viroids are the smallest RNA pathogens that infect and replicate autonomously in higher plants. They exist as circular, single-stranded, non-coding RNA with a length of 246–401 nucleotides. They do not code any peptides; hence they rely entirely on host factors for their replication. Broadly viroids are classified into two families, i.e., the Pospiviroidae family, whose members replicate in the nucleus and contain five conserved structural/functional domains, and the Avsunviroidae family members replicate in chloroplast and exhibit ribozyme activity. Viroid hosts include both herbaceous and woody species. More than 40 viroid diseases have been reported from different plant species. Symptoms of viroid infection include stunting, leaf epinasty and distortion, fruit distortion and color break, stem, and leaf necrosis, and even death of the whole plant, and they are often dependent upon environmental conditions. Viroids were initially identified by their association with specific diseases, but not all viroid-infected plants exhibit obvious signs of disease. Identification of viroid-infected plants can be accomplished by either symptomatology on indicator hosts, a classical method still used in many certification programs. Due to a lack of protein coat, protein-based diagnostics are not valid. Diagnostics thus is limited to: (a) bioassay on sensitive hosts; (b) polyacrylamide gel electrophoresis (PAGE); (c) nucleic acid spot hybridization (NASH); (d) reverse transcription-polymerase chain reaction (RT-PCR). More recently developed methods such as micro- and macroarrays, next-generation sequencing (NGS), enable even faster and more sensitive detection of viroid infections. On-site detection is also an important improvement of detection techniques, since it facilitates the adoption of fast measures to prevent potential disease spread. Isothermal amplification methods like reverse transcriptase-loop mediated isothermal amplification (RT-LAMP), recombinase polymerase amplification (RPA) has also been used for viroid detection.

2.1   INTRODUCTION

Viroids are one of the most intriguing organisms found on the earth and are considered as more plausible living relics of the pre-cellular RNA world by Theodor Otto Diener, who discovered the first viroid causing potato spindle tuber disease in 1971. The viroid was called potato spindle tuber viroid (PSTVd), abbreviated as PSTVd. These are the smallest RNA pathogens having a molecular weight of 25,000–110,000 Daltons, which infect and replicate autonomously in higher plants. They exist as circular, single-stranded, covalently closed, non-encapsulated RNAs with a length of 246–401 nucleotides in size (Flores et al., 1997). These are having extensive internal base-pairing result in rod-like secondary structures that cause diseases in higher plants (Diener, 1983). Their RNAs are non-protein-coding, hence exploiting the transcriptional machinery of the host to replicate and move in the host. Thus, the detection of viroids is not possible using serodiagnostic techniques (Gucek et al., 2016). Presently, sequences of 46 viroid species are available in the genome database of the National Center for Biotechnology Information (NCBI), of which 32 species have been approved by the International Committee on Taxonomy of Viruses (ICTV). These viroid species have been classified into two families, *Pospiviroidae* (five genera) having central conserved region (CCR) in their genome, replicate, and accumulate in host nuclei and *Avsunviroidae* (three genera) is lacking CCR, possessing hammerhead ribozymes, replicating, and accumulating in host plastids, mainly chloroplasts. Among the remaining species, eight viroid species are still unclassified, and six species of the *Pospiviroidae* family have not yet been approved by ICTV.

Viroids are having wide and highly variable host ranges infecting crops like apple, avocado, carnation, chrysanthemum, *Coleus columnea*, citrus, cucumber, coconut, grapevine, hops, oil palm, peach, plum, potato, tea, and tomato plants. Certain viroids like PSTVd and citrus exocortis viroid (CEVd) have a broad host range (Diener, 1991), while others such as grapevine yellow speckle viroid 1 and 2 (GYSVd-1 and GYSVd-2) are highly specific to *Vitis vinifera* (Koltunov et al., 1989). Certain viroids, while causing disease symptoms in one host, replicate in another host without any apparent deleterious effect. Thus, while hop stunt viroid (HSVd) causes economically important disease in hops, it is present in grapevines and citrus in latent state (Puchta et al., 1988). While, many viroids do not produce any symptoms in their hosts (latent infection), others caused symptoms similar to viruses such as dwarfing, stunting, leaf chlorosis, epinasty, and necrosis, vein discoloration, bark cracking, distortion of tubers, flowers, and fruits, delay in flowering, foliation, and ripening, and in some cases, plants may also die (Hammond and Owens, 2006; Kovalskaya and Hammond, 2014). Symptom expression is also influenced by viroid strain and the host, resulting in latent (no symptoms) or mild to severe infection (Kovalskaya and Hammond, 2014). In addition, symptoms of viroid infection also differ due to mixed infection of viroids as well as other pathogens in the hosts. Viroids replicate best at high temperatures (30–33°C), which together with high light intensity are known to induce symptoms rapidly.

Viroids are readily transmitted through vegetative propagation, mechanical transmission through sap, and through seeds and pollen from infected plants. The PSTVd can be transmitted via aphids, if encapsidated with Potato leafroll virus (PLRV) coat protein to escape digestion in the gut of the aphid (Querci et al., 1997). Contaminated agricultural implements still account for the major source of transmission.

Similar to viruses, the management of viroids can be mainly done through eradication of diseased plants, elimination from planting material, host resistance, disinfection of machinery, and vector control (Kovalskaya and Hammond, 2014). Since, exclusion principle is a primary important management strategy in the case of viroids, thus certification and production of healthy planting material through the identification of viroid free geno-types is essential to control the spread of viroids. A reliable and sensitive diagnostic is needed to back most management strategies, and enhancing the demand for constant development of new detection methods (Gucek et al., 2016). Moreover, a cheap and reliable method for simultaneous detection of many viroids may serve the purpose well.

2.2 DETECTION AND DIAGNOSTIC METHODS

The diagnostics for viroids vary considerably from those of viruses due to some of their characteristics viz., present in low concentrations in the plants, localized in the cell nucleolus and plastids, which is also the site of other macromolecules, making it difficult the isolation of viroids form the polysaccharides and polyphenols, and most importantly, lack of protein coat. However, a wide range of detection techniques have been standardized for the identification of viroids whose utility is dependent on the precision, sensitivity, reproducibility rapidness, simplicity, and cost-effectiveness (Narayanasamy, 2011).

2.2.1 BIOLOGICAL METHODS

Biological indexing or bioassay is the most primitive method of viroid detection developed even before physico-chemical nature of viroids was identified. It has been widely used for certification and quarantine programs for viroid detection in vegetatively propagated planting material (Kovalskaya and Hammond, 2014; Narayanasamy, 2011) as well as especially for the fulfillment of Koch's postulates. For a successful biological assay of viroids, the selection of suitable indicator hosts and recognition of characteristic symptoms induced by the viroid in the hosts must be well known (Legrand, 2015). Bioassay plays a vital role in various aspects like resistance screening, epidemiological studies and analysis of plant-pathogen interactions.

Since, majority of the viroids are mechanically transmitted, thus primary as well as secondary hosts can be inoculated by injection or rubbing the inoculum onto leaves treated with carborundum, graft inoculation, slash inoculation and biolistic approaches (Verhoeven et al., 2010). For the pathogens infecting their single respective host like Avocado sunblotch viroid and chrysanthemum chlorotic mottle viroid, their primary host, i.e., avocado and chrysanthemum, respectively, can be used as indicator host for their bioassay (Allen et al., 1981). While for others, causing latent infection not causing visible symptoms in any host such as Iresine viroid 1, hop latent viroid, coleus blumei viroids, Australian grapevine viroid, eggplant latent viroid, and dahlia latent viroid, their detection is not possible through biological indexing (Nie and Singh, 2017). Various factors are needed to be taken into account for bioassay, such as viroid strain, host species, cultivar, mixed infections of viroids and viruses, inoculation method, a variety of symptoms, and time required for symptom development (Gucek et al., 2016). The detection through biological indexing is also dependent on environmental conditions like temperature and light intensity (Verhoeven et al., 2010). Table 2.1 summarizes the primary and indicator hosts for viroids where a bioassay is available.

TABLE 2.1 Detection of Viroids Using Bioassay Approach

Viroid	Primary Hosts	Indicator Hosts	References
Apple dimple fruit viroid (ADFVd)	Apple	*Malus pumila* cvs. "Starkrimson," "Braeburn"	Di Serio et al. (2001)
Apple fruit crinkle viroid (AFCVd)	Apple	*Malus pumila* cvs. "Ohrin," "NY58_22"	Ito and Yoshida (1998)
Apple scar skin viroid (ASSVd)	Apple	*Malus pumila* cvs. "Stark's Earliest," "Sugar Crab"	Howell and Mink (1992)
Australian grapevine viroid (AGVd)	Grapevine	*Cucumis sativus*	Rezaian (1990)
Avocado sunblotch viroid (ASBVd)	Avocado	*Persea americana* cv. "Hass"	Allen et al. (1981)
Chrysanthemum chlorotic mottle viroid (CCMVd)	Chrysanthemum	Chrysanthemum cvs. "Deep Ridge," "Bonnie Jean," "Yellow Delaware"	Horst (1975)
Chrysanthemum stunt viroid (CSVd)	Chrysanthemum	Chrysanthemum cv. "Mistletoe"	Bachelier et al. (1976); Hollings and Stone (1973)
Citrus bark cracking viroid (CBCVd)	Citrus, hop	Etrog citron Arizona 861-S1	Duran-Vila and Semancik (2003); Wang et al. (2013)
Citrus dwarfing viroid (CDVd)	Citrus	Etrog citron Arizona 861-S1	Serra et al. (2008a, b)
Citrus bent leaf viroid (CBLVd)	Citrus	Etrog citron Arizona 861-S1	Serra et al. (2008a, b)
Citrus exocortis viroid (CEVd)	Citrus. tomato	Etrog citron Arizona 861-S1, *Gynura aurantiaca*	Chaffai et al. (2007); Duran-Vila et al. (1988)
Citrus viroid V (CVd)	Citrus	Etrog citron Arizona 861-S1	Serra et al. (2008a, b)
Coconut cadang-cadang viroid (CCCVd)	Coconut	*Cocos nucifera; Corypha elata; Adonidia merrillii; Areca catechu; Elaeis guineensis; Chrysalidocarpus lutescens; Oreodoxa regia*	Imperial et al. (1985)
Columnea latent viroid (CLVd)	Columnea	*Calendula officinalis*	Matsushita and Tsuda (2015)
Grapevine latent viroid (GLVd)	Grapevine	*Vitis vinifera* cv. "Kyoho"	Zhang et al. (2018)

TABLE 2.1 *(Continued)*

Viroid	Primary Hosts	Indicator Hosts	References
Hop stunt viroid (HSVd)	Hop, grape, citrus, cucumber, plum, peach	*Cucumis sativus* L. "Suuyou"; *Luffa aegyptiaca* Mill; Parson's special mandarin	Palacio-Bielsa et al. (2004); Roistacher et al. (1973); Yaguchi and Takahashi (1984)
Mexican papita viroid (MPVd)	*Solanum cardiophyllum*	*Solanum lycopersicum* cv. "Rutgers"; *Nicotiana glutinosa*	Martinez-Soriano et al. (1996); Singh and Ready (2003)
Peach latent mosaic viroid (PLMVd)	Peach	*Prunus persica* cv. "GF-305"	Desvignes (1976)
Pear blister canker viroid (PBCVd)	Pear	*Pyrus communis* cv. "A20," "Fieud 37, "Fieud 110"	Desvignes et al. (1999)
Pepper chat fruit viroid (PCFVd)	Sweet pepper	*Solanum lycopersicum* cv "Money-maker"; *S. tuberosum* cv. "Nicola"	Verhoeven et al. (2009)
Potato spindle tuber viroid (PSTVd)	Potato, tomato	*Solanum lycopersicum* cvs. "Sheyenne," "Rutgers"; *Scopolia sinensis*; *S. berthaultii*; *Datura stramonium*	Matsushita and Tsuda (2015); Singh (1973)
Tomato apical stunt viroid (TASVd)	Tomato	*Solanum lycopersicum*; *Nicotiana glutinosa*; *Nicotiana tabacum* cv. "White Burley"; *Solanum melongena*; *Calendula officinalis*	Matsushita and Tsuda (2015); Verhoeven et al. (2011); Walter (1987)
Tomato chlorotic dwarf viroid (TCDVd)	Tomato	*Solanum lycopersicum* cvs. "Sheyenne," "Rutgers"; *Nicotiana glutionsa*; *Solanum melongena*; *Calendula officinalis*	Matsushita and Tsuda (2015); Singh et al. (1999)
Tomato planta macho viroid (TPMVd)	Tomato	*Solanum lycopersicum* cv. "Rutgers"	Galindo et al. (1982); Verhoeven et al. (2011)

Source: Adapted: Nie and Singh (2017).

2.2.2   NUCLEIC ACID-BASED METHODS

2.2.2.1   ELECTROPHORETIC METHODS

2.2.2.1.1   Polyacrylamide Gel Electrophoresis (PAGE)

PAGE was the first molecular technique standardized for the detection of viroids owing to its non-requirement of nucleotide sequences for detection of viroids; it has been exploited vigorously for detection of unknown viroids (Hanold et al., 2003). It works on the principle of circularity of viroid RNA leading to low electrophoretic mobility under denaturing conditions, whereas the rod-like native structure behaving similar to a dsRNA of comparable length leading to differential migration on gel under denaturing and non-denaturing conditions. Based on this principle, various variants of PAGE have been developed such as sequential-PAGE (S-PAGE), return-PAGE (R-PAGE), two-dimensional denaturing PAGE (2D-PAGE), and temperature gradient gel electrophoresis (TGGE). After electrophoresis, viroids can be visualized by staining with ethidium bromide, toluidine blue, or silver nitrate. PAGE has been utilized for the field detection of CCCVd in the Philippines as a mobile laboratory developed by Randles in 1992.

In S-PAGE, subsequent electrophoresis is carried out, one under normal conditions (urea-free gel) and another under denaturing conditions on separate gels (using the cut gel containing viroid-sized RNAs obtained from first electrophoresis) placed on denaturing gel (urea containing). It was first described by Rivera-Bustamante et al. (1986) and has been used extensively for characterization of viroids complexes of citrus and grapevine (Duran-Vila et al., 1988). It has also been used for the conformation of viroid infectivity. Whereas, R-PAGE is based on temperature modification to achieve denaturation using the same gel first under normal conditions followed by denaturing conditions (increased temperature) by reversing the direction of electrophoresis. This method first described by Schumacher et al. (1983, 1986), has been used for the separation and identification of severe and mild strains (Singh and Boucher, 1987) as well as detection of CSVd and PSTVd (Roenhorst et al., 2000). The 2D-PAGE is based on using denaturing conditions in both the dimensions which allow identification of circular as well as linear RNAs (products of intragel cleavage of their respective circular forms. This method has been used to resolve the complex mixtures of circular and linear RNAs of CCCVd (Vadamalai et al., 2006), etc. TGGE is carried out in a horizontal polyacrylamide gel resting on an electrically insulated metal plate using temperature gradient perpendicular to the direction of electrophoresis for denaturation, which mediates the conversion of rod-like structure into open circular conformation.

PAGE is the only available technique that can be used to identify new viroids and prove the suspected viroid nature of an RNA molecule in combination with other available techniques. It is easy to perform, highly sensitive and specific. Its sensitivity can be further enhanced using silver staining method. The only loophole lies in its susceptibility to impurities, thus care has to be taken while sample preparation.

2.2.2.2   HYBRIDIZATION METHODS

Molecular hybridization is based on the formation of stable hybrid through complementary base pairing between the target nucleic acid sequence of the pathogen and the complementary labeled short sequence, i.e., the probe. As a diagnostic tool, molecular hybridization was in fact first time utilized for the detection of PSTVd (Owens and Diener, 1981). Variants of probes that can be employed are radioactive probes, non-radioactive probes, including biotin and digoxigenin (DIG) probes, single-stranded cDNA probes and riboprobes (Muhlbach et al., 2003). The non-radioactive probes have the ability to detect the target RNA molecules to fentomole level. However, nowadays RT-PCR has been taken up as a routine diagnostic protocol, yet hybridization remains a useful technique as it balances sensitivity with ease of use, time, and cost.

2.2.2.2.1   Dot-Blot Hybridization

Dot-Blot hybridization is based on the direct application of a nucleic acid onto nitrocellulose or nylon membranes, followed by hybridization with viroid-specific probes for detection. The target may either be TNS, total RNAs or purified viroid RNA, depending upon extraction and fractionation protocol, which are fixed to the membrane by baking or by UV cross-linking (5–10 fold increased sensitivity). Hybridization efficiency is dependent on the complexity (length and composition) and concentration of the probe as well as the temperature, salt concentration, base mismatches, and inclusion of hybridization accelerators. Generally, low salt concentration and high temperature increase stringency. However, it is a reliable technique, but use of crude sap extracts or partially purified samples lead to high background interference. The addition of formamide in hybridization solution also leads to enhanced stringency due to increase in correct base pairing, reduction in the background, reduced probe degradation. Good signal/background ratio is usually achieved at 70–72°C in 50% formamide. Recently, nucleic acid spot hybridization (NASH) has been standardized for the detection of PSTVd, along with two viruses at CIP (Muller et al., 2019).

2.2.2.2.2 *Gel-Blot Hybridization*

Gel-Blot, also called northern-blot hybridization (NBH), is performed by separation of RNAs on gel electrophoresis followed by blotting on nylon or nitrocellulose membrane onto which specific probes are hybridized, which can be in turn detected by radioactive, colorimetric, or chemiluminescent signals. Apart from detection, it also reveals the size of the target. Denaturing as well as nondenaturing gels can be used for detection (Murcia et al., 2009). NBH has enabled various studies related to viroid replication, rolling circle amplification (RCA), detection of PSTVd subgenomic RNAs, viroid-derived small RNAs of 21–24 nt in infected tissue (Martinez de Alba et al., 2002). NBH has been useful in discriminating variants of viroids generated due to insertions/deletions in the sequence of RNA (Malfitano et al., 2003). Recently, NBH with cDNA DIG specific probes were designed for the detection of viroids in citrus cultivars (Umana-Castro et al., 2017).

2.2.2.2.3 *Tissue-Print Hybridization*

Tissue printing or imprint hybridization is a faster and cheaper detection technique as it excludes the requirement of nucleic acid extraction. It is a minimal sample preparation technique wherein plant sap from specific tissues are transferred directly onto the membrane followed by hybridization with specific probes for detection similar to dot-blot hybridization. Its main advantage is on-field detection of viroids as well as identification of it localization within the infected tissues (Gucek et al., 2016). However, there are high chances of false negatives either due to probe-protein interaction from sap (WenXing et al., 2009) or due to low viroid titer as well as seasonal fluctuation in its titer as observed in the case of citrus. However, this could be overcome by using citron as an amplification host (Palacio et al., 2000).

2.2.2.2.4   In-Situ Hybridization (ISH)

In-situ hybridization (ISH) assay refers to the combination of molecular hybridization with coupled with electron, light or confocal laser scanning microscopy. It has played a crucial role in studying subcellular localization and trafficking of Pospiviroidae in the nucleus (Zhang et al., 2014) and Avsunviroidae in chloroplasts (Lima et al., 1994) as well as their movement through plasmodesmata (PD) and vascular tissues (Wang and Ding, 2010). *In situ* hybridization has also provided insights on the invasion of PSTVd through ovule or pollen in petunia embryo before embryogenesis (Matsushita and Tsuda, 2014). This method has also highlighted the high efficiency of somatic embryogenesis in eliminating viroids from several infected grapevine cultivars is due to embryos formed in viroid-infected calli remaining viroid-free (Gambino et al., 2011). RNAscope is a novel ISH method wherein double Z probes are allowed to hybridize with the RNA sequence for greater amplification through the formation of scaffold with possess multiple binding sites for the detection enzyme when bind to colorimetric or fluorometric substrate. RNAscope has been standardized for the detection of CEVd and CVd-II in citrus petioles, stem, and root tissues (Stanton et al., 2019). However, ISH provides efficient detection even with the use of a small amount of tissue, but finds difficulty in identification in tissues having low RNA copies.

2.2.2.2.5   Polyvalent Detection Using Polyprobes

Polyprobes (more appropriately riboprobes) for viroids are used in non-isotopic nucleic acid hybridization for simultaneous detection of more than one viroid. This not only saves time and money, but also enables for efficient utilization of resources and labor by using a cocktail of specific single probes or a unique polyprobe for detection. Polyprobes consist of several specific DIG-labeled probes obtained by transcription of cloned partial or full-length cDNA sequences of viroids. Polyprobes are particularly beneficial for large-scale screening and survey, especially for quarantine and certification programs as up to 6–8 viroids can be detected in a single reaction. In recent times, a polyprobe has been designed with the capacity to detect eighteen pathogens, including 13 viruses and 5 viroids in grapevine (Sanchez-Navarro et al., 2018) and also for detection of 12 viruses and 4 viroids infecting tomato (Sanchez-Navarro et al., 2019).

2.2.2.2.6   Ribonuclease Protection Assay (RPA)

RPA is a special type of hybridization technique, i.e., a liquid-hybridization based diagnostic protocol for RNA that employs the use of inhibitors resistant enzymes. It has been standardized for the detection of coconut cadang-cadang viroid (CCCVd) in infected coconut and African oil palm (Vadamalai et al., 2009). It is sensitive and specific technique for detection and quantification of specific RNAs as well as detection of nucleotide variations and genetic heterogeneity in sample of target RNA.

2.2.2.3   POLYMERASE CHAIN REACTION BASED METHODS

2.2.2.3.1   Reverse Transcriptase-PCR (RT-PCR)

RT-PCR is the most commonly used method for the detection of viruses and viroids due to its high sensitivity and specificity which achieved just by optimized primer designing and standardized amplification conditions. RT-PCR also offers advantages of detection of viroid variants using specific primers. RT-PCR can be carried out in a two-step (cDNA synthesis in one tube and amplification in another tube) or one-step reaction. The two steps PCR is less sensitive due to increased risk of contamination as compared to convenient one tube-one step RT-PCR. The size of total nucleic acids (TNS) of infected tissue required for the reaction is just 1–100 pg (Gucek et al., 2016). An essential requirement of RT-PCR is an extraction of highly purified TNS (free of inhibitors, polysaccharides, secondary metabolites, and polyphenols). To overcome this limitation, variants of RT-PCR, such as tissue printing of plant sap followed by RT-PCR (Weidemann and Buchta, 1998), RT-PCR-ELISA (Shamloul and Hadidi, 1999) and microtissue direct RT-PCR (Hosokawa et al., 2006) have been developed. RT-PCR has been regularly employed to prove the etiology of many diseases, including recent reports of rasta disease of tomato caused by PSTVd and tomato apical stunt viroid (TASVd), discovery of expanding host ranges of viroids such as PSTVd in seeds of *S. anguivi, S. coagulans,* and *S. dasyphyllum* (Skelton et al., 2019), reporting viroid presence in different regions such as presence of GYSVd-2 in grapevine in India (Singhal et al., 2019).

Another problem encountered in viroid detection is due to their low concentration in host tissues, to overcome which variants of RT-PCR and molecular hybridization have been developed, such as RT-PCR-DBH, RT-PCR coupled with Southern hybridization, PCR-microplate hybridization (PCR-MPH) and RT-PCR-ELISA (Gucek et al., 2016).

The other variants are *in situ* RT-PCR (Zhao and Niu, 2008) and multiplex RT-PCR (mRT-PCR) for simultaneous detection of several viroids as hosts like citrus and grapevine are known to be infected by mixtures of many viroids at one time. The mRT-PCR is a cost, labor, and time effective technique that combines multiple sets of specific primers for two or more targets in a single reaction enabling amplification and detection of different viroids in a single test (Pallas et al., 2018). The first mRT-PCR involving viroids were standardized by Nie and Singh (2001) for simultaneous detection of five potato viruses and a viroid. Recently, mRT-PCR assays have been developed for simultaneous detection of four viroids infecting hop plants (Gucek et al., 2019) and two viruses and two viroids in grapevine (Kominkova and Kominek, 2019).

2.2.2.3.2   Real-Time RT-PCR

Real-time PCR (qRT-PCR) has gained widespread adoption for the detection of pathogens due to the on-going monitoring and absolute quantification along with amplification of the targets. The qRT-PCR can be performed in two ways, i.e., using non-specific fluorescence dyes which intercalate with dsDNA emitting fluorescence in proportion to the concentration of dsDNA (Wittwer et al., 1997) or using oligonucleotide DNA probe labeled with reporter (fluorescent dye) at 5' end and quencher at 3' end, which when acted upon by *Taq* DNA polymerase are released due to degradation of the probe by 5'-3' exonuclease activity of the polymerase. The release of reporter emits fluorescence which in turn is recorded in real-time for each amplification cycle (Gibson et al., 1996). The second method works on hybridization of the probe with the complementary target sequence. Most commonly used probes for viroid detection are TaqMan and SYBR Green wherein the former enables specific hybridization of probe and target sequence, while the latter binds non-specifically to dsDNA which might lead to false positive results to avoid which melting curve analysis has to be followed after the reaction (Mirmajlessi et al., 2015). Another type of probe is scorpion probe which provide stronger signals in less reaction times enabled by the combination of probe and primer functions. The qRT-PCR is 1000-fold more sensitive as compared to hybridization protocol (Gucek et al., 2016). The main requirements for qRT-PCR include isolation of high quality RNA, proper probe designing, and calibration of thermal cycler.

The main advantage of qRT-PCR based detection of viroids is distinguishing between real replication and contaminations, especially in new host species due to detection of plus as well as minus viroid chains. It has also been employed in the fields of plant resistance, transmission, and genotyping-related study of viroids (Gucek et al., 2016). Recently, qRT-PCR assay has been developed for the detection of CEVd in sensitive and tolerant rootstocks of citrus (Ghobakhloo et al., 2019) as well as for detection of avocado sun blotch viroid (ASBVd) in avocado (Kuhn et al., 2019; Table 2.2).

TABLE 2.2 Comparison of Different Viroid Amplification Methods

Characteristics	RT-PCR	RT-LAMP	RT-RPA
Amplification	Cyclic reaction	Isothermal	Isothermal
Reaction time	~5 hrs, time-consuming	~1.5 hrs, rapid	~1 hr, rapid
Equipment required	expensive thermal cycler requires	not require	not require
Template source	Purified RNA	Purified RNA	Purified RNA as well as crude sap
Primer required	A set of primer	4–6 sets of primers	Single set of primer
Visualization	Through agarose gel electrophoresis	Through agarose gel electrophoresis or by a cheap turbidity-meter or change in color	Through agarose gel electrophoresis or fluorescence detection or lateral flow strips
Application in field diagnosis	No	Yes	Yes

Multiplex qRT-PCR have also been for simultaneous detection of viroids in various crops, especially using TaqMan-based assays, which does not require different size products and offer easier optimization. However, it also comes with a disadvantage of reduced sensitivity and specificity of viroids due to competition between the probes (Mirmajleesi et al., 2015). Recently, a multiplex qRT-PCR has been standardized for the simultaneous detection of pear blister canker viroid (PBCVd) and apple scar skin viroid (ASSVd) along with phytoplasmas in pome trees (Malandraki et al., 2019). The coupling of qRT-PCR with high-resolution melting (HRM) analysis has enabled single-nucleotide polymorphisms (SNPs) detection, which is very much essential in the case of viroids as they are highly variable with similar size and high sequence homology, which makes it difficult to distinguish among them. Using HRM along with EVA green dye or similar has enabled the detection of multiple mutations in a single assay, reducing the time and cost of analysis (Loconsole et al., 2013).

2.2.2.4 POINT-OF-CARE BASED DIAGNOSTIC METHODS

2.2.2.4.1 Reverse Transcriptase-Loop Mediated Isothermal Amplification (RT-LAMP)

RT-LAMP is a fast, specific, sensitive, and robust detection method that can be used for on-site diagnosis (Notomi et al., 2000). The high specificity is the resultant of usage of a set of four to six primers, annealing at different regions on the target genome (six to eight target sequences) via strand displacement DNA synthesis with Bst DNA polymerase giving results in less than 1 hour (Gucek et al., 2016). It works under isothermal conditions at temperatures ranging between 60°C and 65°C. Results can be interpreted via naked eye or UV light or by observing the amplification curves using a portable fluorometer (Gucek et al., 2016).

RT-LAMP has been used for the detection of viroids such as ASSVd (Lee et al., 2018) and columnea latent viroid (CLVd) (Bhuvitarkorn et al., 2019). Multiplex RT-LAMP was standardized by Liu et al. (2014) for simultaneous detection of chrysanthemum stunt viroid (CSVd) and chrysanthemum virus B (CVB). RT-LAMP can be standardized for the on-field diagnosis of viroids facilitating the early detection and quick management as well as prevention of disease spread. The limitation of RT-LAMP lies in the complexity of primer designing, risk of contamination in sample preparation and it cannot be multiplexed for multiple pathogen detection.

2.2.2.4.2 Isothermal and Chimeric Primer-Initiated Amplification of Nucleic Acids (ICAN)

ICAN is an on-site nucleic acid amplification technique using a pair of chimeric 5'-DNA-RNA-3'primers under isothermal conditions at a constant temperature of 55°C along with thermostable RNaseH and a DNA polymerase having strand displacing activity. It can also be used for real-time detection on combination with a cycling probe by measuring the intensity of the fluorescent signal produced due to hybridization and cleavage of the probe that increases in intensity with the increase in the amount of amplified fragment (Owens et al., 2012; Mukai et al., 2007). Till date, ICAN has only been used for the detection of CSVd in chrysanthemum, giving results in 2 hrs (Mukai et al., 2007).

2.2.2.4.3   *Reverse Transcriptase-Recombinase Polymerase Amplification (RT-RPA)*

Recombinase polymerase amplification (RPA) is a single tube, isothermal alternative to the PCR developed by TwistAmp™. RPA is fast, highly specific and can detect single copies of DNA and 10 copies/fewer RNA in complex samples, without the need for prior nucleic acid purification or a thermal cycler. It operates at low temperatures of 37–42°C (room temperature conditions) as compared to the high temperatures required in case of RT-LAMP and also requires only one specific primer set for amplification and thus offers an extremely lucrative approach for on-field diagnostics. Sample preparation is easy by crushing in specific extraction buffers or using purified RNA which can be added to the provided reaction master mix consisting of recombinase, polymerase, and single stranded DNA binding (SSB) protein along with primer followed by incubation in dry bath or room temperature for 30 mins. RPA results can be interpreted either via gel electrophoresis or through real-time by using a fluorescent probe. RPA can also be multiplexed for the simultaneous detection of many viruses and viroids. RT-RPA has been employed for the detection of tomato chlorotic dwarf viroid (TCDVd) (Hammond and Zhang, 2016) and HSVd in hop plants (Kappagantu et al., 2017) while it has been standardized for a large number of viruses including on-field detection. RPA serves as a good alternative detection method in resource constraint laboratories as well as at quarantine stations for rapid detection without the requirement of skilled labor. Table 2.1 summarizes the comparison of various molecular methods for viroid detection.

2.2.2.5   *MICROARRAY AND MACROARRAY*

Microarray and macroarray are multi pathogen diagnostic protocols that work on the principle of hybridization by base pairing between complementary sequences of the target and specific probes previously immobilized on an array which is generally a microscopic slide. The two types of arrays are distinguished on the basis of size of sample spots wherein microarray contains spot sizes of less than 200 μm, while macroarray, contain spot sizes of 300 μm or more. Such arrays are designed using the nucleotide sequence data of viroids available in public databases for synthesizing oligonucleotide probes followed by their printing on the slides allowing the simultaneous detection via fluorescence, colorimetric, or electric currents to detect hybridization with the target. A probe length of 40–50 nt is apt for detection of viroids due to their small size providing consistent hybridization and high sensitivity. Microarrays can also be standardized for the discovery of new viroids of unknown etiology by adding elements obtained from highly conserved regions within viroid families, genera, or species. It is the most useful technique for discriminating and acquiring knowledge on the various pathogens and their strains involved in causing mixed infections in plant. The only limitation in use of microarray for routine diagnosis of pathogens is due to its complexity and costliness.

Macroarray has only been developed for the detection of 11 potato viruses and PSTVd (Agindotan and Perry, 2008) in case of viroids while microarray has been developed for the detection of 37 reported viroids from all the 8 viroid genera (Zhang et al., 2013). The other improved variants of microarray technology have also come up for viroid detection, such as CombiMatrix (Irvine, USA) array which a microchip circuit-based detection system instead of the complex glass slide system, allowing screening of six pospiviroid species, i.e., TASVd, PSTVd, TCDVd, tomato planta macho viroid (TPMVd), citrus exocortis viroid (CEVd), and CLVd and hybridization of multiple samples simultaneously (Tiberini and Barba, 2012).

A microsphere (beads) based flow cytometric system was developed by Luminex (Austin, TX, United States) that incorporates polystyrene microspheres internally dyed with two spectrally distinct fluorochromes precoupled with oligonucelotide sequence. Thus, different microspheres can be distinguished on the basis of their spectral addresses and any number of beads can be combined to form an array for simultaneous detection of up to 150 targets. To determine the presence or absence of a particular pathogen, a threshold or a cut-off value for positive is set within an array. The LUMINEX bead-based 11-plex array has been applied for detection of nine pospiviroids (van Brunschot et al., 2014). Another simplified version of the microarray is using array tubes which are microcentrifuge tubes consisting of 200–400 oligonucleotide based glass array mounted in the bottom. The PCR amplified and biotin-labeled targets are added to array tube and allowed for hybridization followed by interpretation of results through the scanner. It is highly specific, sensitive, and easy approach used for the detection of various pathogens, including viruses, but yet to be used for the detection of viroids (Adams et al., 2015).

Another microarray-based approach enabling the simultaneous detection of both DNA as well as RNA pathogens developed is isothermal nucleic acid sequence-based amplification (NASBA) implemented microarray analysis (NAIMA), which is a fast, reliable, and sensitive detection technique enabling multiplexing. It is another on-site applicable method for detection due to isothermal amplification and no requirement of sophisticated equipments. NASBA was first time developed for detection of HSVd, CEVd and citrus bark cracking viroid (CBCVd) in citrus using Inosine 5'-triphosphate during amplification for stabilization of secondary structure and in turn to increase yield in NASBA (Nakahara et al., 1998). Following this NASBA was combined with electro-chemiluminescence (ECL) detection to develop diagnostic for the ASSVd which proved to be faster and 100-fold more sensitive than RT-PCR for ASSVd detection in apple leaves and also bark tissue on adding polyvinylpyrrolidone in the extraction buffer (Kim et al., 2004). The ASSVd detection using NASBA was further improvised in combination with real-time amplication using molecular beacons (Heo et al., 2019).

2.2.2.6   NEXT GENERATION SEQUENCING (NGS) METHODS

NGS also referred to as deep sequencing or high throughput sequencing (HTS) refers to the method of generation of sequence of nucleotides present in the genome of the target pathogen causing specific disease in the considered sample. It is a robust, reliable, and highly sensitive technique efficient in distinguishing between the strains of pathogens and also identifying the single nucleotide polymorphisms (SNPs) in a pathogen. It has become very useful in the identification of causes of diseases of unknown etiology, especially of viruses and viroids owing to its high accuracy in the sequencing of small genomes of these pathogens. Various NGS platforms have come up since its first technology in 2004, with reduced cost and complexity differing principle and the purpose they serve. The selection of the kind of platform to be used for sequencing is dependent on the size and complexity of the genome to be studied as well as the accuracy and depth of coverage needed. Illumina and SOLiD platforms are considered apt for RNA sequencing projects requiring better coverage with high accuracy at reduced cost (Radford et al., 2012). The other NGS platforms are Roche 454, PacBio, Ion torrent, and nanopore.

NGS offers a fast and cheap approach for multiplexing and identifying how many different viruses and viroids are present in a crop plant which has especially proved useful for crops like citrus and grapevine, which are now known to be simultaneously infected by a large number of viruses and viroids. Plants infected by viroids contain 21–24 nt viroid-derived small RNAs (vd-sRNAs) processed into siRNAs, which can accurately sequenced in single run followed by re-assembling to obtain viroid genome followed by analysis via computational algorithm to perform homology independent discovery of replicating circular RNAs, called PFOR (progressive filtering of overlapping small RNAs) (Wu et al., 2012). Another improvisation of PFOR, led to PFOR2 algorithm which was found useful in the discovery of a new grapevine viroid, i.e., grapevine latent viroid (GLVd) (Zhang et al., 2014). The NGS for characterization of vd-sRNAs was utilized the first time for the detection of two variants of a peach viroid (Di Serio et al., 2009) and two viroids of grapevine (Navarro et al., 2009). Since then, NGS of small RNA libraries has been deeply used for further dissecting the molecular interplay between viroids and their hosts in both herbaceous and woody hosts.

An alternative to vd-sRNAs is the use of replication intermediate of viroids which is a double stranded RNA generated by host RDRs for viroid detection through NGS, such as detection of a new apscaviroid infecting persimmon (Ito et al., 2013), thus confirming the potentiality of this approach in viroid identification as well for detection such as that of CEVd, HSVd, and citrus dwarfing viroid (CDVd) in citrus (Navarro and Di Serio, 2018). Recently, total RNA has been used for the detection of new viroids such as coleus blumei viroid 6 in *Coleus blumei* (Smith et al., 2018) and apple chlorotic fruit spot viroid (ACFSVd) in apple (Leichtfried et al., 2019). Apart from the detection and characterization of viroids, the NGS has been applied in various other fields related to viroids such as their movement within the host, other host interactions, pathogenesis, mutation, evolution, mRNA targeting, extending host range.

NGS is another approach apart from PAGE, which can be used for the identification of unknown viroids and their strains without prior knowledge of the sequence. However, as compared to PAGE, it is more sensitive, specific, and fast approach, efficient even for identification of viroids in woody perennial crops having low-titer of viroids. With the development of new platforms cost incurred can be further reduced for sequencing. NGS can serve as a lucrative diagnostic tool in quarantine and certification programs as well as elimination of viroids from the hosts.

2.3   CONCLUSIONS

Detection and management of plant pathogens go hand in hand, thus fast, efficient, and accurate detection method is of prime importance for adapting apt management approach as well as preventing their spread to other areas. However, the advent of NGS has paved the way for detection and discovery of new pathogens, but their confirmation through conventional techniques such as PAGE in case of viroids is essential. Amongst, all the nucleic acid-based approaches; conventional RT-PCR is the most regularly used technique for the detection of single pathogen. However, in case of detection of large number of pathogens, hybridization techniques prove advantageous. However, qRT-PCR is considered best due to its increased sensitivity with greater detection limit and speedy detection with the closed tube format reduces the risk of contamination. Since, most crops are known to be infected by more than one viroid thus, detection techniques offering simultaneous detection of multiple viroids such as mRT-PCR, mRT-qPCR, microarrays or NAIMA, are not only cost effective but also saves time and resources. The most recent advancement in the field of diagnostics is the isothermal-based point-of-care based approaches such as RT-LAMP, RT-RPA, and ICAN for the detection of viroids. These point-of-care methods are not only fast and sensitive, but also enable the detection to be carried out in resource constrain laboratories as well as on-site detection thus facilitating the prevent of disease spread in the field and quality testing of samples in quarantine and certification programs.

KEYWORDS

- **apple chlorotic fruit spot viroid**
- **avocado sun blotch viroid**
- **central conserved region**
- **polymerase**
- **sequencing**
- **viroids**

REFERENCES

Adams, I., Harrison, C., Tomlinson, J., & Boonham, N., (2015). Microarray platform for the detection of a range of plant viruses and viroids. In: Lacomme, C., (ed.), *Plant Pathology: Methods Mol. Biol.* (p. 1308). Humana Press, New York, NY.

Agindotan, B., & Perry, K. L., (2008). Macroarray detection of eleven potato-infecting viruses and potato spindle tuber viroid. *Plant Dis., 92*, 730–740.

Allen, R. N., Palukaitis, P., & Symons, R. H., (1981). Purified avocado sunblotch viroid causes disease in avocado seedlings. *Australas. Plant Pathol., 10*, 31–32.

Bhuvitarkorn, S., Klinkong, S., & Reanwarakorn, K., (2019). Enhancing columnea latent viroid detection using reverse transcription loop-mediated isothermal amplification (RT-LAMP). *Int. J. Agric. Tech., 15*(2), 215–228.

Di Serio, F., Gisel, A., Navarro, B., Delgado, S., Martı́nez, D. A. A. E., Donvito, G., et al., (2004). Deep sequencing of the small RNAs derived from two symptomatic variants of a chloroplast viroid: Implications for their genesis and for pathogenesis. *PLoS One, 4*, 7539.

Diener, T. O., & Smith, D. R., (1971). Potato spindle tuber viroid. VI. Monodisperse distribution after electrophoresis in 20% polyacrylamide gels. *Virology, 46*, 498, 499.

Diener, T. O., (1983). Viroids. *Adv. Virus Research, 28*, 241–283.

Diener, T. O., (1991). The fronteir of life: The viroids and viroid-like satellite RNAs. In: Maramorosch, K., (ed.), *Viroids and Satellites: Molecular Parasites at the Frontiers of Life*. CRC Press, Boca Raton, Florida, USA.

Duran-Vila, N., Juarez, J., & Arregui, J. M., (1988). Production of viroid-free grapevines by shoot tip culture. *American Journal of Enology and Viticulture, 39*, 217–220.

Flores, R., (1997). Viroids: The non-coding genomes. *Seminars in Virology, 8*, 65–73.

Gambino, G., Navarro, B., Vallania, R., Gribaudo, I., & Di Serio, F., (2011). Somatic embryo-genesis efficiently eliminates viroid infections from grapevines. *Eur. J. Plant Pathol., 130*, 511–519.

Ghobakhloo, M., Dizadji, A., & Yamchi, A., (2019). A real-time PCR assay for detection and absolute quantitation of *citrus exocortis viroid* in two sensitive and tolerant rootstocks. *Crop Prot.*, 27–30.

Gibson, U. E., Heid, C. A., & Williams, P. M., (1996). A novel method for real time quantitative RT-PCR. *Genome Res., 6*, 995–1001.

Gucek, T., Jakse, J., Matousek, J., & Radisek, S., (2019). One-step multiplex RT-PCR for simultaneous detection of four viroids from hop (*Humulus lupulus* L.). *Eur. J. Plant Pathol., 154*, 273.

Guceka, T., Trdanb, S., Jakseb, J., Javornikb, B., Matousekc, J., & Radiseka, S., (2017). Diagnostic technique if viroids. *Plant Pathol., 66*, 339–358.

Hammond, R. W., & Owens, R. A., (2006). Viroids: New and continuing risks for horticultural and agricultural crops. In: *APSnet Features*. https://doi.org./10.1094/APSnetFeature-2006-1106.

Hammond, R. W., & Zhang, S., (2016). Development of a rapid diagnostic assay for the detection of tomato chlorotic dwarf viroid based on isothermal reverse-transcription-recombinase polymerase amplification. *J. Virol. Methods., 236*, 62–67.

Hanold, D., Semancik, J. S., & Owens, R. A., (2003). Polyacrylamide gel electrophoresis. In: Hadidi, A., Flores, R., Randles, J., & Semancik, J., (eds.), *Viroids* (pp. 95–102). CSIRO Publishing. Collingwood, Australia.

Heo, S., Kim, H. R., & Lee, H. J., (2019). Development of a quantitative real-time nucleic acid sequence based amplification (NASBA) assay for early detection of apple scar skin viroid. *Plant Pathol. J., 35*(2), 164–171.

Hosokawa, M., Matsushita, Y., Uchida, H., & Yazawa, S., (2000). Direct RT-PCR method for detecting two chrysanthemum viroids using minimal amounts of plant tissue. *J. Virol. Methods, 131*, 28–333.

Ito, T., Suzaki, K., Nakano, M., & Sato, A., (2013). Characterization of a new apsca viroid from American persimmon. *Arch. Virol., 158*, 2629–2631.

Kappagantu, M., Villamor, E. V., Bullock, J. M., & Eastwell, K. C., (2017). A rapid isothermal assay for the detection of hop stunt viroid in hop plants (*Humulus lupulus*), and its application in disease surveys. *J. Virol. Methods., 245,* 81–85.

Kim, H. R., Lee, S. H., Kim, J. S., Yiem, M. S., Park, J. W., Kim, C. K., Chung, J. D., & Shin, Y. B., (2004). Detection of apple scar skin Viroid in apple trees by the isothermal nucleic acid amplification and electrochemiluminescence. *Acta Hortic., 657,* 361–366.

Koltunow, A. M., & Rezaian, M. A., (1989). Grapevine viroid 1B, a new member of the apple scar skin viroid group contains the left terminal region of tomato planta macho viroid. *Virology, 170,* 575–578.

Kominkova, M., & Kominek, P., (2019). Development and validation of RT-PCR multiplex detection of grapevine viruses and viroids in the Czech republic. *J. Plant Pathol.* https://doi.org/10.1007/s42161-019-00452-x.

Kovalskaya, N., & Hammond, R. W., (2014). Molecular biology of viroid-host interactions and disease control strategies. *Plant Sci., 228,* 48–60.

Kuhn, D. N., Freeman, B., Geering, A., & Chambers, A. H., (2019). A highly sensitive method to detect avocado sunblotch viroid for the maintenance of infection-free avocado germplasm collections. *Viruses, 11*(6), 512.

Lee, S. H., Ahn, G., Kim, M. S., Jeong, O. C., Lee, J. H., Kwon, H. G., Kim, Y. H., & Ahn, J. Y., (2018). Poly-adenine-coupled LAMP barcoding to detect apple scar skin viroid. *ACS Comb. Sci., 20*(8), 472–481.

Legrand, P., (2015). Biological assays for plant viruses and other graft-transmissible pathogens diagnoses: A review. *EPPO Bull., 45,* 240–251.

Leichtfried, T., Dobrovolny, S., Reisenzein, H., et al., (2019). Apple chlorotic fruit spot viroid: A putative new pathogenic viroid on apple characterized by next-generation sequencing. *Arch Virol., 164,* 3137–3140.

Lima, M. I., Fonseca, M. E., Flores, R., & Kitajima, E. W., (1994). Detection of avocado sunblotch viroid in chloroplasts of avocado leaves by *in situ* hybridization. *Arch Virol., 138,* 385–390.

Liu, X. L., Zhao, X. T., Muhammad, I., Ge, B. B., & Hong, B., (2014). Multiplex reverse transcription loop-mediated isothermal amplification for the simultaneous detection of CBV and CSVd in chrysanthemum. *J. Virol. Methods, 210,* 26–31.

Loconsole, G., Onelge, N., Yokomi, R. K., Kubaa, R. A., Savino, V., & Saponari, M., (2013). Rapid differentiation of citrus hop stunt viroid variants by real-time RT-PCR and high-resolution melting analysis. *Mol. Cell Probes., 27,* 221–229.

Malandraki, I., Varveri, C., & Vassilakos, N., (2019) One-step multiplex quantitative RT-PCR for the simultaneous detection of viroids and phytoplasmas. In: Musetti, R., & Pagliari, L., (eds.), *Phytoplasmas*: *Methods Mol. Biol.* (p. 1875). Human Press, New York, NY.

Malfitano, M., Di Serio, F., Covelli, L., Ragozzino, A., Herna´ndez, C., & Flores, R., (2003). Peach latent mosaic viroid variants inducing peach calico (extreme chlorosis) contain a characteristic insertion that is responsible for this symptomatology. *Virology, 313,* 492–501.

Martinez, D. A. A. E., Flores, R., & Hernandez, C., (2002). Two chloroplastic viroids induce the accumulation of small RNAs associated with post transcriptional gene silencing. *J. Virol., 76,* 13094–13096.

Matsushita, Y., & Tsuda, S., (2014). Distribution of potato spindle tuber viroid in reproductive organs of petunia during its developmental stages. *Phytopathology, 104,* 964–969.

Mirmajlessi, S., Loit, E., Mand, M., & Mansouripour, S., (2015). Real-time PCR applied to study on plant pathogens: Potential applications in diagnosis: A review. *Plant Protect. Sci., 51,* 177–190.

Muhlbach, H. P., Weber, U., Gomez, G., Pallas, V., Duran-Vila, N., & Hadidi, A., (2003). Molecular hybridization. In: Hadidi, A., Flores, R., Randles, J. W., & Semancik, J., (eds.), *Viroids* (pp. 103–114). (Collingwood: CSIRO).

Mukai, H., Uemori, T., Takeda, O., et al., (2007). Highly efficient isothermal DNA amplification system using three elements of 5'-DNA-RNA-3' chimeric primers, RNaseH and strand-displacing DNA polymerase. *J. Biochem., 142,* 273–281.

Muller, G., (2019). *HQU-Nucleic Acid Spot Hybridization (NASH) for the Detection of PSTVd, PVT and PYVV* (Vol. 2, pp. 1–18). CIP-SOP020.

Murcia, N., Serra, P., Olmos, A., & Duran-Vila, N., (2009). A novel hybridization approach for detection of citrus viroids. *Mol. Cell Probes., 23,* 95–102.

Nakahara, K., Hataya, T., & Uyeda, I., (1998). Inosine 5'-triphosphate can dramatically increase the yield of NASBA products targeting GC-rich and intramolecular base-paired viroid RNA. *Nucleic Acids Res., 26,* 1854–1856.

Narayanasamy, P., (2011). Diagnosis of viral and viroid diseases of plants. In: *Microbial Plant Pathogens-Detection and Disease Diagnosis: Viral and Viroid Pathogens* (Vol. 3, pp. 295–312).

Navarro, B., & Di Serio, F., (2018). Double-stranded RNA-enriched preparations to identify viroids by next-generation sequencing. *Viral Metagenomics, 37*–43.

Navarro, B., Pantaleo, V., Gisel, A., Moxon, S., Dalmay, T., Bistray, G., et al., (2009). Deep sequencing of viroid-derived small RNAs from grapevine provides new insight on the role of RNA silencing in plant-viroid interaction. *PLoS One, 4,* 7686.

Nie, X., & Singh, R. P., (2001). A novel usage of random primers for multiplex RT-PCR detection of virus and viroid in aphids, leaves, and tubers. *J. Virol. Methods, 91,* 37–49.

Nie, X., & Singh, R. P., (2017). Viroid detection and identification by bioassay. In: Hadidi, A., et al., (eds.), *Viroids and Satellites* (p. 347). Academic Press.

Notomi, T., Okayama, H., Masubuchi, H., Yonekawa, T., Watanabe, K., Amino, N., & Hase, T., (2000). Loop-mediated isothermal amplification of DNA. *Nucleic Acids Res., 28,* 63.

Owens, R. A., & Diener, T. O., (1981). Sensitive and rapid diagnosis of potato spindle tuber viroid disease by nucleic acid hybridization. *Science, 213,* 670–672.

Owens, R. A., Sano, T., & Duran-Vila, N., (2012). Plant viroids: Isolation, characterization/ detection, and analysis. *Methods Mol. Biol., 894,* 253–271.

Palacio, A., Foissac, X., & Duran-Vila, N., (2000). Indexing of citrus viroids by imprint hybridization. *Eur. J. Plant Pathol., 105,* 897–903.

Pallas, V., Sanchez-Navarro, J. A., & James, D., (2018). Recent advances on the multiplex molecular detection of plant viruses and viroids. *From Microbiol., 9,* 2087.

Puchta, H., Ramm, K., & Sänger, H. L., (1988). Nucleotide sequence of hop stunt viroid isolate from the German grapevine cultivar Riesling. *Nucleic Acids Res., 16,* 2730–2730.

Querci, M., Owens, R. A., Bartolini, I., Lazarte, V., & Salazar, L. F., (1997). Evidence for heterologous encapsidation of potato spindle tuber viroid in particles of potato leafroll virus. *J. Gen. Virol., 78,* 1207–1211.

Radford, A. D., Chapman, D., Dixon, L., Chantrey, J., Darby, A. C., & Hall, N., (2012). Application of next generation sequencing technologies in virology. *J. Gen. Virol., 93,* 1853–1868.

Randles, J. W., Hanold, D., Pacumbaba, E. P., & Rodriguez, M. J. B., (1992). Cadang-cadang disease of coconut palm. In: Mukhopadhyay, A. N., Kumar, J., Chaube, H. S., & Singh, U. S., (eds.), *Plant Diseases of International Importance* (Vol. 4, pp. 277–295). Prentice-Hall, Englewood Cliffs, NJ.

Rivera-Bustamante, R. F., Gin, R., & Semancik, J. S., (1986). Enhanced resolution of circular and linear molecular forms of viroid and viroid-like RNA by electrophoresis in a discontinuous pH system. *Anal. Biochem., 156*, 91–95.

Roenhorst, J. W., Butôt, R. P. T., Van, D. H. K. A., Hooftman, M., & Van, Z. A., (2000). Detection of chrysanthemum stunt viroid and potato spindle tuber viroid by return polyacrylamide gel electrophoresis. *EPPO Bull., 30*, 453–456.

Sanchez-Navarro, J. A., Corachán, L., Font, I., et al., (2019). Polyvalent detection of twelve viruses and four viroids affecting tomato by using a unique polyprobe. *Eur. J. Plant Pathol., 155*, 361.

Sanchez-Navarro, J. A., Fiore, N., Zarnorano, A., Fajardo, T. V. M., & Pallás, V., (2018). Simultaneous detection of the 13 viruses and 5 viroids affecting grapevine by molecular hybridization using a unique probe or 'polyprobe'. *Proceedings of the 19th Congress of ICVG*. Santiago, Chile.

Schumacher, J., Meyer, N., Riesner, D., & Weidemann, H. L., (1986). Diagnostic procedure for detection of viroids and viruses with circular RNAs by "return"-gel electrophoresis. *J. Phytopathol., 155*, 332–343.

Schumacher, J., Randles, J. W., & Riesner, D., (1983). A two-dimensional electrophoretic technique for the detection of circular viroids and virusoids. *Anal. Biochem., 135*, 288–295.

Shamloul, A. M., & Hadidi, A., (1999). Sensitive detection of potato spindle tuber and temperate fruit tree viroids by reverse transcription-polymerase chain reaction-probe capture hybridization. *J. Virol. Methods, 80*, 145–155.

Singh, R. P., & Boucher, A., (1987). Electrophoretic separation of a severe from mild strains of potato spindle tuber viroid. *Phytopathology, 77*, 1588–1591.

Singhal, P., Kapoor, R., Saritha, R. K., & Baranwal, V. K., (2019). First report of grapevine yellow speckle viroid-2 infecting grapevine (*Vitis vinifera*) in India. *Plant Dis., 103*, 166.

Skelton, A., Buxton-Kirk, A., Fowkes, A., Harju, V., et al., (2019). Potato spindle tuber viroid detected in seed of uncultivated *Solanum anguivi, S. coagulans*, and *S. dasyphyllum* collected from Ghana, Kenya and Uganda. *New Dis. Rep., 39*, 23.

Smith, R. L., Lawrence, J., Shukla, M., Singh, M., Li, X., Xu, H., Chen, D., Gardner, K., & Nie, N., (*2018*). Occurrence of *Coleus blumei viroid 6* in commercial *Coleus blumei* in Canada: The first report outside of China. *Plant Dis., 103*, 782.

Stanton, D. S., Harper, S. J., & Brlansky, R. H., (2019). *Using RNAscope as a Diagnostic Tool to Identify Two Citrus Viroid in Plant Tissues*. Annual Meeting, Tampa, FL. Society of Integrative and Comparative Biology.

Tiberini, A., & Barba, M., (2012). Optimization and improvement of oligonucleotide microarray-based detection of tomato viruses and pospiviroids. *J. Virol. Methods., 185*, 43–51.

Umana-Castro, R., Pritsch, C., Molina-Bravo, R., & Pagliano, G., (2017). Diagnostic parameters of northern blot hybridization technique for detection of citrus viroids in field-grown plants. *Asian J. Plant Pathol., 11*, 71–80.

Vadamalai, G., Hanold, D., Rezaian, M. A., & Randles, J. W., (2006). Variants of coconut cadang-cadang viroid isolated from an African oil palm (*Elaies guineensis* Jacq.) in Malaysia. *Arch., 151*, 1447–1456.

Vadamalai, G., Perera, A., Hanold, D., Rezaian, M., & Randles, J., (2009). Detection of coconut cadang-cadang viroid sequences in oil and coconut palm by ribonuclease protection assay. *Ann. Appl. Biol., 154*, 117–125.

Van, B. S. L., Bergervoet, J. H. W., Pagendam, D. E., De Weerdt, M., Geering, A. D. W., & Drenth, A., (2014). Development of a multiplexed bead-based suspension array for the detection and discrimination of pospiviroid plant pathogens. *PLoS One*, *9*, 84743. doi: 10.1371/journal.pone.0084743.

Verhoeven, J. T. J., Hüner, L., Virscek, M. M., Mavric, P. I., & Roenhorst, J. W., (2010). Mechanical transmission of potato spindle tuber viroid between plants of *Brugmansia suaveloes*, *Solanum jasminoides* and potatoes and tomatoes. *Eur. J. of Plant Pathol.*, *128*, 417–421.

Wang, Y., & Ding, B., (2010). Viroids: Small probes for exploring the vast universe of RNA trafficking in plants. *J. Integr. Plant Biol.*, *52*, 28–39.

Weidemann, H., & Buchta, U., (1998). A simple and rapid method for the detection of potato spindle tuber viroid (PSTVd) by RT-PCR. *Potato Res.*, *41*, 1–8.

WenXing, X., Ni, H., QiuTing, J., Farooq, A. B. U., ZeQiong, W., & YanSu, S., (2009). Probe binding to host proteins: A cause for false positive signals in viroid detection by tissue hybridization. *Virus Res.*, *145*, 26–30.

Wittwer, C. T., Herrmann, M. G., Moss, A. A., & Rasmussen, R. P., (1997). Continuous fluorescence monitoring of rapid cycle DNA amplification. *BioTechniques*, *22*, 130–138.

Wu, Q., Wang, Y., & Cao, M., (2012). Homology-independent discovery of replicating pathogenic circular RNAs by deep sequencing and a new computational algorithm. *PNAS USA*, *109*, 3938–3943.

Zhang, Y., Yin, J., & Jiang, D., (2013). A universal oligonucleotide microarray with a minimal number of probes for the detection and identification of viroids at the genus level. *PLoS One*, *8*, 64474.

Zhang, Z., Lee, Y., Spetz, C., Clarke, J. L., Wang, Q., & Blystad, D. R., (2014). Invasion of shoot apical meristems by chrysanthemum stunt viroid differs among Argyranthemum cultivars. *Front Plant Sci.*, *6*, 53.

Zhang, Z., Qi, S., Tang, N., Zhang, X., Chen, S., & Zhu, P., (2014). Discovery of replicating circular RNAs by RNA-seq and computational algorithms. *PLoS Pathog.*, *10*, e1004553.

Zhao, Y., & Niu, J., (2008). Apricot is a new host of apple scar skin viroid. *Australas Plant Dis Notes*, *3*, 98–100.

CHAPTER 3

CRISPR/Cas9 Genome Editing in Plants for Virus Resistance: Opportunities and Challenges

VIDYA R. HINGE,[1] SANDEEP. P. KALE,[2] and RAHUL L. CHAVHAN[1]

[1]*Assistant Professor, Department of Plant Biotechnology, Vilasrao Deshmukh College of Agricultural Biotechnology, Latur; Vasantrao Naik Marathwada Krishi Vidyapeeth (VNMKV) Parbhani–431 402, Maharashtra, India, E-mail: rlchavhan@gmail.com (R. L. Chavhan)*

[2]*Senior Research Fellow, USAID-BIRAC International Project, School of Life Sciences, S.R.T.M.U., Nanded, Maharashtra, India*

ABSTRACT

Among the several biotic stresses affecting agricultural crop produce, plant viruses are becoming the foremost concern which causes major crop yield losses globally and illustrate serious threats to the food security. Thus, induction of resistance against plant viruses is becoming crucial for effective management of plant virus diseases. In the present context of the climate change scenario, plant viruses are speedily evolving and overcoming host resistance in a limited period. A newly emerging CRISPR/Cas9 (clustered regularly interspaced palindromic repeats/CRISPR-associated 9) system has been recognized as promising tool for making plant resistance for viruses. This is widely favored due to possessing higher object specificity, efficient, simple, and reproducible. The virus resistance is based upon the principle of cleavage or alteration of plant genome resulting increase in plant innate immunity. In the present book chapter, we have addressed biology of the CRISPR/Cas9 system, plant immunity against plant viruses and Use of CRISPR/Cas9 system to engineer virus resistance in plants. Further, lacunas and promising challenges on utility of CRISPR/Cas9-based plant virus resistance development are also discussed in brief.

3.1   INTRODUCTION

Plant viruses are very tiny, obligatory intracellular parasites made up of nucleic acid, DNA or RNA coated with protective covering, called capsid (coat protein). After fungi, plant viruses are the most prominent plant pathogen causes 60 billion dollars losses per year worldwide in both agricultural and horticultural crops (Karavina et al., 2017). Tobacco mosaic virus (TMV) is the first plant virus reported to cause Tobacco mosaic disease. International Committee on Taxonomy of Viruses (ICTV) has classified viruses into 7 orders, 111 families, 30 subfamilies, 610 genera, and 3705 species. According to this classification around 1407 plant virus species are categorized into 73 genera and 49 families (Adams et al., 2015). Furthermore, Baltimore (1971) has classified these viruses into seven classes based upon the type of viral genome and mRNA synthesis:

1. **Class I: dsDNA Viruses:** mRNA is synthesized normally using –ve strand as template.
2. **Class II: ssDNA Viruses:** mRNA is synthesized by double stranded DNA intermediate.
3. **Class III: dsRNA Viruses:** mRNA is synthesized by complementary strand (template strand).
4. **Class IV: ssRNA Viruses:** RNA directly functions as mRNA.
5. **Class V: Sense (–) ssRNA Viruses:** mRNA is synthesized by synthesis of +ve strand.
6. **Class VI: (+) Strand RNA Viruses:** Reverse transcriptase (RT) enzyme constitutes the viral genome.
7. **Class VII:** Theses are DNA viruses which are reverse transcribed with RNA intermediates.

Moreover, viruses are classified on the basis of genomic composition which pertains double-stranded DNA (dsDNA), single-stranded DNA (ssDNA), ssDNA(–), ssDNA(+), ssDNA(+/-), dsDNA-RT, ssRNA-RT, dsRNA, ssRNA(–), ssRNA(+), ssRNA(–/+) and viroid (Gaur et al., 2016). Economical important plant viruses are belongs to the groups of ssRNA(+), dsDNA, and ssDNA viruses. The ssRNA(+) genome viruses are being designated as most devastating plant viruses which comprises important families namely Bromoviridae [e.g., *Brome mosaic virus* (BMV), *Alfalfa mosaic virus* (AMV), *Cucumber mosaic virus* (CMV), *Tobacco streak virus* (TSV)], Closteroviridae [e.g., *Citrus Tristeza clostero virus* (CTV), *Beet yellow virus* (BYV), *Lettuce infectious yellows virus* (LIYV), *Grapevine leafroll associated virus* 3 (GLRaV-3)], Luteoviridae [e.g., *Barley yellow dwarf virus* (BYD), *Potato leaf roll virus* (PLRV), *Pea enations mosaic virus*-1 (PEMV-1)], Potyviridae [e.g., *plum pox poty virus* (PPV), *Potato virus Y* (PVY), *Blackberry virus Y* (BVY), *Wheat streak mosaic virus* (WSMV), *Barley yellow mosaic virus* (BYMV), *Sweet potato mild mottle virus* (SPMMV), and Tombusviridae [e.g., *Tobacco necrosis virus* (TNV), *Maize chlorotic mottle virus* (MCMV), *Tomato bushy stunt virus* (TBSV), *Carnation mottle virus Necrovirus* (CarMV)]. Whereas, few plant viruses whose genome is made up by negative-sense(–)ssRNA were represented by Bunyaviridae [e.g., *Tomato spotted wilt virus* (TSWV)] is also infectious to the crop plants. The major plant viruses possessing DNA genome are belonging to families Geminiviridae (ssDNA) [e.g., *Bean golden mosaic virus* (BGMV), *Maize streak virus* (MSV)] and Caulimoviridae (dsDNA) [e.g., *Cauliflower mosaic virus* (CaMV), African cassava mosaic virus (ACMV), *Rice tungro bacilliform virus* (RTBV)]. These viruses alone are responsible for causing yield losses ranging from \$ 25 million to \$ 100 million annually (Table 3.1). Among all plant pathogens, viruses contribute 50% of emerging and habitual plant diseases worldwide damaging natural flora as well as cultivated plants (Jones and Naidu, 2019). However, due to the continuous pressure of global climate change and ever-growing human population, management of plant viral diseases is becoming the major challenge in front of global agriculture sector. In addition, agriculture globalization, international trade, advancement, and/or alteration in agricultural practices and evolved cropping pattern are favoring destructive viral disease outbreaks and spreading viruses and their vectors across the boundaries and to the new geographical destinations. This penalizes the unpredicted consequences to the natural ecosystems and worldwide food production and food security. Therefore, it is very much essential to know the plant virus-host interaction, virus symptoms, disease development, plant innate immunity against plant viruses and present plant virus management practices namely Crispr Cas9 genome editing tool in plant virus resistance.

3.2 PLANT VIRUS INTERACTION

Plant viruses easily entered in the cells through cell injury or insect vectors (insects, nematodes, mites, fungi, and plasmodiophoroids) that feeds on or infecting the plants. Further, within infected plant cell virus genome is decapsidated, translated, and replicated which could results into outburst of virus particles (Nicaise, 2014). Initially, virus particles replicated at the site of initial infection foci and further reach to healthy neighboring cells

TABLE 3.1   List of Important Plant Viruses and Their Global Crop Yield Losses

SL. No.	Plant Virus	Crop Infected	Cost of Crop Damage per Year/ Crop Yield Loss Per Year	Location	References
1.	Cassava mosaic begomovirus	Cassava crop	25 million tons	Africa, India, Srilanka	Minato et al. (2019)
2.	Potato leaf roll polerovirus	Potato	$ 100 million	US	Sastry et al. (2014)
3.	Potato leaf roll polerovirus	Potato	$ 30–50 million	UK	Wale et al. (2008)
4.	Barley yellow dwarf luteovirus	Barley, oats, rice, wheat, maize	$ 13.93 million	UK	Ordon et al. (2009)
5.	Rice tungro disease (RTD)	Rice	$ 1.5 billion rice	South-East Asia	Kumar et al. (2019)
6.	Citrus triteza closterovirus	Cacoa trees	200 million trees	Worldwide	Harper (2013)
7.	Cacoa swollen shoot virus	Citrus trees		Togo, Ghana, Nigeria	Dzahini et al. (2010)
8.	Maize streak virus (MSV)	Maize	$480 million	Africa	Rybicki (2015)
9.	Banana bunchy top virus (BBTV)	Banana	$50 million	Australia, Africa, and Asian	Kumar et al. (2015)
10.	Papaya ring spot virus (PRSV)	Papaya	$11 million (80 to 100% yield loss)	Hawaii	Wu et al. (2018)

through plasmodesmata (PD) by viral ribonucleoprotein (vRNP) complexes. Whereas, spread of infection to the other organs of the host plants are adopted by distant or systemic movement mechanism by virus (Dolja et al., 1992; Zwart et al., 2012). The normal function of the plant is interfered due to gathering of viral nucleic acids or proteins molecules insight and which results in induction of a symptomatic defense response by host plant (Pallas and Garcia, 2011; Culver and Padmanabhan, 2007). The commonly occurred viral diseases exhibited symptoms like yellowing or mottling of the leaves, mosaic patterns, dwarfing or developmental abnormalities of the plant and systemic necrotic symptoms, sporadically leading to the death of the plant (Roossinck, 2015).

3.2.1   DEVELOPMENT OF PLANT VIRUS DISEASE SYMPTOMS

A normal plant disease is the outcome of interaction between susceptible host plant, virulent pathogen and the environment. Plant virus disease symptoms are derived from specific interactions between virus and host components. Plant virus-host interactions are of two types: consequential virus-host interactions directly contribute to the establishment of a systemic infection; while inconsequential virus-host interactions do not contribute to the success of the infection but nevertheless disrupt host physiology (Matthews, 2002). Further, virus-induced symptom and disease development is typically explained by two general models (Culver and Padmanbhan, 2007). A *competitive disease model* describes that, plant viruses replicate inside the host cell and usurp a substantial amount of host resources results into adverse effect on growth and development of host and leading to disease development. However, *interaction disease model* is based on the specific interactions between plant virus and components of host that leading to the interruption of physiological process of plants and its metabolism, protein modifications, cell cycle control mechanism, host transport system, and changing the allocation of cellular resources as well as cell-to-cell communication signals (Jameson, 2002; Culver and Padmanabhan, 2007).

3.2.2   MOLECULAR INSIGHTS IN VIRUS-HOST INTERACTION

The virus and host proteins are accountable to acquire infection and further disease development in host plants. The infection process in plant is initiated through disassembling of viral genome, vRNA replication, its movement, and encapsidation. The formation of virions, a mature infectious virus particle, or vRNP complexes and its movement through PD accelerate plant infection process. The process is continued until the vRNPs movement through vascular tissues (xylem and phloem) which could results into systemic infection (Laliberte, 2016). Positive sense RNA viruses synthesize virus protein in the cytoplasm of the invaded host cell. This virus includes the viral enzymes (vRdRp and *helicase)* and proteins (CP protein and MP protein) which are responsible for infection, further replication and movement in host plant (Zaitlin, 1999). However, 'viral factories' are described as virus induced quasi-organelles associated with cellular membranes where vRNA replication takes place. There are two types of viral factories; *Spherule-Shape* viral factories (static, 50–400 nm, associated to the membrane of peroxisome, the mitochondrion and the chloroplast) and *Vesicular-shape* viral factories (motile, 30 to 300 nm, ER-derived, and involved in the intracellular movement of the vRNA). Viral replication complexes (VRCs) are accumulating the vRNA replication to particular location and results into the restriction of host defense functions (Den Boon and Ahlquist, 2010; Laliberte and Sanfacon, 2010). The membrane-associated protein of plant viruses is part of VRCs and which triggers membrane rearrangement (Laliberte, 2016). The host proteins, i.e., endosomal sorting complexes required for transport (ESCRT) factors, host reticulon homology proteins (RHPs) and the early secretary pathway components are involved in the process of virus replication and synthesis of viral factories (Diaz et al., 2015).

DNA viruses such as geminiviruses, replicate inside host cell by hijacking host machineries like DNA polymerase and RNA polymerase II and accelerate the rate of host cell division (Stanley, 1985; Gutierrez, 2002; Hanley-Bowdoin et al., 2004). The host synthesized membrane spanning proteins (TOM1 and TOM3), translation initiation factor (iso) 4E [eIF(iso)4E], the translation elongation factor 1A (eEF1A) and the poly(A)-binding protein (PABP) are assisting the viral replication and their translocation inside host plant (Yamanaka et al., 2002; Laliberte, 2016). A list of various plant viruses, virus proteins, host proteins, and their functions are summarized in Table 3.2.

3.3   INNATE IMMUNITY IN PLANT

Plant pathogen resistance mechanism was explained earlier based on gene for gene theory. The gene for gene theory has successfully demonstrated the host plant resistance against various pathogens of plant (Flor, 1971; Keen, 1990). A hypersensitive response (HR) is activated by reaction between

TABLE 3.2 Plant Virus and Host Protein Regulating the Disease Development in Host Plant

SL. No	Plant Virus Protein	Host Plant Protein/ Component	Resulting Interaction and Disease Development	References
		RNA Replicase Related Proteins		
1.	TMV replicase	Aux/IAA proteins	Alterations in auxin response pathways, developmental symptoms	Padmananbhan et al. (2005, 2006)
2.	TMV replicase	P58 IPK (inhibitor of dsRNA activated PKR)	Regulation of cell death	Bilgin et al. (2003)
3.	RDV P2	ent-Kaurene oxidase	Gibberellin synthesis, dwarfing	Zhu et al. (2006).
4.	Gemini virus Rep proteins	Retinoblastoma protein (pRBR)	Cell cycle reprogramming	Kong et al. (2000)
5.	Nib RNA Replicase (PPV)	—	Virus accumulation and disease development	Guo and García (1997)
6.	p126/p183 (TMV and PPMMoV)	—	Transport protein and its movement	Citovsky and Zambryski (1995)
7.	2a (CMV)	Protein Kinase	Formation of replicase complex	Kim et al. (2002)
8.	PSTVd derived siRNA	Host mRNA	Mis regulation of host mRNA, induction of disease	Wang et al. (2004)
9.	TBSV p19	ALY proteins (nuclear shuttle proteins)	Unknown	Canto et al. (2006); Uhrig et al. (2004)
10.	Geminivirus NSP (nuclear shuttle protein)	AtNSI (Acetyltransferase)	Disruption of AtNSI acetylation activity	Carvalho et al. (2004, 2006); McGarry et al. (2003)
11.	Geminivirus NSP (nuclear shuttle protein)	NIK kinases	Reduce NIK kinase activity, disrupt defense response?	Fontes et al. (2004)
12.	FBNYV 20-kDa protein (F-box protein)	Skp-1 and pRBR	Degradation of pRBR? Cell cycle reprogramming?	Aronson et al. (2000)

TABLE 3.2 *(Continued)*

SL. No	Plant Virus Protein	Host Plant Protein/Component	Resulting Interaction and Disease Development	References
		Coat Protein Gene		
13.	AMV CP	Translation initiation factors eIF4G and eIFiso4G,	interact with the host translation initiation factors eIF4G and eIFiso4G, mimicking the function of host PABP	Krab et al. (2005); Neeleman et al. (2001)
14.	Wheat yellow mosaic virus (WYMV): P2	COPII GTPase Sar1	COPII GTPase Sar1 interacts with the P2 protein of Wheat yellow mosaic virus (WYMV)	Sun et al. (2014)
15.	TuMV 6K2	COPII coatomer Sec24a	The COPII coatomer Sec24a recognizes the N-terminal cytoplasmic tail of the TuMV 6K2 protein, thus facilitating the incorporation of the viral protein into COPII vesicles	Jiang et al. (2015)
16.	CPMV: 60K helicase	ER localized SNARE-like protein VAP27	Interact with the 60K helicase of CPMV	Carette et al. (2002)
17.	TuMV:6K2	VAP27 protein	Binding VAP27, 6K2 associates also with Syp71, which is involved in vesicle fusion	Wei et al. (2013)
		Membrane-Associated Viral Proteins		
18.	TBSV: p33	Peroxisome (switch to ER in the absence of peroxisome)	Upregulates phospholipid biosynthesis; recruits ESCRT factors for VRCs assembly; vRNA recruitment; interacts with the p92pol; binds eEF1A to promote VRCs assembly and (–) vRNA synthesis.	Rajendran and Nagy (2004); Pogany et al. (2005); Jonczyk et al. (2007); Li et al. (2009); Barajas et al. (2014)
19.	Red clover necrotic mosaic virus (RCNMV): p27	GTPase, (Arf1)	The GTPase, such as the Arf1, preferentially binds to the C-terminal region of the viral protein p27.	Hyodo et al. (2013)

TABLE 3.2 *(Continued)*

SL. No	Plant Virus Protein	Host Plant Protein/ Component	Resulting Interaction and Disease Development	References
20.	TBSV: p92pol	Peroxisome	vRdRp: interacts with p33; recruits GAPDH to the VRCs.	Rajendran and Nagy (2004); Huang and Nagy (2011)
21.	BMV: 1a	ER, host reticulon homology proteins (RHPs)	Formation of viral factories; recruits the vRNA to the viral factories; hijacks reticulons for membrane curvature: RHPs involved in viral factory biogenesis	Schwartz et al. (2002); Wang et al. (2005); Liu et al. (2009); Diaz et al. (2010)
22.	BMV: 2apol	ER	vRdRp: interacts with the capsid protein maybe for genome packaging.	Chaturvedi and Rao (2014)
23.	TuMV: 6K2	ER	VRCs assembly; intracellular long distance movement.	Agbeci et al. (2013); Grangeon et al. (2013); Wan et al. (2015)
24.	TuMV: P3	ER	Virus pathogenesis; symptom and avirulence determinant; genome amplification.	Jenner et al. (2003); Cui et al. (2010)
25.	BaMV/PVX: TGBp1	ER	RNA binding; suppresses host gene silencing; virus movement; regulates the size exclusion limit of the PD; induces the formation of X-body.	Howard et al. (2004); Tilsner et al. (2012)
26.	BaMV/PVX: TGBp2	ER	Induces VRCs formation; interacts with TGBp3.	Samuels et al. (2007)
27.	BaMV/PVX: TGBp3	ER	Associates with the virions for virus delivery; interacts with TGBp2.	Samuels et al. (2007); Chou et al. (2013)

single resistant gene (R-gene) of the host and corresponding avirulence (Avr) protein of pathogen (Dangl and Jones, 2001). It was elucidating that R gene (e.g., Kinase protein) from host plant physically interacts with Avr (e.g., AvrPto or AvrPto Bits) for a virulence determination (Martin et al., 1993; Tang et al., 1996). Several R Genes from numerous plant species were classified in two classes namely: (a) NB-LRR proteins; and (b) receptor-like kinase (RLKs)/receptor-like proteins (RLPs) (Rathjen and Moffett, 2003). Further, plant disease resistance was explained by zigzag model which comprises two distinct defense responses level (primary and secondary). The primary defense level is called pathogen or microbe-associated molecular patterns (PAMP/MAMP) induced immunity (PTI), and the subsequent and secondary defense system is called effector-triggered immunity (ETI) (Islam et al., 2017). The reaction process between host cell-surface associated pattern recognition receptors (PRRs) and conserved structural motifs of pathogen, i.e., MAMPs/PAMPs, DAMPs (damage-associated molecular patterns of plant) (Macho and Zipfel, 2014) (Figure 3.1).

In response to the virus infection viral mRNAs are translated into the cytoplasm, producing coat protein (CP), replication protein (Rep), and movement protein (MP) which are required for completion of viral life cycle. Virus genome is synthesized when Rep combines with cellular proteins. These virus genomes and CPs interact together to form new virions or vRNP complexes. Further, MP is involved for movement of virus into neighboring cells. The HR, necrosis, and SAR are results of effector triggered immunity (ETI) by avirulence proteins (Avr), R co-factors (CF) and recognition of cytosolic NB-LR receptors (e.g., R proteins). The interacting complexes (RAR1/SGT1/HSP90 and PAD4/EDS1/SAG101) or MAP Kinases cascades alters the levels of SA, JA, ET, NO, and H_2O_2 and induce defense genes via NPR1, TF, and EIN2. Pathogen triggered immunity (PTI) is activated upon signaling through virus derived PAMPs by the replication of viral RNA genomes and non-self RNA motifs (ssRNA or dsRNA). On the other hand, plant cells when sense viral infection, secretes plant-derived DAMPs which is recognized by PRRs extracellular (Adapted from: Mandadi and Scholthof (2013); Gouveia et al. (2017)).

This results into biosynthesis of specific defense molecules and making plants efficient to respond quickly and effectively against broad array of pathogens (Bigeard et al., 2015; Roux et al., 2014). The second defense response level is triggered when the R gene product directly or indirectly sense specific effectors (Avr factors) produced inside the host by pathogens and activated effector-triggered immunity (ETI) (Howden and Huitema, 2012). ETI activation results in HR, quick death of cells, development of

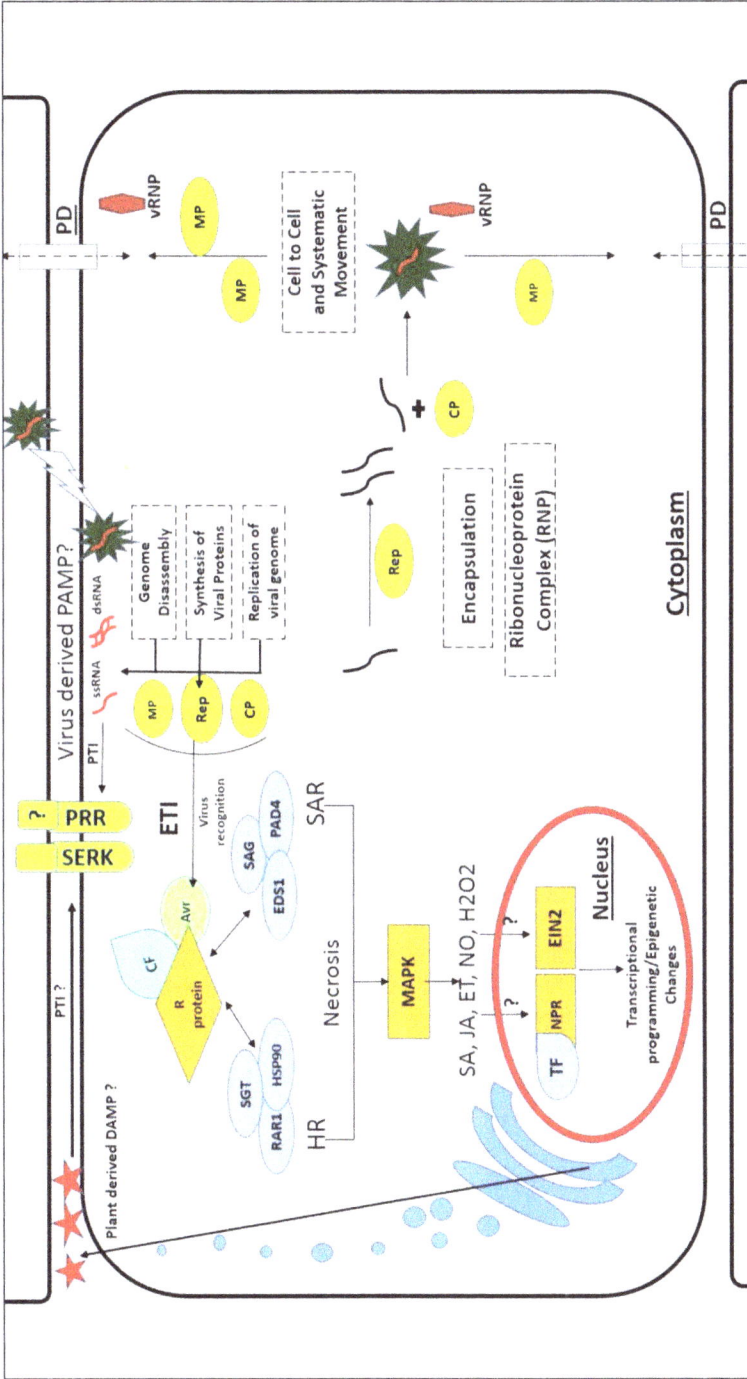

FIGURE 3.1 Plant innate immunity against plant virus infection.
(*Source:* Adapted from Gouveia et al, 2017, which was adapted from Mandadi et al., 2013.)

ROS and salicylic acid (SA) and expression of defense-related genes (Win et al., 2012). The Avr determinants and proteins encoded by virus namely RP, MPs, and CPs, are acting as a vital factor for successful infection (Gouveia et al., 2017; Coll et al., 2011; Reimer-Michalski and Conrath, 2016) (Figure 3.1). Recently natural plant viral resistance is classified as R-gene-mediated resistance, recessive resistance, antiviral RNA silencing, hormone-mediated antiviral defenses and proteasome degradation (Calil and Fontes, 2017; Islam et al., 2017; Gouveis et al., 2017).

3.3.1   R-GENE-MEDIATED RESISTANCE

This is HR mediated resistance causes cell death destroy infected cells and restricts systemic spread of viral invasion. The hyper sensitive response is characterized by mitogen-activated protein kinase (MAPK) signals; jasmonic acid (JA) accumulation, calcium ion influx, SA, membrane permeability modification, defense genes activation, an immediate reactive oxygen species (ROS), nitric oxide (NO) accumulation and callose deposition at the PD (Yang et al., 2001) (Figure 3.1). The majority of plant R genes are nucleotide-binding (NB) and leucine-rich-repeat (LRR) domains encoding genes, Whereas Avr proteins showed very few common characteristics features to each other (Jones and Dang, 2006; Islam et al., 2017). The N gene encoding TIR NB-LRR protein from tobacco was the first characterized viral R Gene conferring resistance to TMV (Whitham et al., 1994). Table 3.3 representing here around identified 20 viral R genes which has dominant inheritance. This class of resistance also comprises an Apaf-1/R protein/CED 4 (ARC) domain, functioning in ATP hydrolysis (Rairdan et al., 2008; Islam et al., 2017). Some non-NB-LRR class of proteins (JAX1, RTM1, RTM2, Ty-1, Ty 3, and Tm-1) though not induces typical ETI-like defense response, such as HR but found to function as sensors of virus infection (Table 3.3). Thus, such non-NB-LRR class of proteins may act as potential target sites in genome editing and induction of virus resistance in the plant.

3.3.2   RECESSIVE RESISTANCE

The recessive R genes are those plant genes which acting as essential factors required to the virus to complete their biological cycle. The resistance governed by these genes were often considered as incompatible virus-host interactions wherein virus infect to plant but further systematic infection is

TABLE 3.3 List of Resistance Genes (D: Dominant and r; Recessive) in Host Plant Against Plant Viruses

SL. No.	Resistance Genes (D; Dominant and r; Recessive)	Plant Virus	Resistance Factors and Features	Avirulence Factor	References
1.	HRT (D)	Turnip crinkle virus	CC-NBS-LRR (HR)	CP	Ren et al. (2000)
2.	JAX1 (D)	*Plantago asiatica* mosaic Virus	Jacalin like lectin (blocks RNA accumulation)	Unknown	Yamaji et al. (2012)
3.	RCY1 (D)	Cucumber mosaic virus	CC-NBS-LRR (HR)	CP	Takahashi et al. (2002)
4.	RTM1 (D)	Tobacco etch virus	Jacalin family (blocking systemic movement)	CP	Chisholm et al. (2000)
5.	RTM2 (D)	Tobacco etch virus	Small heat shock Protein (blocking systemic movement)	CP	Whitham et al. (2000)
6.	RTM3 (D)	Tobacco etch virus	MATH-containing protein (blocking systemic movement)	Unknown	Cosson et al. (2010)
7.	sp1 (r)	Turnip mosaic virus	eIF(iso)4E (mutagenesis)	VPg	Lellis et al. (2002)
8.	cum1(r)	Cucumber mosaic virus	eIF4E (mutagenesis)	Unknown	Yoshii et al. (2004)
9.	cum2 (r)	Cucumber mosaic virus	eIF4E (mutagenesis)	Unknown	Yoshii et al. (2004)
10.	BcTuR3 (D)	Turnip mosaic virus	TIR-NB-LRR (systemic resistance)	Unknown	Cosson et al. (2010)
11.	TuRB07 (D)	Turnip mosaic virus	CC-NBS-LRR (ER)	Unknown	Ma et al. (2010)
12.	L(multi-alleles) (D)	Tobacco mosaic virus	CC-NBS-LRR (HR)	CP	Tomita et al. (2011)
13.	pvr1/pvr2(multi-alleles)(r)	Potato virus Y	eIF4E	VPg	Ruffel et al. (2002)
14.	pvr6 (r)	Pepper veinal mottle virus	eIF(iso)4E	VPg	Ruffel et al. (2006)
15.	Nsv (r)	Melon necrotic spot virus	eIF4E	Unknown	Nieto et al. (2006)
16.	Rsv1 (D)	Soybean mosaic virus	CC-NB-LRR (HR)	P3, HC-Pro	Hayes et al. (2004)
17.	rym4/5(multi-alleles) (r)	Barley yellow mosaic virus	eIF4E	VPg	Stein et al. (2005)
18.	mo1 (multi-alleles) (r)	Lettuce mosaic virus	eIF4E	CI-Cter, VPg	Nicaise et al. (2003)
19.	rymv1 (r)	Rice yellow mottle virus	eIF(iso)	4G, VPg	Albar et al. (2006)

TABLE 3.3   *(Continued)*

SL. No.	Resistance Genes (D; Dominant and r; Recessive)	Plant Virus	Resistance Factors and Features	Avirulence Factor	References
20.	rymv2 (r)	Rice yellow mottle virus	CPR5(H)	unknown	Orjuela et al. (2013)
21.	I (D)	Bean common mosaic virus	TIR–NBS-LRR (HR)	Unknown	Vallejos (2006)
22.	RT4-4 (D)	Cucumber mosaic virus	TIR–NBS-LRR (systemic necrosis)	2a	Seo et al. (2006)
23.	bc3 (r)	Bean common mosaic virus	eIF4E	unknown	Naderpour et al. (2010)
24.	sbm1 (r)	Pea seed-born mosaic virus	eIF4E	VPg	Gao et al. (2004)
25.	Ty1/Ty3 (multi-alleles) (D)	Tomato yellow leaf curl virus	RDR (RNA silencing)	Unknown	Butterbach et al. (2014)
26.	Tm1 (D)	Tomato mosaic virus	TIM-barrel-like domain (blocking replication)	Replication protein	Ishibashi et al. (2007)
27.	pot1 (r)	Potato virus Y	eIF4E	VPg	Ruffel et al. (2005)
28.	Tm2 (multi-alleles) (D)	Tomato mosaic virus	CC-NBS-LRR (HR)	MP	Lanfermeijer et al. (2003)
29.	Sw5b (D)	Tomato spotted wilt virus	CC-NBS-LRR (HR)	MP (NSm)	Brommonschenkel et al. (2000)
30.	Rx (multi-alleles) (D)	Potato virus X	CC-NBS-LRR (blocking replication)	CP	Bendahmane et al. (2002)
31.	Y1 (D)	Potato virus Y	TIR–NBS-LRR (HR)	Unknown	Vidal et al. (2002)
32.	CYR1 (D)	Mung bean yellow mosaic virus	CC_NB_LRR	CP	Maiti et al. (2012)

Note: MATH: meprin and TRAF domain; CP: coat protein; HC Pro: helper component proteinase; MP: movement protein; RDR: RNA-dependent RNA polymerase; ER: extreme resistance without any necrotic local lesion; eIF4E: eukaryotic translation initiation factor 4E; eIF(iso)4E: eukaryotic translation initiation factor iso 4E; Pelo: a messenger RNA surveillance factor.

disrupted by host resistance factors. These include eukaryotic translation initiation factors, namely eIF4E and eIF4G and which induce resistance by functional mutations or modification of these gene products (Revers and Nicaise, 2014). In many plant virus diseases, eukaryotic translation initiation factors eIF4E and eIF4G plays an essential role in successful infection. The natural variation in eIF4E deliberates potyvirus resistance to in crop species. This include in pepper crop pvr1 and pvr2 (Gao et al., 2004), sbml in pea (Nicaise et al., 2003), mol in lettuce (Ling et al., 2009) in barley rym4/5 (Ruffel et al., 2002), in tomato pot1 (Kang et al., 2005a) and in watermelon zym-FL (Wicker et al., 2005), Pelo in resistance ty5 genotype of tomato (Lapidot et al., 2015). Thus, many of plant natural resistance genes functioning as essential host factors for virus infection have been identified and being utilized in genome editing based induction of plant virus resistance.

3.3.3   RNA INTERFERENCE (RNAi) MEDIATED-RESISTANCE

One of the major mechanisms for plant antiviral immunity is RNA interference (RNAi; also called gene silencing) that is caused due to the activities of enzymes which is recognized by double-stranded RNAs (dsRNAs) and being acting as part of conservation process in most of eukaryotes. The dsRNAs are processed by DCL (Dicer-like ribonuclease III-type) enzymes into 21–24 nucleotide sRNAs and further attached to RISC complex. The RISC complex comprises AGO (argonaute) and other proteins. The sRNAs binds to their target-mRNA and degrade it due to the cleavage activity induced by Dicer enzymes. The key components of the RNA silencing pathways are existence of multiple copies of AGO, DRB (dsRNAs binding), RNA dependent RNA polymerase (RDR) and DCL (Dicer-like) genes which are playing an important role in RNAi mechanism against viral pathogens (Qu et al., 2008; Deleris et al., 2006; Garcia-Ruiz et al., 2010; Bologna and Voinnet, 2014). However, RNAi-mediated resistance is regularly hindered by several co-evolving viral suppressors (VSRs) and which could enhance the viral pathogenicity within susceptible hosts (Lewsey et al., 2018).

3.3.4   PLANT HORMONE-MEDIATED RESISTANCE

Plant hormones are playing important roles in regulation of intercellular and systemic signaling networks in viral plant defense mechanism (Lewsey et al., 2018). Plant virus interactions were resulted to the alterations in plant hormone synthesis and signaling in host plant (Culver and Padmanabhan, 2007). The plant hormones facilitates the natural plant defense against viral disease mainly Auxin [IAA; Aux/IAA proteins in TMV infection], Gibberellin [GA3 in RDV], Ethylene [CaMV: P6 expression and disruption of the ethylene response pathway], SA [activation of SAR in TMV, CaLCuV, potato virus Y (PVY) and tomato ringspot virus (ToRSV)], abscisic acid (ABA) [reduce the accumulation of TMV, bamboo mosaic potyvirus (BaMV) and CMV and reduce the cell to cell movement of virus] (Calil and Fontes, 2017; Jovel et al., 2011; Baebler et al., 2014; Alazem et al., 2014). The other plant hormone like JA synthesis genes were found down regulated during the attack of geminivirus (Ascencio-Ibanez et al., 2008). However, brassinosteroids (BRs) was also observed as a positive regulator for inducing a plant defense against viruses such as TMV, TCV, and oilseed rape mosaic virus (ORMV) (Nakashita et al., 2003; Piroux et al., 2007; Korner et al., 2013). Thus, genetic induced alteration in the phytohormone level can restrict virus infection and would offer an opportunity to enhance plant immunity.

3.3.5 PROTEASOME DEGRADATION

The ubiquitin-proteasome pathway (UPS) is also an antiviral defense strategy employed by host plant. It is involved in basic plant processes, pertaining degradation, functional alteration of cellular proteins and signaling in response to abiotic and biotic stimuli. Plant viruses exploit this UPS and induce, inhibits or modify ubiquitin (Ub)-related host proteins enabling successful host infection. However, UPS also plays a role in host defense mechanism to wipe out components of viruses (Alcaide-Loridan and Jupin, 2012). Many viral proteins are encountered to interact with the various subunits of the 20S proteasome or 26S proteasome *viz.*, helper component proteases (HcPro) of Lettuce mosaic virus (LMV), PVY, Papaya ringspot virus (PRSV); bC1 protein of gemini virus (TMV); C2 protein of gemini virus (BSCTV); RNA-dependent RNA polymerase of Turnip yellow mosaic virus (TYMV), factor C4 from Beet severe curly top virus (BSCTV), etc., (Sahana et al., 2012; Zhang et al., 2011; Eini et al., 2009; Camborde et al., 2010; Chenon et al., 2012). In a nutshell, plant viruses are exploiting UPS for control of quality of their own proteins and enhancing their efficacy. However, in parallel, plants are using this pathway as basal resistance and targeting viral proteins for degradation (Dielen et al., 2011; Calil and Fontes, 2017). Thus, the genes which are involved in this pathway can be explored to built-up virus resistance in crop plants.

3.4 PLANT VIRUS MANAGEMENT STRATEGIES

Several strategies are available for control of plant viruses. The chemical, biological, and cultural management practices are mainly based on the control of insect vector for further spread and transmission of virus in the field. Most of the time, these methods could not find effective and associated with environmental hazardous. Thus, breeding of virus resistant varieties is becoming one of the important strategies for control of viral diseases. Unfortunately, due to prolong time requirement and unavailability of resistant gene pool within available germplasms limits its applications. Besides this, further plant virus management strategies are classified into several categories with inclusive of advanced approaches (Figure 3.2; Chauhan et al., 2019). The conventional methods like (meristem tip culture, thermotherapy, cryotherapy, and chemotherapy) have certain limitations such as expensiveness, time, and labor consuming, acclimatization limits, variability, production scheduling, and contamination. Whereas, the advanced approach namely RNAi silencing and cross protection are not durable and induced resistant is also hindered by viral suppressor at field level. The use of transgenic and gene pyramiding approach is time consuming, costly, and have limitations of biosafety concerns. On other side, the emerging genome editing technology based upon homing endonucleases (EMNs), zinc finger nucleases (ZFNs), transcription activator-like effector nuclease (TALENs) and clustered regularly interspaced palindromic repeats (CRISPR/Cas9) repair mechanism have extensive advantages over these methods of management of plant viruses.

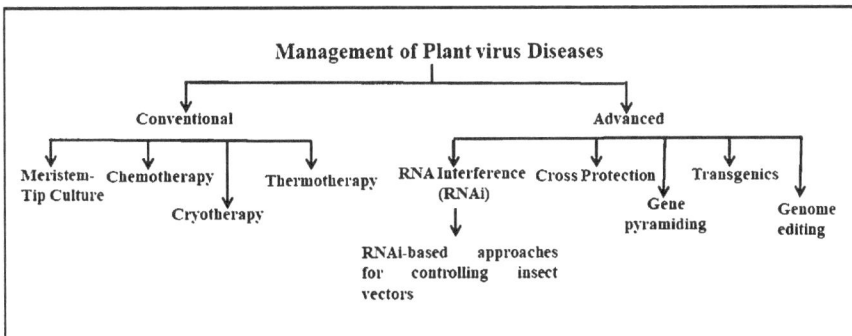

FIGURE 3.2 Plant virus management strategies.

3.5   BIOLOGY OF CRISPR/CAS9

Clustered regularly interspaced palindromic repeats (CRISPR/Cas9) an genome editing tool has been preferred by researchers due to its ease, simplicity, specificity, low off target effect and preciseness (Shan et al., 2013; Araki and Ishii, 2015; Baltes and Voytas, 2015; Kanchiswamy et al., 2015). The clustered regularly interspaced short palindromic repeats (CRISPRs) was initially discovered in *E. coli* in the year 1987 (Ishino et al., 1987). Further, its function as acquired resistance in bacteria and Archea against a bacteriophage was confirmed in 2007 by Barrangou and colleagues. The short DNA segments (20–50 bp) from invading viruses and plasmids have been integrated into their host genomes in between the copies of repeat sequence of 20–50 bp. This type of organization is named as CRISPRs (Jansen et al., 2002). In adjacent with CRISPR array, series of genes encoding Cas9 endo-nuclease and two RNAs, *viz.* CRISPR RNA (crRNA) and trans-activating CRISPR RNA (tracrRNA) are present (Figure 3.3).

Bacterial immunity against the corresponding viruses attacking the host is driven by CRISPR/Cas9 system through acquisition of CRISPR RNA (crRNA) biogenesis, and interference (Figure 3.3). In the acquisition process, foreign DNA are selected, processed, and integrated into the CRISPR array for storage as a memory of infection. A long precursor crRNA (pre-crRNA) was formed during transcription of CRISPR array and which leads to mature crRNAs. The onset of related virus or plasmid infection activates interference machinery mainly cas9 protein which is guided by crRNAs leading to the cleavage of complementary sequences, called protospacers flanked by a protospacer-adjacent motif (PAM), in the incoming foreign nucleic acids (Figure 3.1; Jinek et al., 2012). Based on the presence of Cas genes and the nature of the interference complex, CRISPR-Cas systems have been classified into two classes and subsequently subdivided into six types and several sub-types that each has signature *Cas* genes (Hille et al., 2018). The CRISPR-Cas systems of Class 1 category (types I, III, and IV) comprises multi-Cas protein complexes, whereas in class2 systems (types II, V, and VI), single effector protein undertakes interference (Makarova et al., 2011, 2015). The type II CRISPR/cas9 system with little modification (synthetic single guide RNA (sgRNA) designed for trRNA and crRNA) was first time used by Doudna and Charpentier in 2012 for genome editing (Jinek et al., 2012). Afterward, this system has been extensively exploited by many researchers for genome alteration in life science research. Moreover, various Cas9 nuclease variants are being synthesized and adopted in genome alteration. Conventionally Cas9 nuclease induces site-specific double stranded break

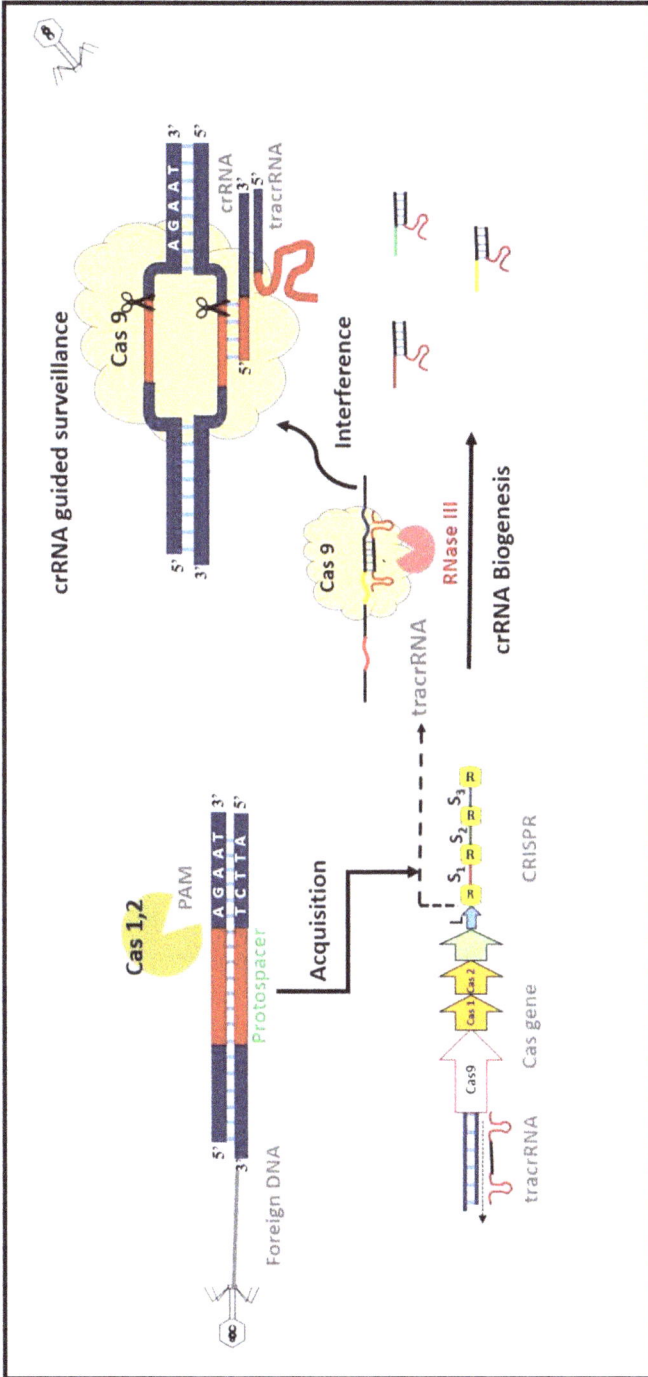

FIGURE 3.3 Biological mechanism of CRISPR/Cas9 nuclease system.

Source: Adapted from: Sorek et al. (2013).

(DSB) which are being repaired by natural host repair systems leading to the induction of genetic mutations. The Cas9 mutant, i.e., Cas9D10A have only nickase activity and cleaves only one DNA strand, whereas dCas9 only possess DNA binding property used for gene konckout (Butler et al., 2018). The DNA repair pathways, non-homologous end joining (NHEJ) and homology directed repair (HDR) are accountable for arbitrary indels and which results into frame shift mutations (Cristea et al., 2013). Whereas, in HDR, exogenously supplied homologs DNA sequences are integrated at targeted site. Therefore, such repair pathways are carefully utilized in specific and targeted genome alteration.

3.6   CRISPR/CAS9 MEDIATED RESISTANCE IN PLANTS AGAINST VIRUS

The RNAi-based virus-resistant transgenic development have showed limited success (Shepherd et al., 2009) and which are facing the barriers of regulatory concerns and public acceptance. Whereas CRISPR/Cas9 technology enabled induction of targeted mutations leading to the control of plant viruses (Zhang et al., 2015; Unniyampurath et al., 2016). The utility of CRISPR/Cas9 system for plant virus resistance was reported independently by several researchers by using model species like Tobacco and Arabidopsis (Table 3.4).

The virus resistance using CRISPR/Cas9 is imparted either by edition of viral genome or host genome (Khatodia et al., 2017). In general, two strategies are commonly employed to develop CRISPR/Cas9 based plant virus resistance. The first strategy is to develop transgenic through engineering and maintenance of Cas9 and sgRNA in plant genome. While second strategy is to develop non transgenic mutants during further segregation of CRISPR/Cas9 machinery and release of non-transgenic mutant lines (Macovei et al., 2018). The procedure for virus resistance development through CRISPR/Cas9 genome editing approach is described in Figure 3.1. In general, CRISPR/cas9 based virus resistant plant developments comprises important steps namely; design of target specific gRNA, mobilization of gRNA-CRISPR/cas9 cassette in plant transformation vector; delivery of gRNA into plant system; Screening of desirable mutant lines; Segregation and selection of desirable mutant lines having effective on-target activity. There are several plant transformation strategies being deployed for the successful delivery of CRISPR/cas9 gene cassettes. The *Agrobacterium*-mediated and biolistic transformation methods are remained the most applied transformation methods, but they possessed

TABLE 3.4 List of CRISPR/Cas9 Mediated Resistance Developed Against Viruses in Various Crop

Virus/Viruses Targeted	Host Plant	Target Gene (Viral/Host)	Gene Function	References
BSCTV	*N. benthamiana* and *A. thaliana*	IR, CP, and Rep	RCA Mechanism	Ji et al. (2015)
BeYDL	*N. benthamiana*	LIR and Rep/RepA	RCA Mechanism	Baltes et al. (2015a)
TYLCV, BCTV, MeMV	*N. benthamiana*	IR, CP, and Rep	RCA Mechanism	Ali et al. (2015)
CLCuKoV, TYLCV, TYLCSV, MeMV, BCTV-Logan, BCTV	*N. benthamiana*	IR, CP, and Rep	RCA Mechanism	Ali et al. (2016)
TuMV	*A. thaliana*	eIF(iso)4E	Host factor for translation of virus genome	Pyott et al. (2016)
CVYV, ZYMV, and PRSMV	*Cucumis sativus*	eIF4E	Host factor for translation of virus genome	Chandrasekaran et al. (2016)
BeYDV	*N. benthamiana*	LIR and Rep/RepA	Transgene free	Baltes et al. (2015b)
BSCTV	*N. benthamiana* and *A. thaliana*	IR, CP, and Rep	Transgene free	Ji et al. (2015)
TuMV	*N. benthamiana*	GFP1,2,3, CP, and 3'UTR	Replication mechanism	Aman et al. (2018)
CMV, TMV	*N. benthamiana* and *A. thaliana*	ORF,1,2,3, CP, and 3'UTR	Replication mechanism	Zhang et al. (2018)
RTSV	*Oryza sativa* L. *japonica*	eIF4G	Host factor for translation of virus genome	Macovei et al. (2018)

Note: Short guide RNA (sgRNA); beet severe curly top virus (BSCTV); bean yellow dwarf virus (BeYDV); tomato yellow leaf curl virus (TYLCV); beet curly top virus (BCTV); merremia mosaic virus (MeMV); cotton leaf curl Kokhran virus (CLCuKoV); tomato yellow leaf curl Sardinian virus (TYLCSV); turnip mosaic virus (TuMV); cucumber vein yellowing virus (CVYV); zucchini yellow mosaic virus (ZYMV); papaya ring spot mosaic virus (PRSMV); intergenic region (IR); coat protein (CP); replication associated protein (Rep); long intergenic region (LIR); replicase protein A (Rep/RepA); eukaryotic translation initiation factor 4E (eIF4E); green fluorescent protein (GFP); 3' untranslated region (3'UTR); translation initiation factor 4 gamma gene (eIF4G);

Source: Modified from: Zaidi et al. (2016); Khatodia et al. (2017); Makarova et al. (2018).

certain limitations. Moreover, several other transformation methods are being practiced namely; inflorescence dip, *Agrobacterium*-infiltration, virus based transformation, pollen magnetofaction, mesoporous silica nanoparticles (MSN) method, and DNA free reagent delivery method, i.e., ribonucleopro-tein (RNP), etc., (Vats et al., 2019).

3.6.1   CRISPR-CAS9-MEDIATED RESISTANCE FOR DNA VIRUSES

The power of the CRISPR/Cas9 system to impart resistance against Gemini viruses in plants was efficiently demonstrated in model plants *N. Benthamiana* and *Arabidopsis* against *Tomato yellow leaf curl virus* (TYLCV), *Beet curly top virus* (BCTV), *Merremia mosaic virus* (MeMV) (Ali et al., 2015), *Bean yellow dwarf virus* (BeYDV) (Baltes et al., 2015) and BSCTV (Ji et al., 2015).

Ji et al. (2015) transiently expressed *Streptococcus pyogenes Cas9* gene and 43 candidate sgRNA target sites within coding and non-coding regions of the BSCTV genome into *Nicotiana benthamiana* and *Arabidopsis thaliana* leaves. Further, these Agro-infiltrated leaves were challenged by BSCTV after two days of infiltration. The challenged plants after 10 days of post-infection were examined towards symptom development and presence of viral load by qPCR analysis. Control plants were exhibited typical shoot tip leaf curling symptoms. Whereas, infiltrated plants did not show virus symptoms and qPCR revealed more than decline in the growth of BSCTV. Further, sequencing of virus DNA exhibited 1 to 10 nucleotide long deletions in the target region. The higher virus resistance was found in proportion of higher intensity of expression of Cas9 endonuclease gene and vis-a-vis.

The BeYDV resistant transgenic *N. benthamiana* was generated by Baltes et al. (2015a). They exploited 11 sgRNAs and targeted Rep motifs, Rep-binding sites, hairpin, and the nonnucleotide sequence of BeYDV, the constitutively expressed Cas9 and sgRNAs (gBRBS+ or gBM3+) into N. benthami showed reduced symptoms and copy number of BeYDV genome.

The transgenic *N. benthamiana* plants showed 87% reduction in virus load and symptom development due to induction of mutations in the BeYDV genome.

Ali et al. (2015a) engineered the CRISPR/Cas9 cassettes into *N. benthamiana* plants which exhibited resistance against TYLCV. They selected specific sgRNAs to multiple protein of TYLCV namely capsid protein (CP), RCR-II motif of the Rep and intergenic region (IR). Among all these tested sgRNAs, the targeted stem-loop invariant IR sequence interfered and reduced virus replication. Further, they tested these gRNA against BCTV and MeMV and induced mixed infection immunity by single sgRNA targeting multiple viral strains.

Further Ali et al. (2016) stated the emerging trend of mutant viruses having potential to survive, replicate, and further movement in systemic manner. They reported that, the sgRNA/Cas9 designed on target sequences of viruses specially, Cotton leaf curl Kokhran virus (CLCuKoV), MeMV, and strains of TYLCV resulted into generation of viral variants and which were capable to replicate and escape from genome editing machinery. In addition, they reported the *N. benthamiana* plants carrying gRNAs targeting the IR sequences of viral genome could not produce novel variants. They proved IR-repaired variants of virus was unable to replicate and thus provided a better interference to the virus multiplication.

Zaidi et al. (2016) described that, sgRNA/Cas9 designed on targets of non-coding intergenic sequences was competent, and produced high levels of virus interference in geminiviruses (Figure 3.4) than of coding regions.

3.6.2 CRISPR-CAS-MEDIATED RESISTANCE FOR RNA VIRUSES

Beside use of CRISPR-Cas system to develop immunity against DNA viruses, it is also exploited to target plant RNA viruses. However, its applicability in mammalian cells against RNA has been successfully demonstrated by engineering *Franciscella novicida* Cas endonuclease (FnCas9) (Price et al., 2015). The important host factors of plants being utilized by RNA virus are eukaryotic translation initiation factors or host susceptibility genes, i.e., eIF4E, eIF(iso)4E and eIF4G (Sanfacon, 2015) which are acting as potential target site for inducing mutation through CRISPR-Cas9.

Chandrasekaran et al. (2016) has raised mutant cucumber line in cucumber (*Cucumis sativus*) by targeting translation initiation factor eIF4E gene by CRISPR/Cas9 and developed non-transgenic Cucumis eIF4E mutant. During testing of homozygous Cucumis mutant against a member of Potyviridae family they revealed resistance against cucumber vein yellowing virus (CVYV), Zucchini yellow mosaic virus (ZYMV), and Papaya ring spot mosaic virus-W (PRSV-W).

Moreover, in case of *Arabidopsis thaliana* isoform of eIF4E gene locus was targeted by Pyott et al. (2016) and introduced site-specific mutation by 1 bp insertion as well as 1 bp deletions within this gene. The edited plants showed complete resistance to Potyvirus, turnip mosaic virus (TuMV) without any significant off-target effect with its wild.

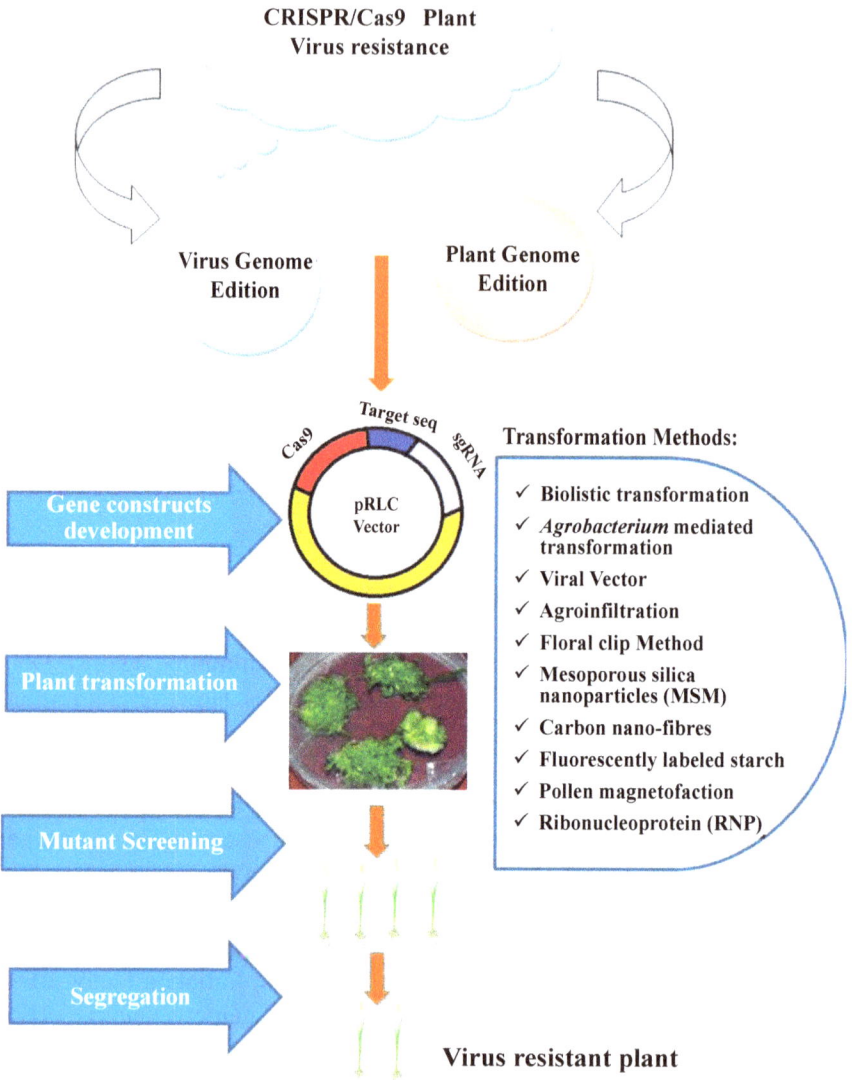

FIGURE 3.4   Flow chart representing CRISPR Cas9 based virus resistance development in crop plants.

The CRISPR technology and control of RNA viruses was also demonstrated by Zhang et al. (2018). They expressed FnCas9 and targeted CMV and TMV by using specific sgRNAs in *N. benthamiana* and *Arabidopsis* plants. They reported 40–80% reduced accumulation of viruses and obtained resistance stability of sgRNA-FnCas9 against CMV and TMV up to T6 generation. Surprisingly, Zhang et al. (2018) reported the RNA-binding activity of FnCas9 is responsible to make interference with the CMV genome. This would help to restrict the escape mechanism of mutated viral variants to the CRISPR/Cas9 machinery. In addition, FnCas9 system interferes RNA viruses into the cytoplasm itself and did not require nuclear localization and interference like geminiviruses.

Aman et al. (2018) utilized RNA-guided ribonuclease Cas13a for editing TuMV RNA genome and reported minimal replication and extend of TuMV in tobacco leaves. They have targeted four different viral genes namely, two targets on green fluorescent protein (GFP) and each of target site in the helper component proteinase silencing suppressor (HC-Pro), and coat protein (CP) gene. The effective virus interference was obtained in the attempt of edited HC-Pro and GFP2 genes and which could result into declining in multiplication of TuMV in tobacco.

A translation initiation factor 4 gamma gene (eIF4G) enabled to control of a recessive trait-based RTSV resistance in rice (*Oryza sativa*). Three gRNA's showing resistance reaction to RTSV were developed at SNPs nucleotide positions of 4387 and 4390 of eIF4G and induced CRISPR/Cas9 based mutation in the eIF4G alleles of rice (Macovei et al., 2018). Further, *Agrobacterium*-mediated transformation was attempted into IR-64 cultivar and progenies were advanced up to T2 and T3 generation. They converted IR-64 a susceptible cultivar of rice to the resistant for tungro spherical virus (RTSV) without finding any off-target activity.

3.7   OBSTACLES AND POSSIBLE CHALLENGES AHEAD

The occurrence of off-target effect is crucial challenge in CRISPR/ Cas9 technology. The mismatches in the sequence of sgRNA, maximum expression of Cas9 nuclease (Tsai and Joung, 2016), the location and accessibility of target site inside the host or target genome of organism, choice of promoter for sgRNA and Cas9 gene expression (Pattanayak et al., 2013) also influences the off-target effect. Moreover, high homology with the desired target in the genome, choice of endonuclease, lack of availability of genomic data, presence of cultivar to cultivar polymorphism within same species (Kadam et al., 2018) and polyploidy genome of plants makes difficult to induce target specific gene mutation. The chances of off-target activity found increased in the case of GC content of sgRNA are lower than 30%. In addition, the off-target activity could be minimized by reducing sgRNA length from 5' end up to 17–18 nucleotides (Fu et al., 2014).

Therefore appropriate selection of target site(s) rendering high efficiency of mutagenesis and prediction of less chances of off-target effect can be achieved by use of several online computational tools and servers towards designing of sgRNAs, gene constructs and data analysis (Hsu et al., 2013; Montague et al., 2014; Vats et al., 2019). Also, the use of web tool 'CRISPR-P' for designing sgRNAs (Lei et al., 2014), careful choice of the sgRNA sequence and various experimental techniques would help to avoid mismatch and may help to increase on-target genome editing efficiency (Mohanta et al., 2017; Cho et al., 2014).

3.8   CONCLUSION AND FUTURE PERSPECTIVES

In conclusion, CRISPR/Cas9 based advanced genome editing technique is being exploited for developing DNA and RNA virus resistance in crop plants. Beauty of this technology is to develop public acceptable virus resistant non-transgenic plants which can be escaped from regulatory concerns. Beside this, use of this technology can be explored in variety of way in agriculture systems by resolving the issue of its off-target activity, accuracy, and increase in the efficiency of the customized enzymes. Moreover, development of highly sophisticated bioinformatics programs, modification in endonucleases, gRNA modifications, and sensitive NGS-based detection methods are to be needed to increase the on-target genome editing efficiency and prevent the interruption of native gene function.

KEYWORDS

- bamboo mosaic potyvirus
- capsid protein
- carnation mottle virus necro virus
- Cas9
- genome editing
- plant innate immunity
- plant virus

REFERENCES

Adams, M. J., Hendrickson, R. C., Dempsey, D. M., & Lefkowitz, E. J., (2015). Tracking the changes in virus taxonomy. *Archives of Virology, 160*, 1375–1383.

Agbeci, M., Grangeon, R., Nelson, R. S., Zheng, H., & Laliberte, J. F., (2013). Contribution of host intracellular transport machineries to intercellular movement of turnip mosaic virus. *PLoS Pathogenes, 9*(10), e1003683.

Alazem, M., & Lin, N. S., (2015). Roles of plant hormones in the regulation of host-virus interactions. *Molecular Plant Pathology., 16*, 529–540.

Albar, L., Bangratz-Reyser, M., Hébrard, E., Ndjiondjop, M. N., Jones, M., & Ghesquière, A., (2006). Mutations in the eIF (iso) 4G translation initiation factor confer high resistance of rice to rice yellow mottle virus. *The Plant Journal, 47*(3), 417–426.

Alcaide-Loridan, C., & Jupin, I., (2012). Ubiquitin and plant viruses, let's play together. *Plant Physiology, 160*, 72–82.

Ali, Z., Abulfaraj, A., Idris, A., Ali, S., Tashkandi, M., & Mahfouz, M. M., (2015). CRISPR/Cas9-mediated viral interference in plants. *Genome Biology, 16*, 238.

Ali, Z., Ali, S., Tashkandi, M., Zaidi, S. S., & Mahfouz, M. M., (2016). CRISPR/Cas9-mediated immunity to Gemini viruses: Differential interference and evasion. *Science Reporter, 6*(26), 9–12.

Aman, R., Ali, Z., Butt, H., Mahas, A., Aljedaani, F., Zuhaib, K. M., Ding, S., & Mahfouz, M., (2018). RNA virus interference via CRISPR/Cas13a system in plants. *Genome Biology, 19*, 1.

Araki, M., & Ishii, T., (2015). Towards social acceptance of plant breeding by genome editing. *Trends in Plant Science, 1, 20*(3), 145–149.

Aronson, M. N., Meyer, A. D., Gyorgyey, J., Katul, L., Vetten, H. J., Gronenborn, B., & Timchenko, T., (2000). Clink, a nanovirus-encoded protein, binds both pRB and SKP1. *Journal of Virology, 74*, 2967–2972.

Ascencio-Ibáñez, J. T., Sozzani, R., Lee, T. J., Chu, T. M., Wolfinger, R. D., Cella, R., & Hanley-Bowdoin, L., (2008). Global analysis of *Arabidopsis* gene expression uncovers a complex array of changes impacting pathogen response and cell cycle during Gemini virus infection. *Plant Physiology, 148*(1), 436–454.

Baebler, Š., Witek, K., Petek, M., Stare, K., Tušek-Žnidarič, M., Pompe-Novak, M., Renaut, J., et al., (2014). Salicylic acid is an indispensable component of the Ny-1 resistance-gene-mediated response against potato virus Y infection in potato. *Journal of Experimental Botany, 65*(4), 1095–1109.

Baltes, N. J., & Voytas, D. F., (2015b). Enabling plant synthetic biology through genome engineering. *Trends in Biotechnology, 1, 33*(2), 120–131.

Baltes, N. J., Hummel, A. W., Konecna, E., Cegan, R., Bruns, A. N., & Bisaro, D. M., (2015a). Conferring resistance to Gemini viruses with the CRISPR-Cas prokaryotic immune system. *Nature Plants, 1*, 1–4.

Baltimore, D., (1971). Expression of animal virus genomes. *Bacteriol. Rev., 35*, 235–241.

Barajas, D., Martin, I. F., Pogany, J., Risco, C., & Nagy, P. D., (2014). Noncanonical role for the host Vps4 AAA+ ATPase ESCRT protein in the formation of tomato bushy stunt virus replicase. *PLoS Pathogens, 10*(4), e1004087.

Barrangou, R., Fremaux, C., Deveau, H., Richards, M., Boyaval, P., Moineau, S., Romero, D. A., & Horvath, P., (2007). CRISPR provides acquired resistance against viruses in prokaryotes. *Science, 315*, 1709–1712.

Bendahmane, A., Farnham, G., Moffett, P., & Baulcombe, D. C., (2002). Constitutive gain-of-function mutants in a nucleotide binding site-leucine rich repeat protein encoded at the Rx locus of potato. *Plant J., 32,* 195–204.

Bigeard, J., Colcombet, J., & Hirt, H., (2015). Signaling mechanisms in pattern-triggered immunity (PTI). *Molecular Plant, 8,* 521–539. doi: 10.1016/j.molp.2014.12.022.

Bilgin, D. D., Liu, Y., Schiff, M., & Dinesh-Kumar, S. P., (2003). P58IPK, a plant ortholog of double-stranded RNA-dependent protein kinase PKR inhibitor, functions in viral pathogenesis. *Developmental Cell, 4*(5), 651–661.

Bologna, N. G., & Voinnet, O., (2014). The diversity, biogenesis and activities of endogenous silencing small RNAs in *Arabidopsis. Annual Review Plant Biology, 65,* 473–503.

Brommonschenkel, S. H., Fray, A., & Tanksley, S. D., (2000). The broad-spectrum tospo virus resistance gene Sw-5 of tomato is a homolog of the root-knot nematode resistance gene. *Mol. Plant-Microbe Interact, 13,* 1130–1138.

Butler, N. M., Jiang, J., & Stupar, R. M., (2018). Crop improvement using genome editing. *Plant Breeding Reviews, 55*–101.

Butterbach, P., Verlaan, M. G., Dullemans, A. D., Lohuis, R. G., Visser, Y., & Bai, R. K., (2014). Tomato yellow leaf curl virus resistance by Ty-1 involves increased cytosine methylation of viral genomes and is compromised by cucumber mosaic virus infection. *Proceedings of the National Academy of Sciences USA, 11,* 12942–12947.

Calil, I. P., & Fontes, E. P. B., (2017). Plant immunity against viruses: Antiviral immune receptors in focus. *Annals of Botany, 119*(5), 711–723. https://doi.org/10.1093/aob/mcw200.

Camborde, L., Planchais, S., Tournier, V., Jakubiec, A., Drugeon, G., Lacassagne, E., Pflieger, S., Chenon, M., & Jupin, I., (2010). The ubiquitin-proteasome system regulates the accumulation of Turnip yellow mosaic virus RNA-dependent RNA polymerase during viral infection. *The Plant Cell, 22*(9), 3142–3152.

Canto, T., Uhrig, J. F., Swanson, M., Wright, K. M., & MacFarlane, S. A., (2006). Translocation of Tomato bushy stunt virus P19 protein into the nucleus by ALY proteins compromises its silencing suppressor activity. *Journal of Virology, 80,* 9064–9072.

Carette, J. E., Stuiver, M., Van, L. J., Wellink, J., & Van, K. A., (2000). Cowpea mosaic virus infection induces a massive proliferation of endoplasmic reticulum but not Golgi membranes and is dependent on De Novo membrane synthesis. *Journal of Virology, 74*(14), 6556–6563.

Carvalho, M. F., & Lazarowitz, S. G., (2004). Interaction of the movement protein NSP and the Arabidopsis acetyltransferase AtNSI is necessary for cabbage leaf curl Gemini virus infection and pathogenicity. *Journal of Virology, 78,* 11161–11171.

Carvalho, M. F., Turgeon, R., & Lazarowitz, S. G., (2006). The Gemini virus nuclear shuttle protein NSP inhibits the activity of AtNSI, a vascular-expressed *Arabidopsis* acetyltransferase regulated with the sink-to-source transition. *Plant Physiol., 140,* 1317–1330.

Chandrasekaran, J., Brumin, M., Wolf, D., Leibman, D., Klap, C., & Pearlsman, M., (2016). Development of broad virus resistance in non-transgenic cucumber using CRISPR/Cas9 technology. *Molecular Plant Pathology, 17,* 1140–1153.

Chaturvedi, S., & Rao, A. L., (2014). Live cell imaging of interactions between replicase and capsid protein of brome mosaic virus using bimolecular fluorescence complementation: Implications for replication and genome packaging. *Virology, 464,* 67–75.

Chauhan, P., Singla, K., Rajbhar, M., Singh, A., Das, N., & Kumar, K., (2019). A systematic review of conventional and advanced approaches for the control of plant viruses. *Journal of Applied Biology and Biotechnology, 7*(4), 89–98. https://doi.org/10.7324/jabb.2019.70414.

Chenon, M., Camborde, L., Cheminant, S., & Jupin, I., (2012). A viral deubiquitylating enzyme targets viral RNA-dependent RNA polymerase and affects viral infectivity. *EMBO Journal, 31*, 741–775.

Chisholm, S. T., Mahajan, S. K., Whitham, S. A., Yamamoto, M. L., & Carrington, J. C., (2000). Cloning of the *Arabidopsis* RTM1 gene, which controls restriction of long-distance movement of tobacco etch virus. *Proceedings of the National Academy of Sciences, USA, 97*, 489–494.

Cho, K., Van, M. B., Bahdanau, D., & Bengio, Y., (2014). *On the Properties of Neural Machine Translation: Encoder-Decoder Approaches, arXiv, 3*, 1409.1259.

Chou, Y. L., Hung, Y. J., Tseng, Y. H., Hsu, H. T., Yang, J. Y., Wung, C. H., Lin, N. S., et al., (2013). The stable association of virion with the triple-gene-block protein 3-based complex of Bamboo mosaic virus. *PLoS Pathogenes, 9*(6), e1003405.

Citovsky, V., & Zambryski, P., (1995). Transport of protein-nucleic acid complexes within and between plant cells. In: *Membrane Protein Transport* (Vol. 1, pp. 39–57).

Coll, N. S., Epple, P., & Dangl, J. L., (2011). Programmed cell death in the plant immune system. *Cell Death and Differentiation, 18*(8), 1247–1256.

Cosson, P. L., Sofer, Q. H., Le, V. L., Schurdi-Levraud, V., Whitham, S. A., Yamamoto, M. L., Gopalan, S., et al., (2010). RTM3, which controls long-distance movement of potyviruses, is a member of a new plant gene family encoding a meprin and TRAF homology domain-containing protein. *Plant Physiol., 154*, 222–232.

Cristea, S., Freyvert, Y., Santiago, Y., Holmes, M. C., Urnov, F. D., Gregory, P. D., & Cost, G. J., (2013). *In vivo* cleavage of transgene donors promotes nuclease-mediated targeted integration. *Biotechnology and Bioengineering, 110*(3), 871–880.

Cui, X., Wei, T., Chowda-Reddy, R. V., Sun, G., & Wang, A., (2010). The tobacco etch virus P3 protein forms mobile inclusions via the early secretory pathway and traffics along actin microfilaments. *Virology, 397*(1), 56–63.

Culver, J. N., & Padmanabhan, M. S., (2007). Virus-induced disease: Altering host physiology one interaction at a time. *Annual Review Phytopathology, 45*, 221–243.

Dangl, J. L., & Jones, D. J., (2001). Plant pathogens and integrated defense responses to infection. *Nature, 411*(6839), 826–833.

Deleris, A., Gallego-Bartolome, J., Bao, J., Kasschau, K. D., Carrington, J. C., & Voinnet, O., (2006). Hierarchical action and inhibition of plant Dicer-like proteins in antiviral defense. *Science, 313*, 68–71.

Den, B., Johan, A., & Ahlquist, P., (2010). Organelle-like membrane compartmentalization of positive-strand RNA virus replication factories. *Annual Review of Microbiology, 64*, 241–256.

Diaz, A., Wang, X., & Ahlquist, P., (2010). Membrane-shaping host reticulon proteins play crucial roles in viral RNA replication compartment formation and function. *Proceeding of Natural Academic Science, 107*(37), 16291–16296.

Diaz, A., Zhang, J., Ollwerther, A., Wang, X., & Ahlquist, P., (2015). Host ESCRT proteins are required for bromovirus RNA replication compartment assembly and function. *PLoS Pathogenes, 11*(3), e1004742.

Dielen, A. S., Sassaki, F. T., Walter, J., Michon, T., & Menard, G., (2011). The 20S proteasome alpha-5 subunit of *Arabidopsis thaliana* carries an RNase activity and interacts in planta with the lettuce mosaic potyvirus HcPro protein. *Molecular Plant Pathology, 12*, 137–150.

Dolja, V. V., McBride, H. J., & Carrington, J. C., (1992). Tagging of plant potyvirus replication and movement by insertion of beta-glucuronidase into the viral polyprotein. *Proceedings of the National Academy of Sciences of the United States of America, 89*, 10208–10212.

Dzahini-Obiatey, H. K., Domfeh, O., & Amoah, F. M., (2010). Over seventy years of a viral disease of cocoa in Ghana: From researchers' perspective. *African J. Agric. Res., 5*(7), 476–485.

Eini, O., Dogra, S., Selth, L. A., Dry, I. B., Randles, J. W., & Rezaian, M. A., (2009). Interaction with a host ubiquitin-conjugating enzyme is required for the pathogenicity of a geminiviral DNA β satellite. *Molecular Plant-Microbe. Interactions, 22*(6), 737–746.

Flor, H., (1971). Current status of the gene-for-gene concept. *Annual Review of Phytopathology, 9*(1), 275–296.

Fontes, E. P., Santos, A. A., Luz, D. F., Waclawovsky, A. J., & Chory, J., (2004). The Gemini virus nuclear shuttle protein is a virulence factor that suppresses transmembrane receptor kinase activity. *Genes Dev., 18,* 2545–2556.

Fu, Y., Sander, J. D., Reyon, D., Cascio, V. M., & Joung, J. K., (2014). Improving CRISPR-Cas nuclease specificity using truncated guide RNAs. *Nature Biotechnology, 32*(3), 279.

Gao, Z., Johansen, E., Eyers, S., Thomas, C. L., Noel, E. T. H., & Maule, A. J., (2004). The potyvirus recessive resistance gene, sbm1, identifies a novel role for translation initiation factor eIF4E in cell-to-cell trafficking. *The Plant Journal, 40*(3), 376–385.

Garcia-Ruiz, H., Takeda, A., Chapman, E. J., Sullivan, C. M., Fahlgren, N., Brempelis, K. J., & Carrington, J. C., (2010). Arabidopsis RNA-dependent RNA polymerases and dicer-like proteins in antiviral defense and small interfering RNA biogenesis during turnip mosaic virus infection. *The Plant Cell, 1, 22*(2), 481–496.

Gaur, R. K., Petrov, N. M., Patil, B. L., & Stoyanova, M. I., (2016). *Plant Viruses: Evolution and Management.* Springer.

Gouveia, B. C., Calil, I. P., Machado, J. P. B., Santos, A. A., & Fontes, E. P., (2017). Immune receptors and co-receptors in antiviral innate immunity in plants. *Frontiers in Microbiology, 7,* 2139.

Grangeon, R., Agbeci, M., Chen, J., Grondin, G., Zheng, H., & Laliberte, J. F., (2012). Impact on the endoplasmic reticulum and Golgi apparatus of turnip mosaic virus infection. *Journal of Virology, 86*(17), 9255–9265.

Guo, H. S., & García, J. A., (1997). Delayed resistance to plum pox potyvirus mediated by a mutated RNA replicase gene: Involvement of a gene-silencing mechanism. *Molecular Plant-Microbe Interactions, 10*(2), 160–170.

Gutierrez, C., (2002). Strategies for Gemini virus DNA replication and cell cycle interference. *Physiology Molecular Plant Pathology, 60,* 219–230.

Hanley, B., Linda, S., Settlage, B., & Robertson, B., (2004). Reprogramming plant gene expression: A prerequisite to Gemini virus DNA replication. *Molecular Plant Pathology, 5*(2), 149–156.

Harper, S. J., (2013). *Citrus tristeza virus*: Evolution of complex and varied genotypic groups. *Front Microbiology, 4,* 93. Available at: http://www.ncbi.nlm.nih.gov/pubmed/23630519 (accessed on 8 February 2021).

Hayes, A. J., Jeong, S. C., Gore, M. A., Yu, Y. G., Buss, G. R., & Tolin, S. A., (2004). Recombination within a nucleotide-binding-site/leucine-rich-repeat gene cluster produces new variants conditioning resistance to soybean mosaic virus in soybeans. *Genetics, 166,* 493–503.

Hille, F., Richter, H., Wong, S. P., Bratovič, M., Ressel, S., & Charpentier, E., (2018). The biology of CRISPR-Cas: Backward and forward. *Cell, 172*(6), 1239–1259. https://doi. org/10.1016/j.cell.2017.11.032.

Howard, A. R., Heppler, M. L., Ju, H. J., Krishnamurthy, K., Payton, M. E., & Verchot-Lubicz, J., (2004). Potato virus X TGBp1 induces plasmodesmata gating and moves between cells in several host species whereas CP moves only in *N. benthamiana* leaves. *Virology, 328*(2), 185–197.

Howden, A. J., & Huitema, E., (2012). Effector-triggered post-translational modifications and their role in suppression of plant immunity. *Frontiers in Plant Science, 16*, 3–160.

Hsu, P. D., Scott, D. A., Weinstein, J. A., Ran, F. A., Konermann, S., Agarwala, V., Li, Y., et al., (2013). DNA targeting specificity of RNA-guided Cas9 nucleases. *Nature Biotechnology, 31*(9), 827.

Huang, T. S., & Nagy, P. D., (2011). Direct inhibition of tombusvirus plus-strand RNA synthesis by a dominant negative mutant of a host metabolic enzyme, glyceraldehyde-3-phosphate dehydrogenase, in yeast and plants. *Journal of Virology, 85*(17), 9090–9102.

Hyodo, K., Mine, A., Taniguchi, T., Kaido, M., Mise, K., Taniguchi, H., & Okuno, T., (2013). ADP ribosylation factor 1 plays an essential role in the replication of a plant RNA virus. *Journal of Virology, 87*(1), 163–176.

Ishibashi, K., Masuda, K., Naito, S., Meshi, T., & Ishikawa, M., (2007). An inhibitor of viral RNA replication is encoded by a plant resistance gene. *Proceedings of the National Academy of Sciences of USA, 104*, 3833–13838.

Ishino, Y., Shinagawa, H., Makino, K., Amemura, M., & Nakata, A., (1987). Nucleotide sequence of the iap gene, responsible for alkaline phosphatase isozyme conversion in *E coli* and identification of the gene product. *J. Bacteriol., 169*, 5429–5433.

Islam, W., Zaynab, M., & M. Q. Z. W., (2017). Plant-virus interactions: Disease resistance in focus. *Hosts and Viruses, 4*(1), 5–20. https://doi.org/10.17582/journal.bjv/2017/4.1.5.20.

Jameson, P. C. S., (2002). Hormone-virus interaction in plants. *Critical Review of Plant Science, 21*, 205–228.

Jansen, R., Embden, J. D. A. V., Gaastra, W., & Schouls, L. M., (2002). Identification of genes that are associated with DNA repeats in prokaryotes. *Molecular Microbiology, 43*, 1565–1575.

Jenner, C. E., Wang, X., Tomimura, K., Ohshima, K., Ponz, F., & Walsh, J. A., (2003). The dual role of the potyvirus P3 protein of Turnip mosaic virus as a symptom and a virulence determinant in brassicas. *Mol. Plant Microbe. Interact., 16*(9), 777–784.

Ji, X., Zhang, H., Zhang, Y., Wang, Y., & Gao, C., (2015). Establishing a CRISPR-Cas-like immune system conferring DNA virus resistance in plants. *Nature Plants, 1*(10), 1–4.

Jiang, J., Patarroyo, C., Cabanillas, D. G., Zheng, H., & Laliberte, J. F., (2015). The vesicle-forming 6K2protein of turnip mosaic virus interacts with the COPII coatomer sec24a for viral systemic infection. *Virology, 89*(13), 6695–6710.

Jinek, M., Chylinski, K., Fonfara, I., Hauer, M., Doudna, J. A., & Charpentier, E., (2012). A programmable dual-RNA-guided DNA endonuclease in adaptive bacterial immunity. *Science, 337*, 816–821.

Jonczyk, M., Pathak, K. B., Sharma, M., & Nagy, P. D., (2007). Exploiting alternative subcellular location for replication: Tombusvirus replication switches to the endoplasmic reticulum in the absence of peroxisomes. *Virology, 362*(2), 320–330.

Jones, J. D., & Dangl, J. L., (2006). The plant immune system. *Nature, 444*, 323–329. https://doi. org/10.1038/nature05286.

Jones, R. A. C., & Naidu, R. A., (2019). Global dimensions of plant virus diseases: Current status and future perspectives. *Annual Review of Virology, 6*(1), 387–409. https://doi.org/10.1146/annurev-virology-092818-015606.

Jovel, J., Walker, M., & Sanfacon, H., (2011). Salicylic acid-dependent restriction of tomato ringspot virus spread in tobacco is accompanied by a hypersensitive response, local RNA silencing, and moderate systemic resistance. *Molecular Plant-Microbe Interactions, 24,* 706–718.

Kadam, U. S., Shelake, R. M., Chavhan, R. L., & Suprasanna, P., (2018). Concerns regarding 'off-target' activity of genome editing endonucleases. *Plant Physiology and Biochemistry, 131,* 22–30.

Kanchiswamy, C. N., Malnoy, M., & Maffei, M. E., (2015). Chemical diversity of microbial volatiles and their potential for plant growth and productivity. *Frontiers in Plant Science, 13*(6), 151.

Kang, B. C., Yeam, I., Frantz, J. D., Murphy, J. F., & Jahn, M. M., (2005). The pvr1 locus in capsicum encodes a translation initiation factor eIF4E that interacts with tobacco etch virus VPg. *Plant Journal, 42,* 392–405. https://doi.org/10.1111/ j.1365-313X.2005.02381.x.

Karavina, C., & Gubba, A., (2017). Iris yellow spot virus in Zimbabwe: Incidence, severity and characterization of allium-infecting isolates. *Crop Protection, 94,* 69–76.

Keen, N. T., (1990). Gene-for-gene complementarity in plant-pathogen interactions. *Annual Review of Genetics, 24*(1), 447–463.

Khatodia, S., Bhatotia, K., & Tuteja, N., (2017). Development of CRISPR/Cas9 mediated virus resistance in agriculturally important crops. *Bioengineered, 8*(3), 274–279.

Kim, S. H., Palukaitis, P., & Park, Y. I., (2002). Phosphorylation of cucumber mosaic virus RNA polymerase 2a protein inhibits formation of replicase complex. *The EMBO Journal, 21*(9), 2292–2300.

Kong, L. J., Orozco, B. M., Roe, J. L., Nagar, S., Ou, S., Feiler, H. S., Durfee, T., et al., (2000). A gemini virus replication protein interacts with the retinoblastoma protein through a novel domain to determine symptoms and tissue specificity of infection in plants. *The EMBO Journal., 19*(13), 3485–3495.

Kørner, C. J., Klauser, D., Niehl, A., Domínguez-Ferreras, A., Chinchilla, D., Boller, T., Heinlein, M., & Hann, D. R., (2013). The immunity regulator BAK1 contributes to resistance against diverse RNA viruses. *Molecular Plant-Microbe Interactions., 26*(11), 1271–1280.

Krab, I. M. C., Caldwell, D. R., & Gallie, J. F., (2005). BolCoat protein enhances translational efficiency of alfalfa mosaic virus RNAs and interacts with the eIF4G component of initiation factor eIF4FJ. *General Virology, 86,* 1841–1849.

Kumar, G., Jyothsna, M., Valarmathi, P., Roy, S., Banerjee, A., Tarafdar, J., Senapati, B. K., et al., (2019). Assessment of resistance to rice tungro disease in popular rice varieties in India by introgression of a transgene against rice tungro bacilliform virus. *Archives of Virology, 3, 164*(4), 1005–1013.

Kumar, P. L., Selvarajan, R., Iskra-Caruana, M. L., Chabannes, M., & Hanna, R., (2015). Biology, etiology, and control of virus diseases of banana and plantain. In: *Advances in Virus Research* (Vol. 91, pp. 229–269).

Laliberte, J. J., (2016). Membrane association for plant virus replication and movement. In: Zhou, X., & Wang, A., (eds.), *Current Research Topics in Plant Virology* (pp. 1–335). https://doi.org/10.1007/978-3-319-32919-2.

Laliberté, J., & Sanfaçon, H., (2010). Cellular remodeling during plant virus infection. *Annual Review of Phytopathology., 8*(48), 69–91.

Lanfermeijer, F. C., Dijkhuis, J., Sturre, M. J., De Haan, P., & Hille, J., (2003). Cloning and characterization of the durable tomato mosaic virus resistance gene Tm-2(2) from *Lycopersicon* esculentum. *Plant Mol. Biol., 52,* 1037–1049.

Lapidot, M., Karniel, U., Gelbart, D., Fogel, D., Evenor, D., Kutsher, Y., Makhbash, Z., et al., (2015). A novel route controlling begomovirus resistance by the messenger RNA surveillance factor pelota. *PLoS Genetics*, *11*, 1005538. https://doi.org/10.1371/ journal. pgen.1005538.

Lei, Y., Lu, L., Liu, H. Y., Li, S., Xing, F., & Chen, L. L., (2014). CRISPR-P: A web tool for synthetic single-guide RNA design of CRISPR-system in plants. *Molecular Plant Pathology*, *7*, 1494–1496. doi: 10.1093/mp/ssu044.

Lellis, A. D., Kasschau, K. D., Whitham, S. A., & Carrington, J. C., (2002). Loss-of-susceptibility mutants of *Arabidopsis thaliana* reveal an essential role for eIF(iso) 4E during potyvirus infection. *Current Biol.*, *12*, 1046–1051.

Lewsey, M., Palukaitis, P., & Carr, J. P., (2018). Plant-virus interactions: Defense and counter-defense. *Annual Plant Reviews Online*, 134–176. https://doi.org/10.1002/9781119312994. apr0366.

Li, Z., Pogany, J., Panavas, T., Xu, K., Esposito, A. M., Kinzy, T. G., & Nagy, P. D., (2009). Translation elongation factor 1A is a component of the tombusvirus replicase complex and affects the stability of the p33 replication co-factor. *Virology*, *385*(1), 245–260.

Ling, K. S., Harris, K. R., Meyer, J. D., Levi, A., Guner, N., Wehner, T. C., Bendahmane, A., & Havey, M. J., (2009). Non-synonymous single nucleotide polymorphisms in the watermelon eIF4E gene are closely associated with resistance to zucchini yellow mosaic virus. *Theory Applied Genetics*, *120*, 191–200. https://doi.org/10.1007/s00122009-1169-0.

Liu, L., Westler, W. M., Den, B. J. A., Wang, X., Diaz, A., Steinberg, H. A., & Ahlquist, P., (2009). An amphipathic alpha-helix controls multiple roles of brome mosaic virus protein 1a in RNA replication complex assembly and function. *PLoS Pathogenes*, *5*(3), e1000351.

Ma, J. F., Hou, X. L., Xiao, D., Qi, L., Wang, F., Sun, F. F., & Wang, Q., (2010). Cloning and characterization of the BcTuR3 gene related to resistance to turnip mosaic virus (TuMV) from non-heading Chinese cabbage. *Plant Mol. Biol. Rep.*, *28*, 588–596.

Macho, A. P., & Zipfel, C., (2014). Plant PRRs and the activation of innate immune signaling. *Molecular Cell*, *24*, *54*(2), 263–272.

Macovei, A., Sevilla, N. R., Cantos, C., Jonson, G. B., Slamet-Loedin, I., Čermák, T., Voytas, D. F., et al., (2018). Novel alleles of rice eIF4G generated by CRISPR/Cas9-targeted mutagenesis confer resistance to rice tungro spherical virus. *Plant Biotechnol Journal*, *16*(11), 1918–1927.

Maiti, S., Paul, S., & Pal, A., (2012). Isolation, characterization, and structure analysis of a nonTIR-NBS-LRR encoding candidate gene from MYMIV-resistant Vigna mungo. *Mol. Biotechnol.*, *52*, 217–233.

Makarova, K. S., Haft, D. H., Barrangou, R., Brouns, S. J., Charpentier, E., Horvath, P., Moineau, S., et al., (2011). Evolution and classification of the CRISPR-Cas systems. *Nature Reviews Microbiology*, *9*(6), 467–477.

Makarova, K. S., Wolf, Y. I., Alkhnbashi, O. S., Costa, F., Shah, S. A., Saunders, S. J., Barrangou, R., et al., (2015). An updated evolutionary classification of CRISPR-Cas systems. *Nature Reviews Microbiology*, *13*(11), 722–736.

Mandadi, K. K., & Scholthof, K. B. G., (2013). Plant immune responses against viruses: How does a virus cause disease?. *The Plant Cell*, *25*(5), 1489–1505.

Martin, G. B., Brommonschenkel, S. H., Chunwongse, J., Frary, A., Ganal, M. W., Spivey, R., Wu, T., et al., (1993). Map-based cloning of a protein kinase gene conferring disease resistance in tomato. *Science*, *262*(5138), 1432–1436.

Matthews, H. R., (2002). *Plant Virology* (4th edn.). Academic Press, New York, NY, USA.

McGarry, R. C., Barron, Y. D., Carvalho, M. F., Hill, J. E., Gold, D., Cheung, E., Kraus, W. L., & Lazarowitz, S. G., (2003). A novel *Arabidopsis* acetyl transferase interacts with the Gemini virus movement protein NSP. *Plant Cell*, *15*, 1605–1618.

Minato, N., Sok, S., Chen, S., Delaquis, E., Phirun, I., Le, V. X., Burra, D. D., et al., (2019). Surveillance for Sri Lankan cassava mosaic virus (SLCMV) in Cambodia and Vietnam one year after its initial detection in a single plantation in 2015. *PloS One, 14*(2).

Mohanta, Y. K., Panda, S. K., Bastia, A. K., & Mohanta, T. K., (2017). Biosynthesis of silver nanoparticles from protium serratum and investigation of their potential impacts on food safety and control. *Frontiers in Microbiology, 8,* 626.

Montague, T. G., Cruz, J. M., Gagnon, J. A., Church, G. M., & Valen, E., (2014). CHOPCHOP: A CRISPR/Cas9 and TALEN web tool for genome editing. *Nucleic Acids Research, 42*(W1), W401–W407.

Naderpour, M., Lund, O. S., Larsen, R., & Johansen, E., (2010). Potyviral resistance derived from cultivars of *Phaseolus vulgaris* carrying bc-3 is associated with the homozygotic presence of a mutated IF4E allele. *Mol. Plant Pathol., 11,* 255–263.

Nakashita, H., Yasuda, M., Nitta, T., Asami, T., Fujioka, S., Arai, Y., Sekimata, K., et al., (2003). Brassinosteroid functions in a broad range of disease resistance in tobacco and rice. *The Plant Journal, 33*(5), 887–898.

Neeleman, L., Olsthoorn, R. C., Linthorst, H. J., & Bol, J. F., (2001). Translation of a non-polyadenylated viral RNA is enhanced by binding of viral coat protein or polyadenylation of the RNA. *Proc. Nat. Acad. Sci. USA., 98,* 14286–14291.

Nicaise, V., (2014). Crop immunity against viruses: Outcomes and future challenges. *Frontiers in Plant Science, 5,* 1–18. https://doi.org/10.3389/fpls.2014.00660.

Nicaise, V., German-Retana, S., Sanjuan, R., Dubrana, M. P., Mazier, M., Maisonneuve, B., Candresse, T., et al., (2003). The eukaryotic translation initiation factor 4E controls lettuce susceptibility to the potyvirus lettuce mosaic virus. *Plant Physiol., 132,* 1272–1282. https://doi.org/10.1104/ pp.102.017855.

Nieto, C., Morales, M., Orjeda, G., Clepet, C., Monfort, A., Sturbois, B., Puigdomènech, P., et al., (2006). An eIF4E allele confers resistance to an uncapped and nonpolyadenylated RNA virus in melon. *Plant J., 48,* 452–462.

Ordon, F., Habekuss, A., Kastirr, U., Rabenstein, F., & Kühne, T., (2009). Virus resistance in cereals: Sources of resistance, genetics and breeding. *Journal of Phytopathology, 157*(9), 535–545. Available at: http://doi.wiley.com/10.1111/j.1439-0434.2009.01540.x.

Orjuela, J., Deless, E. F., Kolade, O., Chéron, S., Ghesquière, A., & Albar, L., (2013). A recessive resistance to rice yellow mottle virus is associated with a rice homolog of the CPR5 gene, a regulator of active defense mechanisms. *Mol. Plant Microbe. Interact, 26,* 1455–6143.

Padmanabhan, M. S., Goregaoker, S. P., Golem, S., Shiferaw, H., & Culver, J. N., (2005). Interaction of the tobacco mosaic virus replicase protein with the Aux/IAA protein PAP1/IAA26 is associated with disease development. *Journal of Virology, 79*(4), 2549–2558.

Padmanabhan, M. S., Shiferaw, H., & Culver, J. N., (2006). The Tobacco mosaic virus replicase protein disrupts the localization and function of interacting Aux/IAA proteins. *Molecular Plant-Microbe Interactions, 19*(8), 864–873.

Pallas, V., & Garcia, A. J., (2015). Viral factors involved in plant pathogenesis. *Current Opinion in Virology, 11,* 21–30.

Pattanayak, V., Lin, S., Guilinger, J. P., Ma, E., Doudna, J. A., & Liu, D. R., (2013). High-throughput profiling of off-target DNA cleavage reveals RNA-programmed Cas9 nuclease specificity. *Nature Biotechnology, 31*(9), 839.

Piroux, N., Saunders, K., Page, A., & Stanley, J., (2007). Gemini virus pathogenicity protein C4 interacts with *Arabidopsis* thaliana shaggy-related protein kinase AtSKg, a component of the brassinosteroid signaling pathway. *Virology, 362,* 428–440.

Pogany, J., White, K. A., & Nagy, P. D., (2005). Specific binding of tombus virus replication protein p33 to an internal replication element in the viral RNA is essential for replication. *Journal Virology, 79*(8), 4859–4869.

Price, A. A., Sampson, T. R., Ratner, H. K., Grakoui, A., & Weiss, D. S., (2015). Cas9-mediated targeting of viral RNA in eukaryotic cells. *Proc. Natl. Acad. Sci. USA, 112*, 6164–6169.

Pyott, D. E., Sheehan, E., & Molnar, A., (2016). Engineering of CRISPR/Cas9-mediated potyvirus resistance in transgene-free *Arabidopsis* plants. *Molecular Plant Pathology, 17*, 1276–1288.

Qu, F., Ye, X., & Morris, T. J., (2008). Arabidopsis DRB4, AGO1, AGO7, and RDR6 participate in a DCL4-initiated antiviral RNA silencing pathway negatively regulated by DCL1. *Proceedings of the National Academy of Sciences, USA, 105*, 14732–14737.

Rairdan, G. J., Collier, S. M., Sacco, M. A., Baldwin, T. T., Boettrich, T., & Moffett, P., (2008). The coiled-coil and nucleotide binding domains of the Potato Rx disease resistance protein function in pathogen recognition and signaling. *The Plant Cell, 20*, 739–751.

Rajendran, K. S., & Nagy, P. D., (2004). Interaction between the replicase proteins of tomato bushy stunt virus *in vitro* and *in vivo*. *Virology, 326*(2), 250–261.

Rathjen, J. P., & Moffett, P., (2003). Early signal transduction events in specific plant disease resistance. *Current Opinion Plant Biology, 6*, 300–306. https://doi.org/10.1016/S1369-5266(03)00057-8.

Reff, M., (1991). *Plant Virology* (p. 835). San Diego: Academic.

Reimer-Michalski, E. M., & Conrath, U., (2016). Innate immune memory in plants. In: *Seminars in Immunology* (Vol. 28, No. 4, pp. 319–327). Academic Press.

Ren, T., Qu, F., & Morris, T. J., (2000). HRT gene function requires interaction between a NAC protein and viral capsid protein to confer resistance to turnip crinkle virus. *Plant Cell, 12*, 1917–1926.

Revers, F., & Nicaise, V., (2001). Plant resistance to infection by viruses. *eLS, 30*.

Roossinck, M. J., (2010). Lifestyles of plant viruses. *Philosophical Transactions of the Royal Society B, 365*, 1899–1905.

Roossinck, M. J., (2015). Plants, viruses and the environment: Ecology and mutualism. *Virology, 479*, 271–277. https://doi.org/10.1016/j.virol.2015.03.041.

Roux, F., Voisin, D., Badet, T., Balagué, C., Barlet, X., Huard, C. C., Roby, D., & Raffaele, S., (2014). Resistance to phytopathogens every single one: Placing plant quantitative disease resistance on the map. *Molecular Plant Pathology, 15*(5), 427–432.

Ruffel, S., Dussault, M. H., Palloix, A., Moury, B., Bendahmane, A., Robaglia, C., & Caranta, C., (2002). A natural recessive resistance gene against Potato virus Y in pepper corresponds to the eukaryotic initiation factor 4E (eIF4E). *Plant Journal, 32*, 1067–1075.

Ruffel, S., Gallois, J. L., Lesage, M. L., & Caranta, C., (2005). The recessive potyvirus resistance gene pot-1 is the tomato orthologue of the pepper pvr2eIF4E gene. *Molecular Genetics and Genomics, 274*, 346–353.

Rybicki, E. P., (2015). A top ten list for economically important plant viruses. *Archives of Virology, 1, 160*(1), 17–20.

Sahana, N., Kaur, H., Basavaraj, F. T., Jain, R. K., Palukaitis, P., Canto, T., & Praveen, S., (2012). Inhibition of the host proteasome facilitates papaya ringspot virus accumulation and proteosomal catalytic activity is modulated by viral factor HcPro. *PLoS One, 7*(12).

Samuels, T. D., Ju, H. J., Ye, C. M., Motes, C. M., Blancaflor, E. B., & Verchot-Lubicz, J., (2007). Subcellular targeting and interactions among the potato virus X TGB proteins. *Virology, 367*(2), 375–389.

Sanfaçon, H., (2015). Plant translation factors and virus resistance. *Viruses, 7*(7), 3392–3419.

Sasaya, T., Nakazono-Nagaoka, E., Saika, H., Aoki, H., Hiraguri, A., Netsu, O., Ichiki, T. U., et al., (2014). Transgenic strategies to confer resistance against viruses in rice plants. *Front Microbiol., 4*, 409. Available at: http://www.ncbi.nlm.nih.gov/pubmed/24454308 (accessed on 8 February 2021).

Sastry, K. S., & Zitter, T. A., (2014). Management of virus and viroid diseases of crops in the tropics. In: *Plant Virus and Viroid Diseases in the Tropics* (pp. 149–480). Springer Netherlands, Dordrecht, The Netherlands.

Schwartz, M., Chen, J., Janda, M., Sullivan, M., Den, B. J., & Ahlquist, P., (2002). A positive-strand RNA virus replication complex parallels form and function of retrovirus capsids. *Molecular Cell, 9*(3), 505–514.

Seo, Y. S., Rojas, M. R., Lee, J. Y., Lee, S. W., Jeon, J. S., Ronald, P., Lucas, W. J., & Gilbertson, R. L., (2006). A viral resistance gene from common bean functions across plant families and is up-regulated in a non-virus-specific manner. *Proceedings of the National Academy of Sciences of USA, 103*, 11856–11861.

Shan, Q., Wang, Y., Li, J., Zhang, Y., Chen, K., Liang, Z., Zhang, K., et al., (2013). Targeted genome modification of crop plants using a CRISPR-Cas system. *Nature Biotechnology, 31*(8), 686–688.

Shepherd, A. R., Chatterton, J. E., Clark, A. F., & Wax, M. B., (2009). Alcon, Inc. *RNAi-Mediated Inhibition of Ocular Targets*. U.S. Patent., 7,592,324.

Soitamo, A. J., Jada, B., & Lehto, K., (2012). Expression of geminiviral AC2 RNA silencing suppressor changes sugar and jasmonate responsive gene expression in transgenic tobacco plants. *BMC Plant Biology, 12*, 204.

Sorek, R., Lawrence, C. M., & Wiedenheft, B., (2013). CRISPR-mediated adaptive immune systems in bacteria and archaea. *Annual Review of Biochemistry, 82*, 237–266.

Stanley, J., (1985). The molecular biology of Gemini viruses. *Adv. Virus Res., 30*, 139–177.

Stein, N., Perovic, D., Kumlehn, J., Pellio, B., Stracke, S., Streng, S., Ordon, F., & Graner, A., (2005). The eukaryotic translation initiation factor 4E confers multiallelic recessive bymovirus resistance in *Hordeum vulgare* (L.). *Plant J., 42*, 912922.

Sun, L., Andika, I. B., Shen, J., Yang, D., & Chen, J., (2014). The P2 of Wheat yellow mosaic virus rearranges the endoplasmic reticulum and recruits other viral proteins into replication-associated inclusion bodies. *Mol. Plant Pathol., 15*(5), 466–478.

Takahashi, H., Miller, J., Nozaki, Y., Takeda, M., Shah, J., Hase, S., Ikegami, M., et al., (2002). RCY1, an *Arabidopsis thaliana* RPP8/HRT family resistance gene, conferring resistance to cucumber mosaic virus requires salicylic acid, ethylene and a novel signal transduction mechanism. *Plant J., 32*, 655–667.

Tang, X., Frederick, R. D., Zhou, J., Halterman, D. A., Jia, Y., & Martin, G. B., (1996). Initiation of plant disease resistance by physical interaction of AvrPto and Pto kinase. *Science, 274*(5295), 2060–2063.

Taxonomy, (2015). *Archives of Virology, 160*, 1375–1383.

Tilsner, J., Linnik, O., Wright, K. M., Bell, K., Roberts, A. G., Lacomme, C., Santa, C. S., & Oparka, K. J., (2012). The TGB1 movement protein of potato virus X reorganizes actin and endomembranes into the X-body, a viral replication factory. *Plant Physiol., 158*(3), 1359–1370.

Tomita, R., Sekine, K. T., Mizumoto, H., Sakamoto, M., Murai, J., Kiba, A., Hikichi, Y., et al., (2011). Genetic basis for the hierarchical interaction between *Tobamovirus* spp. and L resistance gene alleles from different pepper species. *Mol. Plant Microbe Interaction, 24*, 108–117.

Tsai, S. Q., & Joung, J. K., (2016). Defining and improving the genome-wide specificities of CRISPR-Cas9 nucleases. *Nature Reviews Genetics, 17*(5), 300.

Uhrig, J. F., Canto, T., Marshall, D., & MacFarlane, S. A., (2004). Relocalization of nuclear ALY proteins to the cytoplasm by the tomato bushy stunt virus P19 pathogenicity protein. *Plant Physiol., 135,* 2411–2423.

Unniyampurath, U., Pilankatta, R., & Krishnan, M. N., (2016). RNA interference in the age of CRISPR: Will CRISPR interfere with RNAi? *International Journal of Molecular Sciences, 17*(3), 291.

Vallejos, C. E., Astua-Monge, G., Jones, V., Plyler, T. R., Sakiyama, N. S., & Mackenzie, S. A., (2006). Genetic and molecular characterization of the I locus of *Phaseolus vulgaris. Genet., 172,* 1229–1242.

Vats, S. L., Kumawat, S., Kumar, V., Patil, G. B., Joshi, T., Sonah, H., Sharma, T. R., & Deshmukh, R., (2019). Genome editing in plants: Exploration of technological advancements and challenges. *Cells, 8,* 1386.

Vidal, S., Cabrera, H., Andersson, R. A., Fredriksson, A., & Valkonen, J. P., (2002). Potato gene Y-1 is an N gene homolog that confers cell death upon infection with potato virus Y. *Mol. Plant Microb. Interact., 15,* 717–727.

Wale, S., Platt, B., & Cattlin, N. D., (2008). *Diseases, Pests and Disorders of Potatoes: A Color Handbook.* CRC Press, Boca Raton, FL. Available at: https://bsppjournals.onlinelibrary. wiley.com/toc/13653059/2008/57/5 (accessed on 8 February 2021).

Wan, J., Cabanillas, D. G., Zheng, H., & Laliberte, J. F., (2015). Turnip mosaic virus moves systemically through both phloem and xylem as membrane-associated complexes. *Plant Physiol., 167*(4), 1374–1388.

Wang, M. B., Bian, X. Y., Wu, L. M., Liu, L. X., Smith, N. A., Isenegger, D., Wu, R. M., et al., (2004). On the role of RNA silencing in the pathogenicity and evolution of viroids and viral satellites. *Proceedings of the National Academy of Sciences, 101*(9), 3275–3280.

Wang, X., Lee, W. M., Watanabe, T., Schwartz, M., Janda, M., & Ahlquist, P., (2005). Brome mosaic virus 1a nucleoside triphosphatase/helicase domain plays crucial roles in recruiting RNA replication templates. *Journal of Virology, 79*(21), 13747–13758.

Wei, T., Zhang, C., Hou, X., Sanfacon, H., & Wang, A., (2013). The SNARE protein Syp71 is essential for turnip mosaic virus infection by mediating fusion of virus-induced vesicles with chloroplasts. *PLoS Pathog., 9*(5), e1003378.

Whitham, S. A., Anderberg, R. J., Chisholm, S. T., & Carrington, J. C., (2000). Arabidopsis RTM2 gene is necessary for specific restriction of tobacco etch virus and encodes an unusual small heat shock-like protein. *Plant Cell, 12,* 569–582.

Whitham, S., Dinesh-Kumar, S. P., Choi, D., Hehl, R., Corr, C., & Baker, B., (1994). The product of the tobacco mosaic virus resistance gene N: Similarity to toll and the interleukin-1 receptor. *Cell, 78,* 1101–1115. https://doi.org/10.1016/00928674(94)90283-6.

Wicker, T., Zimmermann, W., Perovic, D., Paterson, A. H., Ganal, M., Graner, A., & Stein, N., (2005). A detailed look at 7 million years of genome evolution in a 439 kb contiguous sequence at the barley Hv-eIF4E locus: Recombination, rearrangements and repeats. *Plant Journal, 41,* 184–194. https://doi.org/10.1111/j.1365313X.2004.02285.x.

Win, J., Chaparro-Garcia, A., Belhaj, K., Saunders, D. G. O., Yoshida, K., Dong, S., Schornack, S., et al., (2012). Effector biology of plant-associated organisms: Concepts and perspectives. In: *Cold Spring Harbor Symposia on Quantitative Biology* (Vol. 77, pp. 235–247). doi: 10.1101/ sqb.2012.77.015933.

Wu, Z., Mo, C., Zhang, S., & Li, H., (2018). Characterization of papaya ringspot virus isolates infecting transgenic papaya 'Huanong No. 1'in South China. *Scientific Reports*, *8*(1), 1–11.

Yamaji, Y., Maejima, K., Ozeki, J., Komatsu, K., Shiraishi, T., Okano, Y., Himeno, M., et al., (2012). Lectin-mediated resistance impairs plant virus infection at the cellular level. *Plant Cell, 24*, 778–793.

Yamanaka, T., Imai, T., Satoh, R., & Kawashima, A. T. M., (2002). Complete inhibition of tobamo virus multiplication by simultaneous mutations in two homologous host genes. *Journal of Virology*, *76*, 2491–2497.

Yang, K. Y., Liu, Y., & Zhang, S., (2001). Activation of a mitogen-activated protein kinase pathway is involved in disease resistance in tobacco. *Proceedings of the National Academy of Sciences of USA*, *98*, 741–746. https://doi.org/10.1073/ pnas.98.2.741.

Yoshii, M., Nishikiori, M., Tomita, K., Yoshioka, N., Kozuka, R., Naito, S., & Ishikawa, M., (2004). The *Arabidopsis* cucumovirus multiplication 1 and 2 loci encode translation initiation factors 4E and 4G. *J. Virol., 78*, 6102–6111.

Zaidi, S. S. E. A., Tashkandi, M., & Mahfouz, M. M., (2017). Engineering molecular immunity against plant viruses. In: *Progress in Molecular Biology and Translational Science* (Vol. 149, pp. 167–186).

Zaitlin, M., (1999). Elucidation of the genome organization of tobacco mosaic virus. *Philosophical Transactions of the Royal Society of London. Series B: Biological Sciences, 354*(1383), 587–591.

Zhang, D., Li, Z., & Li, J. F., (2015). Genome editing: New antiviral weapon for plants. *Nature Plants, 1*(10), 1–2.

Zhang, T., Zheng, Q., Yi, X., An, H., Zhao, Y., Ma, S., & Zhou, G., (2018). Establishing RNA virus resistance in plants by harnessing CRISPR immune system. *Plant Biotechnology Journal, 16*(8), 1415–1423.

Zhang, X., Zhang, X., Singh, J., Li, D., & Qu, F., (2012). Temperature-dependent survival of Turnip crinkle virus-infected *Arabidopsis* plants relies on an RNA silencing-based defense that requires dcl2, AGO2, and HEN1. *Journal of Virology, 15, 86*(12), 6847–6854.

Zhu, S., Gao, F., Cao, X., Chen, M., Ye, G., Wei, C., & Li, Y., (2005). The rice dwarf virus P2 protein interacts with Ent-kaurene oxidases *in vivo*, leading to reduced biosynthesis of gibberellins and rice dwarf symptoms. *Plant Physiology, 139*(4), 1935–1945.

Zwart, M. P., Daròs, J. A., & Elena, S. F., (2012). Effects of potyvirus effective population size in inoculated leaves on viral accumulation and the onset of symptoms. *Journal of Virology, 86*, 9737–9747.

CHAPTER 4

Recent Insights in Detection and Diagnosis of Plant Viruses Using Next-Generation Sequencing Technologies

SAJAD UN NABI,[1] NIDA YOUSUF,[1] MANOJ K. YADAV,[2]
DEVENDRA K. CHOUDHARY,[3] DILSHAD AHMAD,[3]
SHOAIB N. KIRMANI,[1] PARVEEZ SHEIKH,[4] and IQBAL AHMED[3]

[1]ICAR-Central Institute of Temperate Horticulture, Srinagar–191132, Jammu and Kashmir, India, E-mail: sajad_patho@rediffmail.com (S. Un Nabi)

[2]ICAR-National Rice Research Institute, Cuttack–753006, Odisha, India

[3]ICAR-Indian Agricultural Research Institute, New Delhi–110012, India

[4]Sher-e-Kashmir University of Agricultural Science and Technology, Srinagar, Jammu and Kashmir, India

ABSTRACT

Traditional approaches like immunodiagnostics or PCR targets the viruses that have historically been associated with known diseases. These tests may not result in an accurate reflection of the etiological status of the tested plant. However, new and robust techniques could help in knowing the proper etiology of disease complexes, identification of new viruses and give an idea about the frequency of viruses present in either infected or uninfected plant material. These technologies widely branded as next-generation/deep sequencing technologies (NGS) has broadened the contours in diagnostics and improved our knowledge of the viruses infecting plants. These technologies have become available at the inception of the 21st century and provided a highly efficient, robust, cheap, and fast sequencing platform ahead of the standard Sanger sequencing technologies. In plant virology, these technologies found universal scope in numerous areas, viz., detection, genome organization, characterization, discovery of novel viruses, replication, and transcription. Numerous sequencing technologies have widespread uses, capable of reading thousands to billions of DNA sequences per run. These technologies were new, more competitive and robust were commercialized in large scale. From these recent advances of major importance, it is anticipated that NGS technologies along with CRISPR-CAS will play a noteworthy role in the diagnosis and management of viral diseases in near future.

4.1   INTRODUCTION

Viruses are ubiquitous pathogens, containing small genome either DNA or RNA enveloped by capsid proteins (CPs) with or without a lipid envelope and ranging from 20 to 2000 nm in size (Sastry, 2013). Viruses being obligate in nature hence are difficult to detect and identify as compared to other cellular organisms. As in viruses, no single gene being common, which could be used as a general diagnostic test, hence it is imperative, that these viruses before becoming established and causing potential harm by spreading could be detected and prevented at entry level. The first technique developed was biological indexing, in which virus infected suspected samples were inoculated onto a sensitive indicator host plant, which was further observed for symptom development (Thompson et al., 2011). Symptomology is characteristic feature for virus species/indicator host species combinations, however, despite its effectiveness; it is labor intensive, less robust and time-consuming (Van et al., 1987). To make a robust detection, a choice of methods have been employed, including electron microscopy (EM), Immunosorbant EM, enzyme linked immunosorbant assays (ELISA) and nucleic acid based methods viz., polymerase chain reaction (PCR) with improved specificity, robustness, and sensitivity (Nabi et al., 2018). These techniques provided cheap and rapid detection for the known viruses (Yadav and Khurana, 2016). However, all these techniques dependent on antibodies and primer sequences developed from the characterized viruses and were proved ineffective in diseases with unknown etiology or diseases caused by a mixture of viruses or viroids that share little or no sequence similarity with previously described ones.

4.2   NEXT GENERATION SEQUENCING (NGS)

With demand of producers for access to the latest plant varieties, the efficient, rapid, cheap methods of detection to facilitate the movement of plants across borders while maintaining phytosanitary standards are need of the hour. These methods could replace many of the existing tools currently used by quarantine facilities (Varvara et al., 2018). An accurate and rapid method to fully index viruses in a sample by inspection and quarantine services has been a long sought, but unattainable goal (Jo et al., 2016). Advances in sequencing technology from last two decades have led to the development of new approaches for the detection, identification, and characterization of viruses. In the early 21st century, efforts were made towards the development of new and novel sequencing methods to substitute the first-generation technology, i.e., Sanger sequencing (Voelkerding et al., 2009; Niedringhaus et al., 2011). The new approaches are referred to as next-generation sequencing (NGS) which has changed the scientific approaches in both applied and basic research in many disciplines, including plant virology (Barzon et al., 2011; Quingfa et al., 2015; Jones et al., 2017). It is sequencing of total nucleic acid content in diseased samples for subsequent identification of pathogen(s) by bioinformatics tools. The metagenomics approach does not require prior information of the pathogens and can potentially identify both the known, novel, and divergent viruses in a sample. These technologies have effectively lowered the costs of sequencing, beyond what is possible with the standard dye-terminator methods. The major advance in NGS is the production of large voluminous data (1 billion and more short reads per run) with lengths between 25 bp and has advanced to produce reads longer than 1000 bp. It has ability to deliver robust, fast, and accurate genome information (Varvara et al., 2018). The VirFind and Virus Detect are specific workflows for plant virus detection and discovery from NGS data. The traditional diagnostic techniques, viz., ELISA, multiplex PCR (M-PCR), and bioassays could be eliminated in favor of a single NGS assay, followed up with individual validation for positive identification. This would reduce the effort, time, and cost to a great extent associated with importing foreign planting material, enabling industry to more rapidly respond to market demands for new plant cultivars developed around the world (Miller et al., 2010).

4.3 HISTORICAL DEVELOPMENTS

- In 2000, the first NGS technology, massively parallel signature sequencing (MPSS) was launched by Lynx Therapeutics (USA) Company.
- In 2004, parallel version pyrosequencing was marketed by 454 Life Sciences (Branford, CT, USA) as second new generation sequencing

technology. This sequencing has reduced 6-fold costs and provides intermediate read length.

- In 2005, Roche sequencing platform introduced 454 GS 20, produced 20 million bases. Also, in 2005, Solexa released the genome analyzer (GA), which was based on sequencing by synthesis (SBS). This sequencing generates up to 50 billion bases with latest model up to 85 billion bases per run (Elizabeth et al., 2014).
- In 2007, 454 GS 20 Roche was replaced by the GS FLX model, capable of producing over 100 million bp of sequence in just a few (4) hours. This model was further upgraded to the 454 GS-FLX+ Titanium sequencing platform, producing over 600 Mbp of sequence data in a single run.
- In 2007, SOLiD technology employed sequencing by oligo ligation detection. The quantity and length of sequences were comparable to Illumina sequencing. The sequencing technology was later purchased by applied biosystems (AB) and released the first SOLiD sequencing system and in 2010, released SOLiD 5500 w and 5500 xlw systems. The SOLiD has read lengths of 85 bp, with 99.99% accuracy and 30 Gbp per run (Simone et al., 2017).
- Illumina has developed the series of HiSeq sequencing platforms (HiSeq1000, HiSeq 1500, HiSeq 2000, and HiSeq 2500) which varies in their run times, outputs, cluster generations, read length, and paired end reads. In 2011, it has also released a bench top platform MiSeq, generating 1.5 Gbp per run in about 10 h (Nicholas et al., 2013).
- Other recently developed sequencing technologies include: Helicos sequencer (2009), Ion Torrent sequencer (2011), Pacific Biosciences (Pac Bio), single molecule real-time (SMRT) sequencer (2011) and Oxford Technologies Nanopore (2013) (Marina et al., 2014; Rhoads et al., 2015).

4.4 STEPS OF NEXT-GENERATION SEQUENCING TECHNOLOGIES

Typically identification and detection of viruses and viroids through metagenomics technique includes sample preparation, DNA sequencing through NGS platform followed by bioinformatics analysis (Qingfa et al., 2015).

4.4.1 ENRICHMENT OF PLANT VIRAL SEQUENCES

Numerous strategies have been employed to enrich the sequencing reads which are specific for particular viruses for their better diagnosis. Owing to vast abundance of the host rRNAs, as well as their reduction from total RNA preparations will enhance the quantum of specific reads of virus and viroid as compared to that of host and non-host origin. Likewise, rRNA reduction could also be achieved through sequencing the polyadenylated transcripts (poly-A RNAs) only and further enrichment of poly-A RNAs for NGS led to the detection of novel RNA and DNA viruses. Further, the enrichment will also remove the reads from several RNA virus families which do not produce poly-A RNAs, including, *Cucumber mosaic virus* (CMV), *Tobacco mosaic virus* (TMV), and *Turnip yellow mosaic virus*. Therefore, enrichment strategy for the subsequent molecules has been more widespread in the viral metagenomics studies.

4.4.1.1   DOUBLE STRANDED RNAs

RNA viruses and viroids synthesized a double stranded RNA (dsRNA) as replicative intermediates and in normal conditions dsRNA are not produced by plants. The total dsRNA sequencing will greatly enhance the proportion of viruses and viroids reads specific to them. It was reported after performing dsRNA enrichment, the deep sequencing of dsRNA the viral reads has increased from 2% to 50%. On the other hand, six of the seven novel viruses discovered through dsRNA sequencing possess RNA genomes, probably due to the fact that plant DNA viruses do not produce adequate long dsRNA in their life cycle.

4.4.1.2   VIRUS-LIKE PARTICLES (VLPs)

The genetic material, of virus RNA or DNA enclosed in viral particles get protection from DNase and RNase treatments. Thus, VLPs enrichment through homogenization, filtration, and ultracentrifugation has been comprehensively utilized for the discovery of virus in plentiful plant samples. The VLP preparations included bacteria as well as contaminating mitochondria, so VLPs are frequently treated with chloroform which causes the disruption of mitochondrial and bacterial membranes prior to nuclease digestion and the VLP-associated nucleic acids extraction for deep sequencing. Unfortunately, enveloped viruses showed sensitiveness toward chloroform treatment. Furthermore, since it needs development of specific protocols for successful virion purification of several plants it is very difficult to capture all viruses through the same protocol used for VLP enrichment. Nonetheless, next generation sequencing (NGS) of the TNS from the VLPs obtained from plants led to the detection and recognition of DNA and RNA viruses (Table 4.2). Since, total nucleic acid preparation obtained from both VLP and dsRNA is very low, an additional step needs to be performed to amplify the extracted nucleic acids through PCR or reverse transcription PCR (RT-PCR) in a sequence-independent manner is essential prior to the library construction for deep sequencing. An improved rolling-circle amplification technique to amplify circular dsDNA viral genomes is available (Hunter et al., 2012).

4.4.1.3   SMALL RNAs

Replication of viruses (RNA and DNA) as well as subviral agents (viroids and satellite RNAs) in plants, induced enormously buildup of the pathogen-specific siRNAs, which correspond to nearly 30% of total small RNAs sequenced from the infected plants (Fatima and Aline, 2019). The earliest deep-sequencing report of viral siRNAs has revealed that viral siRNAs actually overlap each other extensively, in spite of being just 21 to 24 nt long. Ubiquitous production in different eukaryotic hosts and the overlapping property of viral siRNAs have been used separately by two groups to employ a novel strategy for the discovery of virus through enrichment and small RNAs sequencing. In this approach, small RNAs are enriched from infected plant tissues using NGS platforms and assembled into large sequence contigs/fragments that are further used in virus discovery since these sequences are obtained from dsRNAs and VLPs (Jones et al., 2017).

4.4.2   BIOINFORMATIC ANALYSIS

Before the sequence assembly and pathogen identification, the generated raw data through NGS platforms should be preprocessed to eliminate adaptors as well as low-quality sequences. The threshold value and standard parameters are generally provided by the manufacturer. Now, it is a standardized procedure on "older" technologies like 454/Roche pyrosequencing or Illumina SBS (Rhoades et al., 2015). For multiplex sequencing of diverse libraries in a single lane, an additional step of de-multiplexing by means of barcodes built-in the PCR primers is essential prior to sequence assembly. When the host genome sequence is available, in silico removal of host-specific sequences prior to assembly will expedite the downstream bioinformatics analysis. It should be noted that the removal of adaptor and computational filtering of host sequences can guide to other artifacts (Menzel et al., 2016). The assembly of the preprocessed reads can be performed using a number of mainstream algorithms which are publicly available, including Velvet, Oases, and VCAKE (Miller et al., 2010; Haas et al., 2013; Souvorov et al., 2018). The parameters used in sequence assembly are alike to those for genome assembly and defined by the employed algorithm. Consequently, the assembled contig sequences are queried with the help of homology search tools against previously reported sequences stored in local database or public databases for example, GenBank. A familiar way is to evaluate the assembled sequences with the non-redundant nucleotide database of GenBank with the help of a BLAST package. The two most frequently used programs for comparison of nucleotide and amino acid sequences are BLASTn and BLASTx, respectively. Novel and known viruses are easily identified when contigs illustrate high resemblance (>90% similarity and 85% coverage) with an identified virus. When a contig represents distant homology with an identified virus, specially only at the protein level, often, the contig represents a novel virus that can be assigned taxonomically only at the virus family level (Qingfa et al., 2015).

4.5 TYPES AND ADVANTAGES OF NEXT-GENERATION SEQUENCING

The narration of different types of NGS technologies, their applications are presented in Tables 4.1 and 4.2. The various advantages of NGS technologies are:

- The sequence can be obtained from a single strand, whereas multiple reads are required in Sanger sequencing.
- It is quicker as the chemical reaction may be combined with the signal detection and cheaper, as it requires less reagents and manpower.
- It is more precise and accurate, as each segment of DNA or RNA is sequenced multiple times.

4.6 APPLICATIONS OF NEXT-GENERATION SEQUENCING IN PLANT VIRUS DETECTION

Since 2009–2010, the NGS technologies along with complicated bioinformatics tools were effectively employed and have revolutionized the field of

TABLE 4.1 Narration of Different NGS Technologies Along with Their Read Length, Amplification Method, Accuracy, and Sequencing Chemistry

Sequencing Platform	Amplification Method	Sequencing Chemistry	Run Time	Read Length (bp)	Sequencing Speed/h	Maximum Output per run	Accuracy (%)
454 (Roche)	Emulsion PCR	Pyrosequencing	10–23 h	400–700	13 Mbp	700 Mbp	99.9%
Illumina	Bridge PCR	Reversible terminators	8–14 days	100–300	25 Mbp	600 Gbp	99.9%
SOLiD (life technologies)	Emulsion PCR	Ligation	7–12 days	75–85	21–28 Mbp	80–360 Gbp	99.9%
PacBio (pacific biosciences)	No amplification single molecule real-time (or SMRT)	Fluorescently labeled nucleotides	30 min	4,000–5,000	50–115 Mbp	200 Mb–1 Gbp	95%
Helicos (helicos biosciences)	No amplification single molecule	Reversible terminators	8 days	25–55	83 Mbp	35 Gbp	97%
Ion torrent (life technologies)	Emulsion PCR	Detection of released H	4.5 hrs	100–400	25 Mb–16 Gbp	100 Mb–64 Gbp	99%
Nanopore (Oxford technologies)	No amplification single molecule	–	60 min	Very long reads up to 50 kbp	150 Mbp	Tens of Gbp	96%

Source: Adapted from Qingfa et al. (2014); Varvara et al. (2018).

TABLE 4.2 Major NGS Platforms and Their Applications

Platform Systems	Applications
454 GS FLX	DNA sequencing: whole genome sequencing, *de novo* and re-sequencing of large genes in a single run with up to 1 kbp read length. RNA sequencing, sequencing capture metagenomics.
Illumina HiSeq systems	DNA sequencing: Candidate region targeted sequencing; RNA sequencing: transcriptome analysis, gene expression profiling, small RNA and mRNA sequencing, Epigenetic sequencing: chromatin immune-precipitation sequencing (ChIP-Seq), methylation analysis through sequencing.
SOLiD	DNA sequencing: whole genome and exome, RNA sequencing, Epigenetic sequencing
PacBio RS II	DNA sequencing through single molecule real-time (SMRT) system with the long read lengths. Characterization of genetic variation, methylation, InDels, structural variants analysis, haplotypes, and phasing, understanding gene expression through base modification detection,
Helicos genetic analyzer system	DNA sequencing and RNA sequencing.
Ion torrent ion PGM system	Semiconductor sequencing with 400-bp length-best for sequencing of small genes and genomes. DNA sequencing for microbial: genes and genomes, amplicons, and exomes (unrevealed disease-causing variants), viral typing and other microbial typing, targeted sequencing. RNA sequencing
Nanopore GridION system	DNA sequencing; Epigenetic sequencing; genetic variation characterization. RNA sequencing: to investigate the original sample RNA directly, without undergoing conversion to cDNA.

Source: Adapted from Qingfa et al. (2014); Varvara et al. (2018).

plant virology, especially in the domain of identification, detection, genome sequencing, replication, transcriptomics, epidemiology, and ecology. Identified and novel plant RNA as well as DNA viruses (their satellites) as well as phytoplasmas and viroids from various infected host plants were successfully employed using these NGS technologies. Presently, RNA-seq has been used in identification of large number of plant viruses and viroids from infected plant tissues. Moreover, total nucleic acid or whole double-stranded RNA (dsRNA) was isolated from infected plant tissue and the plant virus or viroid was identified through NGS technology. However, in some cases, total nucleic acid from infected host plant was partly removed by hybridization with nucleic acid from healthy plant tissue helps in enriching the virus sequences proceeding to sequencing. The viruses and/or viroids identified and detected in infected host plants or in viruliferous vectors using deep sequencing are known as "virome." The deep sequencing of siRNAs tender excellent opportunities to recognize viruses or viroids infecting host plants species, even at very low titers, in symptomless infections, including previously unidentified viruses or viroids. Next-generation sequencing technology provides millions of siRNA sequences from infected host plant species, which can be further employed in reconstruction of genome of virus or viroid. The siRNAs are 21–24 nt, their sequences can be directly utilized as a primer sequences to amplify viral or viroid fragments by PCR or RT-PCR. Table 4.3 summarize the use of deep sequencing technologies for detection and identification of virome in infected host plant species (Radford et al., 2012; Ho et al., 2014).

4.7   LIMITATIONS OF NGS

The major limitation of deep sequencing in the clinical setting is the required infrastructure, like computer capacity and storage, and requisite expertise needed to comprehensively evaluate and subsequent interpretation of data. Additionally, it needed skillful management of huge data for mining the pathologically vital information in a comprehensive and robust interface. Yet, to formulate NGS as more cost effective, there is need to carry out large batches samples run that may need supra-regional centralization. After the initial capital investment, the deep sequencing technologies would offer sequencing service at national level likely contributing economic benefits as well as improvement in plant protection care.

TABLE 4.3 Viruses Identified from Various Host Plants Using Next-Generation Sequencing Platforms

Host	Study Finding/Virome	Target	Sequencing Platform
Arabidopsis thaliana	*Oilseed rape mosaic virus, cauliflower mosaic virus, cabbage leaf curl virus*	Total RNAs and siRNAs	Illumina genome analyzer
Apple	Apple stem pitting virus, apple chlorotic leaf spot virus, apple necrotic mosaic virus, apple stem grooving virus, apricot latent virus, apple green crinkle associated virus	siRNAs	Illumina
Black pepper	*Piper yellow mosaic virus*	Viral and plant DNA	Roche 454 GS-FLX
Citrus	*Citrus yellow vein clearing viru, citrus chlorotic dwarf-associated virus*	siRNAs and total DNA	Illumina
Cassava	*Cassava brown streak virus*	Total RNA	Roche 454 GS-FLX
Cotton	*Cotton leafroll dwarf virus*	siRNAs	Illumina
Fig	*Fig mosaic virus, Fig latent virus-1*	dsRNAs	Illumina
Grapevine	Grapevine leaf roll associated virus-3, grapevine red blotch-associated virus, grapevine rupestris stem pitting associated virus, *grapevine Syrah 1 virus, grapevine virus E, grapevine virus A, grapevine virus F, Australian grapevine viroid, grapevine yellow speckle viroid, Hop stunt viroid, citrus exocortis Yucatan viroid, citrus exocortis viroid*	Total RNA or dsRNA, siRNAs	Roche 454
Gomphrena globosa	Gay feather mild mottle virus	Total RNA	Roche 454 GS-FLX
Maize	*Maize chlorotic mottle virus and Sugarcane mosaic virus*	Total RNA	Roche 454 GS-FLX+
Nicotiana benthamiana	*Cucumber mosaic virus, cymbidium ringspot virus, tobacco rattle virus, pepper mild mosaic virus, potato virus X*	siRNAs	Roche 454 and Illumina
Passion fruit	*Passion fruit woodiness virus*	Polyadenylated RNA	Illumina Solexa
Plum	Plum pox virus, prune necrotic ringspot virus	dsRNA	Roche 454
Pepper, and eggplant	*Pepper yellow leaf curl virus, eggplant mild leaf mottle virus*	Purified virons viral RNA	SOLiD
Peach	*Peach latent mosaic viroid*	siRNAs	Illumina

TABLE 4.3   *(Continued)*

Host	Study Finding/Virome	Target	Sequencing Platform
Rice	*Rice stripe virus*	siRNAs	Illumina Solexa
Raspberry	*Raspberry latent virus*	dsRNA	Illumina
Sweet potato	*Sweet potato feathery mottle virus, sweet potato chlorotic stunt virus, sweet potato virus C, sweet potato leaf curl Georgia virus, sweet potato pakakuy virus B*	siRNAs	Illumina
Tomato	*Tomato spotted wIt virus, potato spindle tuber viroid,* tomato necrotic stunt virus, *Pepino mosaic virus,*	siRNAs	Illumina genome analyzer IIx
Wild cocksfoot	*Cereal yellow dwarf virus and Cocksfoot streak virus* (Potyvirus)	siRNAs	Roche 454

Source: Adapted from Varvara et al. (2018); Villamor et al. (2019).

4.8 CONCLUSION

The next-generation technologies have been existing for several years. These technologies and bioinformatics offer speedy and low cost DNA sequencing for biological samples, including plant viruses and viroid, full genome sequencing, RNA sequencing, metagenomics, *etc*. The technologies are in incessant development and advancement. Sequencing chemistries are appropriately established, maturing, and growing, fidelity, and read lengths and are improving, allowing us to explore comprehensively the viral genome diversity, virus epidemiology, virus diagnosis, virus-host interaction, and its elimination, *etc*. Most of the sequencing platforms are being designed to be simple and cost effective. Recently, Illumina, Oxford Nanopore sequencing technology, Life Technologies Ion Torrent companies have developed and released cost effective priced bench top platforms to make NGS technologies readily accessible to most of the researchers and diagnostic laboratories. Owing to varying biochemistries and sequencing protocols, NGS platforms generated large amounts of DNA sequence of short read length (typically millions to billions) whose length varied from 25 bp to 400 bp. As compared to Sanger sequencing, the reads generated are shorter. Since its inception, the NGS market has expanded with numerous companies launching different sequencing platforms models, like the bench top platforms launched recently. With the continuous improvement in the sequencing platforms w.r.t speed, efficiency, and cost effectiveness NGS has reached to many laboratories to explore new research in the area of genomics and diagnostics. With approximately 1500 plant viruses listed in the latest report of the International Committee for the Taxonomy of Viruses. The employment of NGS technology for plant viruses would certainly enhance this figure very significantly since new viruses are being reported and characterized in several plant species, as well as insect vectors. The potential use of NGS technologies in quarantine programs as well as plant certification can significantly improve the affectivity and consistency of these programs and restricting the movement of plant virus and viroid diseases at international and national levels.

4.9 FUTURE PROSPECTS

Many constraints remain in the implementation of high throughput technologies to the discovery of plant pathogen. First, new computational algorithms must be established which is competent of discovering novel plant viruses from NGS data in a homology-independent algorithms, as demonstrated in viroid discovery by PFOR. Furthermore, these algorithms may also aid to recognize the origins of the huge quantity of sequences of unknown nature are generated through deep sequencing which does not share any sequence similarity to any GenBank entry. Second, improvement of easy to use software interfaces for instance, Search. Small RNA that are publicly available and need little informatics exercise will help in more extensive way to utilize the NGS technologies for the diagnosis, characterization of new and novel plant viruses and viroids. Important barriers in developing diagnostic tools for virus detection from RNA-seq data includes the uploading of large raw data files, processing of raw and intensive data (assembly, mapping, and alignment), dependence on pre-computed custom databases and availability of tools. Additionally, the difficulty in detection and identification of novel viruses is not directly solved via any bioinformatics tool, as it needs extensive mapping and assembly by a user with expertise in viral genomics.

ACKNOWLEDGMENT

The corresponding author is highly thankful to all co-authors for their contribution and valuable time.

CONFLICT OF INTEREST STATEMENT

The authors declare that they have no conflict of interest.

KEYWORDS

- **diagnostics**
- **double-stranded RNA**
- **enzyme-linked immunosorbent assays**
- **massively parallel signature sequencing**
- **next-generation sequencing**
- **plant viruses**

REFERENCES

Andika, I. B., Kondo, H., & Sun, L., (2016). Interplays between soil-borne plant viruses and RNA silencing-mediated antiviral defense in roots. *Front. Microbial., 7*, 1458.

Barzon, L., Lavezzo, E., Militello, V., Toppo, S., & Palu, G., (2011). Applications of next-generation sequencing technologies to diagnostic virology. *Int. J. Mol. Sci., 12*, 7861–84.

Elizabeth, A. T., & Cherie, L. H., (2013). Next-generation HLA sequencing using the 454 GS FLX system. *Methods Mol Biol., 1034*, 197–219.

Fatima, Y. G., & Aline, K., (2019). Catch me if you can! RNA silencing-based improvement of antiviral plant immunity. *Viruses, 11*.

Haas, B. J., (2013). De novo transcript sequence reconstruction from RNA-seq using the trinity platform for reference generation and analysis. *Nature Protocols, 8*(8), 1494.

Ho, T., & Tzanetakis, I. E., (2014). Development of a virus detection and discovery pipeline using next generation sequencing. *Virology, 471–473*, 54–60.

Jo, Y., Choi, H., Kim, S. M., Kim, S. L., Lee, B. C., & Cho, W. K., (2016). Integrated analyses using RNA-Seq data reveal viral genomes, single nucleotide variations, the phylogenetic relationship, and recombination for apple stem grooving virus. *BMC Genomics, 17*, 57.

Jo, Y., Choi, H., Kyong, C. J., Yoon, J. Y., Choi, S. K., & Kyong, C. W., (2015). In silico approach to reveal viral populations in grapevine cultivar tannat using transcriptome data. *Sci. Rep., 5*, 1584.

Jones, S., Baizan-Edge, A., MacFarlane, S., & Torrance, L., (2017). Viral diagnostics in plants using next generation sequencing: Computational analysis in practice. *Front. Plant Sci., 8*, 1770.

Marina, B., Henryk, C., & Ahmed, H., (2014). Historical perspective, development and applications of next-generation sequencing in plant virology. *Viruses, 6*, 106–136.

Menzel, P., Kim, L. N., & Anders, K., (2016). Fast and sensitive taxonomic classification for metagenomics with kaiju. *Nature Communications*, 711257.

Miller, J. R., Koren, S., & Sutton, G., (2010). Assembly algorithms for next-generation sequencing data. *Genomics., 95*, 315–327.

Nabi, S. U., Mir, J. I., Sharma, O. C., Singh, D. B., Zaffer, S., & Sheikh, M. A., (2018). Optimization of tissue and time for rapid serological and molecular detection of Apple stem pitting virus and apple stem grooving virus in apple. *Phytoparasitica, 46*(5), 705–713.

Nicholas, A. B., Sathish, S. J., Faith, D., Gevers, J., Gordon, R., Knight, D. A., Mills, J., & Gregory, C., (2013). Quality-filtering vastly improves diversity estimates from illumina amplicon sequencing. *Nat. Methods, 10*(1), 57–59.

Niedringhaus, T. P., Milanova, D., Kerby, M. B., Snyder, M. P., & Barron, A. E., (2011). Landscape of next-generation sequencing technologies. *Anal Chem., 83*, 4327–4341.

Qingfa, W., Shou, W. D., Yongjiang, Z., & Shuifang, Z., (2015). Identification of viruses and viroids by next-generation sequencing and homology-dependent and homology-independent algorithms. *Annu. Rev. Phytopathol., 53*, 425–444.

Radford, A. D., Chapman, D., Dixon, L., Chantrey, J., Darby, A. C., & Hall, N., (2012). Application of next-generation sequencing technologies in virology. *J. Gen. Virol., 93*, 1853–1868.

Rhoads, A., & Au, K. F., (2015). PacBio sequencing and its applications. *Genomics, Proteomics Bioinforma., 13*, 278–289.

Sastry, K. S., (2013). diagnosis and detection of plant virus and viroid diseases. In: *Plant Virus and Viroid Diseases in the Tropics.* Springer, Dordrecht.

Simone, R., Peter, W., Monika, R., Jan, M., Elisabeth, B., & Mira, J., (2017). Sequencing on the solid 5500×l system-in-depth characterization of the GC bias. *Nucleus, 8*(4), 370–380.

Souvorov, A., Richa, A., & David, J. L., (2018). SKESA: Strategic K-Mer extension for scrupulous assemblies. *Genome Biology, 19*(1), 153.

Sydney, B., Maria, J., John, B., George, G., & David, H., (2000). Gene expression analysis by massively parallel signature sequencing (MPSS) on microbead arrays. *Nature Biotechnology.*

Van, D. P., Van, D. M., & Piron, P. G. M., (1987). *Netherlands Journal of Plant Pathology, 93,* 73. https://doi.org/10.1007/BF01998093.

Varvara, I., Maliogka, Angelantonio, M., Pasquale, S., & Ana, B., (2018). Recent advances on detection and characterization of fruit tree viruses using high-throughput sequencing technologies. *Viruses, 10,* 436.

Villamor, D. E. V., Ho, T., Rwahnih, Martin, R. R., & Tzanetakis, I. E., (2019). High throughput sequencing for plant virus detection and discovery. *Phytopathology, 109,* 716–725.

Voelkerding, K. V., Dames, S. A., & Durtschi, J. D., (2009). Next-generation sequencing: From basic research to diagnostics. *Clin. Chem., 55,* 641–658.

Yadav, N., & Khurana, S. M. P., (2016). Plant virus detection and diagnosis: Progress and challenges. In: Shukla, P., (ed.), *Frontier Discoveries and Innovations in Interdisciplinary Microbiology.* Springer, New Delhi.

CHAPTER 5

Innovative Diagnostic Tools for Plant Pathogenic Virus

SUNIL KUMAR and SHIVAM MAURYA

Department of Plant Pathology, SKN College of Agriculture, Jobner, Jaipur–303329, Rajasthan, India, E-mail: Khaliasunil1987@gmail.com (S. Kumar)

ABSTRACT

Plant pathogenic infections are answerable for expanding monetary misfortunes worldwide and still turned into a significant worry in present-day farming. It has been evaluated that plant infections can cause as much as 50 billion Euros misfortune around the world, every year. Exact distinguishing proof and analysis of plant viral sicknesses are significant in the period of environmental change and the related changes in infection the study of disease transmission and globalization for food security just as anticipation of the spread of Plant pathogenic infections. Dependable and early location techniques are, as yet, one of the principle and best activities to create control procedures for plant viral sicknesses. Discovery and ID of plant pathogenic infections are one of the most significant techniques for economical plant sicknesses on the board. Likewise, for a productive and efficient administration of plant maladies exact, delicate, and explicit determination is fundamental. The study of plant sickness finding has advanced from visual investigation and ID of plant viral illnesses to identify with high-throughput serological strategies like chemical connected immunosorbent measure (ELISA) and sub-atomic techniques, for example, polymerase chain response (PCR). Laboratory-based procedures, for example, polymerase chain response (PCR), immunofluorescence (IF), fluorescence in-situ hybridization (FISH), compound connected immunosorbent test (ELISA), stream cytometry (FCM), and gas chromatography-mass spectrometry (GC-MS) are a portion of the immediate identification techniques. Aberrant strategies incorporate thermography, fluorescence imaging, and hyperspectral procedures. Serological and sub-atomic methods are right now the most proper when high quantities of tests should be dissected. Atomic techniques can be applied for determination of numerous viral ailments when hereditary data of infections is accessible. Numerous serological techniques including Antigens and antibodies, chemical connected insusceptible sorbent measure (ELISA), tissue smudge immunoassay (TBIA), and immuno electron microscopy (EM) are utilized for Detection and analysis of plant viral maladies.

Nucleic corrosive based identification procedures are, for example, Karyotyping, Hybridization, and Polymerase chain response particularly appropriate for this reason. Polyvalent identification has permitted the location of numerous plant infections at the family level. There are numerous kinds of PCR procedures, for example, RT-PCR, continuous PCR, settled PCR, multiplex PCR (M-PCR) are the most widely recognized DNA enhancement innovation with high affectability for the location of one or a few pathogens in a solitary test. The most recent advancement in the investigation of nucleic acids is microarray innovation, yet it requires nonexclusive DNA/RNA extraction and pre-enhancement techniques to build location affectability. The advances in research that will result from the sequencing of many plant viral pathogen genomes, particularly now in the period of proteomics, speak to another wellspring of data for the future improvement of touchy and explicit location strategies for these microorganisms.

The recently rose proteomic innovation is additionally a promising instrument for giving data about pathogenicity and destructiveness factors that will open up additional opportunities for plant illness finding and fitting security measures.

5.1   INTRODUCTION

Plant pathogenic infections are one of the reasons for low farming efficiency around the world. Principle reasons are new, old and developing plant viral ailments. Plant infections cause genuine sicknesses in field, plant, and different harvests and for the most part lead to huge financial misfortunes. Plant infections alone involve about 47% of the irresistible maladies revealed in plants. Notwithstanding the spread through vector bearers, they are additionally transmitted principally through contaminated proliferating materials (seeds, cuttings, bulbs, bulb lets, rhizomes, suckers, bud woods, tissue culture determined plantlets and so on.). Plants show various indications on leaves, stems, and organic products because of plant viral ailment contaminations. These side effects are especially valuable for visual perception as a traditional initial step for plant sickness determination; however, it falls flat in distinguishing the nearness of pathogen in early disease stages when plant contaminations are symptomless. The nature and assorted variety of plant infections make it hard to utilize a general strategy for their location and portrayal. Numerous harvests would thus be able to be tainted by unidentified infections, significantly constraining the opportunities for the appropriate conclusion and controlling these specialists. Location of destructive infections in plant material, vector, or common stores is basic to guarantee manageable horticulture. Plant infection research has been grouped into three times, for example, "Old-style Discovery Period" from 1883–1951, "Early Molecular Era" from 1952–1983, and "Late Period" from 1983 onwards (Zaitlin and Palukaitis, 2000). During early sub-atomic time and late period, identification methods accessible for plant infections have developed altogether. With the headway of exceptionally touchy recognition instruments and nano innovation, it is currently conceivable to complete infection discovery from tests disguised by a high number of microorganisms, and with low infection titer. As of now, the recognition of phytopathogenic infections is a changing, dynamic, and developing world. Presently a day, strategies can be adjusted or improved just months subsequent to having been created. Every one of these strategies has its own favorable circumstances and impediments. By and by, the novel location procedures can be extensively characterized into two classifications, serological measure and nucleic corrosive based test.

Location and analysis of viral operators related with various sicknesses, are either founded on natural measures, serological tests, nucleic corrosive based examines or mix of various techniques. Decision of identification strategy relies upon various measures: affectability (capacity to distinguish least amount of infection), precision, and reproducibility, number of tests that can be tried at once, starting speculation for mechanical assembly and running expense per test, understanding, preparing, and specialized skill required. Since the primary depiction of infections with plant maladies, the discovery techniques were generally founded on the portrayal of manifestations on the hosts, bio-tests on the symptomatic hosts and transmission electron microscopy (TEM) in rough leaf sap arrangements. These two strategies were at first utilized for a considerable length of time for distinguishing proof of related etiological viral operators with new uncharacterized sicknesses; be that as it may, these techniques were a greater amount of emotional nature and didn't portray the related infection at species or strain level. Infection diagnostics were changed by the disclosure of compound connected immunosorbent examine (ELISA) (Clark and Adams, 1977) which notwithstanding rearrange the discovery techniques can screen enormous number of tests one after another. The outcomes got are of semi-quantitative arrangements, in this way giving a sign of infection titer in the tainted plant tests. After this the strategies like nucleic acid spot hybridization (NASH), polymerase chain reaction (PCR) and Reverse Transcription-PCR (RT-PCR) were produced for increasingly powerful identification of plant infections which recognize viral nucleic corrosive (DNA or RNA whatever is the situation). Most recent shut cylinder based demonstrative tests, for example, quantitative-PCR (qPCR) and continuous PCR (rt-PCR) are the most delicate measures which lessen the danger of post-PCR defilement permitting the positive and touchy recognition of plant infections. In the most recent decade, different techniques like macroarray, horizontal stream measures, circle interceded isothermal intensification (LAMP), and cutting edge sequencing (NGS) were created which have explicit favorable circumstances of either affectability, appropriateness for on location recognition or disclosure of novel diseases.

5.2　DETECTION METHODS BEFORE SEROLOGICAL ERA

5.2.1　BIOASSAY

These strategies depend on the organic properties of related plant infections. These techniques are otherwise called organic ordering where research center measure has are utilized for assurance of natural response of infection as far as run of the mill articulation of manifestations. Neighborhood injury examine has been utilized for estimation of infection fixation and its recognition. Just because Holmes (1939) demonstrated that the quantities of nearby sores created on the mechanical immunization of Tobacco mosaic infection (TMV) on the leaves of *Nicotiana glutinosa* are legitimately corresponding to the infection titer in the sap. Various measures has like *Nicotiana benthamiana, N. tabacum, N. glutinosa, N. sylvestris, Chenopodium amaranticolor* and so forth are routinely utilized as measure has for the sap transmitted infections. The infections which are not sap transmitted, tainted plant parts are joined on the pointer have species on which symptomatic side effects are delivered. Marker has are as yet utilized for discovery or ordering of join transmitted infections of apple, citrus, and other woody organic product plants. Vector transmission is likewise utilized if there should arise an occurrence of vector transmitted infections. In any case, the use of organic ordering depends of the accessibility of appropriate marker plants and indication articulation is subject to natural conditions and may take a while or years.

5.2.2 ELECTRON MICROSCOPY (EM)

Electron microscopy (EM) is utilized to watch the infection particles or infection initiated proteinaceous structures (incorporation bodies) in the plant cell. Rough sap from tainted leaf tests is covered on carbon matrices, which are adversely stressed and seen under electron magnifying instrument. TEM is the most generally utilized strategy in the representation of virions in the tainted examples. This is the main technique where infections can really be believed to be related with the sickness. Be that as it may, the consequences of EM rely upon the centralization of infection in the example and infections can be recognized morphologically up to sort level as it were. Subsequently, EM results should be enhanced with other corroborative tests. What's more, EM is utilized to recognize the infection prompted consideration bodies (for example in potyviruses) in the host cell consequently giving the roundabout proof of the infection disease. Both bio-measures and EM are conventional strategies for wide screening of tests for nearness of infection contamination EM is utilized to watch the infection particles or infection actuated proteinaceous structures (incorporation bodies) in the plant cell. Rough sap from tainted leaf tests is covered on carbon lattices, which are contrarily stressed and seen under electron magnifying instrument. TEM is the most broadly utilized technique in the perception of virions in the contaminated examples. This is the main strategy where infections can really be believed to be related with the illness. Be that as it may, the aftereffects of EM rely upon the grouping of infection in the example and infections can be distinguished morphologically up to class level as it were. Along these lines, EM results should be enhanced with other corroborative tests. In addition, EM is utilized to recognize the infection prompted incorporation bodies (e.g., in potyviruses) in the host cell along these lines giving the roundabout proof of the infection disease. Both bio-tests and EM are conventional strategies for wide screening of tests for nearness of infection contamination.

5.3 MODERN DIAGNOSIS METHODS

5.3.1 SEROLOGICAL TECHNIQUES FOR DETECTION AND DIAGNOSIS OF PLANT VIRUS

Serology is one of the most significant parts of plant virological investigations. Serological properties being unmistakable and stable element are considered as a significant taxonomical measure for separating and grouping plant infections. Immunogenic properties of plant infection have been used to build up an assortment of serological strategies, which are applied for the discovery and analysis of plant infections. In this section, fundamental highlights of viral antigen and antibodies, steps include in creating serological devices and brief depiction of the serological test methods have been portrayed.

5.3.1.1   INTRODUCTION

Recognizable proof of infection is an essential prerequisite in all parts of virological examinations. There are a few methods utilized for discovery and determination of plant infections, which are basically founded on the properties of infection. Serological properties are one of the inherent elements of viral proteins that have been abused to create assortments of procedures for the discovery of plant infections. The term Serology implies ('Sero' got from serum) concentrates on the connections of infection and the cognet counter acting agent. In immunological term, the entire infection molecule speaks to as antigen. The molecule protein (auxiliary protein) that characterizes the external structure of infection (capsid) is overwhelmingly utilized as viral antigen nonetheless; a couple non-basic proteins encoded by the infection qualities are additionally utilized as viral antigen.

Serological strategies are broadly utilized for estimation of field frequency of known infections, affirmation of seeds, and other planting materials for the opportunity from infection, assessment of reproducing or transgenic lines for powerlessness to infections, portrayal, and order of obscure infection or strain. Likewise, serological methods are basic for central investigations of infections properties, for example, replication, development, and quality articulation.

5.3.1.2   BASIC REAGENTS AND PRINCIPLES OF SEROLOGICAL REACTIONS

For a plant virologist, knowledge on the whole intricacy of Immunology is impractical, nevertheless, some basic concept of Immunology is useful as it is the foundation of serological techniques.

5.3.1.2.1   Reagents

Antigen and counter acting agent are the fundamental reagents in a serological response. The idea of antigen and counter acting agent has a hitting similarity with a couple. As a man can't be spouse without a wife and the other way around. An atom can't be antigen without a comparing counter acting agent and the other way around. An antigen can be characterized as a particle with the two natural properties, immunogenicity, and antigenicity. In this manner, the antigen ought to produce immunresponse (immunogenic) and it ought to respond with antibodies independent of its capacity to create them. The plant infections are commonly compelling in inspiring creation of a particular counter acting agent. An antigen associates with a neutralizer at a particular surface site assigned as epitope. The size of epitopes may shift, for instance, if the antigen is a straight peptide the size of epitope is around five to six amino acid deposits, if the epitope is a globular protein, the size of the epitope might be 16 or so amino corrosive buildups (Van Regenmortel, 1988).

Counter acting agent is a globular protein atom known as immunoglobulin (Ig), which is delivered in a vertebrate creature because of reaction to an outside antigen. Counter acting agent particle is comprised of two indistinguishable substantial and light polypeptide chains held together by disulfide bonds. Antibodies or Ig are partitioned into five classes-IgG, IgM, IgA, IgD, and IgE. The major or serum immunoglobulin is IgG. IgG is a Y-formed pivoted atom with a solitary antigen joining locales (paratope) toward the finish of every one of the two arms of Y. As one IgG particle can join two epitopes, its valancy is two.

The pivot area of Ig, which gives adaptability, is helpless against proteolytic cleavage. Absorption by papain at the pivot district of IgG results into three parts (F)-two indistinguishable sections each with a solitary antigen official (abdominal muscle) site assigned as Fab, and the third piece that does not have the antigen gorging limit, assigned as Fc (section crystallizable). Pepsin processes the IgG at an alternate focuses, which results into two pieces one enormous part assigned as F(ab')2 as it holds the divalency of the parent atom and a little section assigned as Fc.' The N-terminal succession of around 100 to 110 amino acids on both substantial and light chains shift extensively in the diverse IgG particles. In this manner, this district is assigned as a variable area of an IgG. The grouping of the rest of the piece of the IgG atom is amazingly same, which is assigned as steady district. In the variable areas, there are three sections on the light chain and three fragments on the substantial chain, strikingly assorted called hypervariable locales, which represent the particularity of immune response.

The heterologous populaces of immunizer with variable particularity delivered against more than one epitopes present on the outside of infection molecule are alluded as polyclonal counter acting agent. The homologous populace of counter acting agent, which is delivered against a solitary epitope on the outside of infection molecule, is alluded as monoclonal immune response. The polyclonal and monoclonal antibodies (MAbs) offer an extraordinary adaptability in the utilization of serological strategies for infection sort/gathering or infection species explicit recognition.

5.3.1.2.2 *Principles*

Interactions of antigen-antibody are the basis of serological reaction. An antibody recognizes an antigen through complementary shapes such as protuberance and depressions (like lock and key) on paratope and epitope, respectively (Sheriff et al., 1987). The forces that hold antigen-antibody complex is not by covalent bonding, but by weak forces such as electrostatic, hydrogen bonding, hydrophobic, and van der Waal's forces. Thus, this interaction is reversible. A large number of serological assay techniques has been developed and utilized for the study of plant viruses. All these techniques are based on some basic principles such as: neutralization, precipitation, agglutination, reaction augmentation, and ultra-magnification:

1. **Neutralization:** Virus balance quantifies the loss of infectivity because of official of the infection with the counter acting agent. The official of immune response maybe hinders the early procedure of contamination in the host cells. The system of this disease restraint isn't completely comprehended. The balance tests have not been generally utilized in Plant Virology.

2. **Precipitation:** Virus particles are multivalents concerning authoritative with counter acting agent, while, antibodies are bivalents. Enormous number of viral antigens and antibodies tie to shape cross section bringing about insolubilization of infection immunizer complex.

3. **Agglutination:** It is the clustering of cells or particles of comparative size bringing about representation of serological responses. This is finished utilizing red platelets and latex particles covered with antigen or immune response. Less counter acting agent particles are expected to deliver an obvious agglutination than that in the event of precipitation response.

4. **Reaction Augmentation:** When a little amount of immunizer ties to the viral antigen, the response can't be noticeable by unaided eye as found if there should be an occurrence of precipitation and agglutination responses. The sign of explicit authoritative of immune response

with the infection can be increased to the edge of perceivability by marking the neutralizer with compound, fluorescent specialist, radioactive operator and chemiluminescent specialist. Expansion of antigen-immunizer restricting sign has prompted create numerous amazing serological strategies.

5. **Ultra-Magnification:** Electron minuscule procedures are applied for the representation of the virion morphology after immunological official of infection and counter acting agent.

5.3.1.3 STEPS IN DEVELOPING SEROLOGICAL TOOLS

There are many steps in developing serological tools against a specific virus. The major steps are preparation of pure viral antigen, immunization, and purification and characterization of the antibody for its desired specificity.

5.3.1.3.1 Viral Antigen

The molecule protein (structural protein) that characterizes infection structure (capsid) has been overwhelmingly utilized for serology of plant infection. Other viral coded nonstructural proteins, for example, NSs protein of Tomato spotted shrivel infection has been utilized to deliver immune response for distinguishing proof of viruliferous thrips (Bundla, 1994). The incorporation body protein, another non-capsid protein (CP) is a possible viral antigen (Heibert et al., 1984).

Refinement of infection molecule protein can be accomplished from contaminated plant materials. For this, the means include are: (a) foundation of infection culture; (b) affirmation of personality of the infection; (c) mass engendering of infection; (d) collect and capacity of contaminated tissues; (e) extraction and explanation; (f) focus by polyethelene glycol and ultracentrifugation lastly; (g) decontamination by thickness slope ultracentrifugation.

Numerous a periods, refinement of infections particularly those are phloem borne is troublesome. The filtered arrangement with low centralization of virion and debased plant cell segments frequently results into low quality of antisera. To defeat these downsides, another way to deal with acquire viral antigen is by cloning and communicating viral qualities in *Escherichia coli*. Nikolaeva et al. (1995) utilized recombinant coat protein of Citrus tristeza infection (CTV) to deliver polyclonal antisera, which recognized a wide scope of CTV separates in catalyst connected immunosorbent measure.

5.3.1.3.2   *Inoculation and Counter Acting Agent Creation*

A wide range of creatures can be utilized for infusing viral antigen. Bunnies are normally utilized creature as it is anything but difficult to back and deal with, and it delivers high-titer antisera against moderately modest quantity of antigen utilized. The viral antigen is managed to a bunny by both intramuscular and intravenous infusions. For intramuscular infusion, antigen is emulsified with adjuvant, (Freund's finished and inadequate adjuvants), which helps in moderate arrival of antigen invigorating more counter acting agent creation. By and large, two intramuscular and one intravenous infusions of 1–2 mg of antigen at a week-by-week span results into creation of sufficient measure of counter acting agent. By 5 to 7 weeks after the primary infusion, immunizer ventures into top titer. Subtleties of conventions of inoculation, blood assortment, serum handling, IgG readiness, and portrayal are given in Hampton et al. (1990) and Dijkstra and de Jager (1998).

5.3.1.4   *ASSAY TOOLS AND THEIR APPLICATIONS*

5.3.1.4.1   *Infectivity Balance Tests (INT)*

INT is finished by serological hindering of natural action of infection. One method of doing INT is infection immunizer blend is taken care of to its bug vector and afterward it is kept an eye on a host whether the vector transmission is conceivable or not. INT has been demonstrated valuable to consider the connection between the individuals from luteoviurs gathering (Rochow and Duffus, 1978). INT is seldom utilized in Plant Virology.

5.3.1.4.2   *Uninvolved Hemagglutination Test (UHT)*

UHT is a roundabout hemagglutination. Erythrocytes are combined with either infection particles or neutralizer by settling and coupling operators. At the point when the particular counter acting agent or infection is added to the infection or immune response covered red platelets, agglutination will be noticeable. UHT has been utilized to recognize modest quantities of infection in unrefined and explained plant sap (Abu Salih et al., 1986). UHT is 1000 time more delicate than the traditional precipitation test.

5.3.1.4.3   Latex Test (LT)

In the LT, polystyrene latex particles are caught up with antigen or immunizer. Utilizing counter acting agent covered latex molecule, limited quantity of infection could be distinguished. This procedure is 100 to 1000 creases more delicate than the cylinder precipitation test. LT can be applied to distinguish both lengthened and isometric plant infections in unrefined sap. LT has been effectively used to consider the connections inside potexvirus and tymoviurs gatherings (Bercks and Querfurth, 1971).

5.3.1.4.4   Microprecipitation Test (MPT)

MPT is a simple serological technique used widely by Plant Virologists for indexing of plants for virus infection. A small drop of antigen and antibody is mixed on a glass slide, and precipitation is observed under a dark-field microscope. The mixed drop can be covered with a layer of mineral oil to prevent evaporation.

5.3.1.4.5   Tube Precipitation Test (TPT)

TPT is based on precipitation of antigen-antibody complex in liquid medium. The test is done in the tube mixing 0.5 volumes of antigen and antisera. A series of dilution of antibody can be considered against a fixed amount of antigen for determining the titer of the antisera. This test can also be performed carefully layering antigen and antibody on one another; in that case, the precipitation will occur as a ring at the interface of antigen and antibody. This procedure can be employed for quantifying the extent of serological cross-reaction between strains of elongated virus. However, this technique has many drawbacks and is no longer used for routine test.

5.3.1.4.6   Agar Double Diffusion Test (ADDT)

In the ADDT, antigen, and counter acting agent diffuse toward one another in a gel and structure a white band or line at a spot where the two reactants meet and accelerate. Various kinds of gel, for example, respectable agar, particle agar, bacto agar and agarose have been utilized to consider a wide assortment of plant infections. The grouping of gel can fluctuate from 0.4

to 1.0%. Sodium azide (0.02 to 1.0%) ought to be added to the gel as an additive. There are numerous boundaries, for example, size, and relative convergence of the reactants, sort of gel, electrolyte focuses in the gel, sort of cushion utilized, temperature, denaturing operator and so on can impact dissemination procedure and arrangement of precipitation line.

Agar dispersion reads are appropriate for unrefined plant sap, explained separate and exceptionally cleaned antigen. ADDT can be performed with different alterations in gel in cylinder or plate. Twofold dissemination examine in plate is frequently alluded as "Ouchterlony twofold dispersion test" (ODDT). Different examples of well size and plan can be made in the agar gel for performing ODDT. The example of precipitation lines, for example, incomplete combination, blend, and intersection of lines show the nearness or nonappearance of normal or related serological connections.

ODDT is likewise appropriate for dissecting the antigenic parts of separated virion or incorporation bodies. For this, 0.5% sodium dodecyl sulfate (SDS) is utilized in agar gel or SDS can be included the test antigen for better separation. This method was produced for the extended infections yet can be utilized for infections with other morphology. ODDT has been broadly utilized in Plant Virology to decide the serological connections of infection and their strains. It has been additionally utilized for testing field tests for the overview of infections.

5.3.1.4.7   Immunoelectrophoresis (IE)

A blend of electrophoretic development of infection molecule and agar dispersion is used in IE strategy. Viral antigen is applied in wells on one side of the gel and electrophoresis is accomplished for the development of the infection from cathode ( ) to anode (+), which assists with isolating the blend of antigens in gel then antisera is applied in the troughs and immuno-diffusion assists with imagining the precipitation response. IE has been used to concentrate for the most part the isometric infections.

5.3.1.4.8   Immuno-Osmophoresis (IO)

IO is otherwise called counter insusceptible electrophoresis or electrosyneresis. By and large, popular antigen moves electrophoretically toward the anode (+) at pH7, while, antibodies toward the cathode (−) due to electro-endosmotic stream. This method is more fast and delicate than immunodiffusion in gel, and can be utilized with lengthened infections that diffuse gradually in agar gel (Regetli and Weintraub, 1964). The wells for antigen and immunizer are cut two by two about 1.4 cm separated and adjust toward current stream. Antigen is stacked in the well nearest to the cathode and antisera to the well nearest to the anode. The electrophoresis cell can be set over a dark paper and the improvement of hasten can be seen with white light through the agar bed.

5.3.1.4.9   Rocket Immunoelectrophoresis (RIE)

RIE is like insusceptible electrophoresis. Antigen is put in the wells of immune response containing agarose gel. The viral antigen relocates toward anode (+) and structures a rocket-molded precipitation design. RIE is a quantitative strategy, where the zone of the rocket is straightforwardly relative to the centralization of the antigen. The method has been utilized to gauge the grouping of virion and capsid subunit of CMV and to separate strains of TMV (Hvranek, 1978).

5.3.1.4.10   Immunofluorescence Examine (IFA)

IFA is commonly utilized for cell restriction of viral antigen. The antibodies are marked with fluorescent color, fluorescin isothiocyanate. The antigen and counter acting agent complex can be seen under UV-magnifying instrument. There are two strategies for leading IFA: (1) Direct recoloring procedure, where antibodies are legitimately conjugated with fluorescent color and utilized for immediate and explicit recoloring of the test viral antigen; (2) roundabout recoloring method, where in the initial step antigen and immunizer are permitted to tie and afterward in the subsequent advance, antiglobulin counter acting agent conjugate with fluorescent color is utilized to distinguish the antigen-neutralizer complex. In this technique, a solitary fluorescent immune response conjugate can be utilized for identification of an enormous number of infections, and it is more delicate than the immediate one, however liable to deliver vague recoloring (Emmons and Riggs, 1977).

5.3.1.4.11   Radioimmuno Measure (RIA)

RIA is as touchy as ELISA. RIA isn't generally utilized in Plant Virology as ELISA is progressively advantageous to utilize. In RIA, viral antigen is radioactively marked. The particular immune response is connected with the mass of a cylinder. The measure of marked infection that is bound to the immune response covered cylinder is then estimated (Ball, 1973).

5.3.1.4.12   Enzyme Linked Immunosorbent Assay (ELISA)

Catalyst connected immunosorbent measure (ELISA), otherwise called a chemical immunoassay (EIA), is a biochemical method utilized for the most part in immunology to identify the nearness of a neutralizer or an antigen in an example. The ELISA has been utilized as a symptomatic apparatus in medication and plant pathology, just as a quality-control check in different enterprises, for example, ELISA application in food industry. In basic terms, in ELISA, an obscure measure of antigen is appended to a surface, and afterward a particular immune response is applied over the surface with the goal that it can tie to the antigen. This counter acting agent is connected to a catalyst, and in the last stage a substance is included that the compound can change over to some discernible sign, most usually a shading change in a synthetic substrate.

In the event that a protein with various epitopes is being recognized, a sandwich test is a decent decision. It normally requires two antibodies that respond with various epitopes. Be that as it may, if the particle has numerous rehashing epitopes, it is conceivable in a sandwich examine to utilize a similar immunizer for both catch and identification. On the other hand, if there is a gracefully of the analyte to be recognized in unadulterated structure that can retain successfully to a microwell, at that point one can set up a serious measure in which the cleansed analyte is immobilized and analyte in the example contends with the immobilized analyte for authoritative to named counter acting agent. For this situation it is basic to titrate the immune response with the goal that it is constraining, or, more than likely the measure affectability will be.

The most significant reagent of ELISA is conjugate (immunoprobe), which is set up by covalently connecting an Ig or a protein particle that is equipped for restricting immunizer, for example, protein An or avidin with a chemical. The essential rule of ELISA depends on the two properties of the conjugate, immuno particularity of the counter acting agent and synergist movement of the chemical. The test antigen is immobilized on the outside of strong plate and antigen-immune response is expanded by the enzymatic breakdown of a substrate to a chromogenic item; which can be seen in unaided eyes or can be estimated by an ELISA peruser.

ELISA was first applied for the recognition of infection by Avraneas in 1969. Engvall and Perlmann (1971); and Voller et al. (1976) showed its likely application in clinical science. Clark and Adams (1977) showed the use of ELISA for the identification of plant infections, and from that point forward ELISA has been used for the recognition and quantitative estimation of various plant pathogens, for example, infections, microbes, growths, and phytoplasma.

There are various forms of ELISA utilized in Plant Virology. The type of ELISA changes dependent on the different blends of layers of infection, immune response, conjugate, and protein A. There are two chief type of ELISA, for example, immediate, and roundabout ELISA. At the point when a conjugate straightforwardly ties to the test antigen, the ELISA is alluded as an immediate ELISA and when the conjugate ties the test antigen by means of another layer of counter acting agent that is explicitly bound with the test antigen is alluded as a backhanded ELISA. Inside the immediate and roundabout structure, there are numerous adaptations of ELISA that are utilized by Virologists. Viral antigen can be legitimately adsorbed on the plate and afterward-different reagents can be layered progressively, this strategy is known as immediate antigen covered ELISA (DAC-ELISA). The particularity of ELISA can be upgraded by covering the plate with infection explicit counter acting agent and afterward the viral antigen can be explicitly caught on the plate. For this situation, antigen is sandwiched between a catching and recognizing immune response. This type of ELISA is alluded as twofold immune response sandwich ELISA (DAS-ELISA). DAS-ELISA can be proceeded as a direct or as a circuitous technique. Direct DAS-ELISA is progressively explicit though, backhanded DAC-ELISA is increasingly valuable for expansive range identification of infections. For the general recognition of infections, backhanded type of ELISA has a bit of leeway is that the conjugate is monetarily accessible and can be utilized for the discovery of a wide scope of infections, while, in direct ELISA, explicit conjugate should be set up for the location of every infection.

ELISA systems include multi steps and require different reagents. Normalization of ELISA for the conclusion of an infection requires creation of neutralizer, purging of IgG, planning of conjugate, readiness of test, and recording the force of shading created because of enzymatic breakdown of the substrate. For leading ELISA, polystyrene plates with various wells are utilized as the strong stage. Soluble phosphatase (AIP) and horseradish peroxidase (HRP) are the normal catalysts used to get ready conjugate. p-nitrophenyl phosphate (PNPP) and tetramethylbenzidine (TMB) are the substrates utilized for ELISA with AIP and HRP, separately. Yellow and blue shading responses demonstrate positive outcomes with AIP and HRP, separately. It is critical to incorporate known positive and negative controls while directing ELISA. The ELISA esteem 2 to multiple times higher than the normal of the negative control is commonly considered as positive response.

ELISA additionally called neutralizer based identification procedures. So Serological examines for plant pathogens, for example, infections, can't be developed specially appointed and henceforth-serological measures were created to recognize them. So unquestionably we have to have various innovations for those specific organisms that causes plant infections like infections which are not culturable and that is the reason serological techniques is likewise useful or valuable to recognize certain pathogens of that sort. In excess of a thousand different pathogens, similar to microorganisms and parasites, would now be able to be distinguished utilizing polyclonal just as monoclonal antisera and procedures, for example, ELISA western smears, immunostrip tests, dab smudge safe restricting measures, just as serologically explicit electron microscopy (SSEM). Therefore, these are for the most part strategies that are serology based, and these procedures are sent for distinguishing proof of different pathogenic infections. Among them, ELISA was first utilized, 1970s and till now it is the most well-known strategy utilized for high throughput recognition of certain plant infections. The significant guideline behind ELISA strategy where the epitope or the antigen is made to tie with antibodies, which is connected with a protein and afterward, when the substrate is included this compound can change shade of the arrangement, however regularly, it transforms into yellow shading. The discovery can be imagined dependent on shading changes, coming about because of the collaboration between the substrate and immobilized catalyst. So here, this is a case of PCR plate, where you can see that the yellow shading wells, giving the positive outcome for the specific pathogen and the darker the alludes to the more number antigens present in the example. Therefore, the lighter shading wells demonstrate lesser measure of antigens nearness in the well. The rule how ELISA functions, in light of this—there are four distinct sorts of ELISA that are planned that is immediate ELISA, aberrant ELISA, sandwich ELISA just as serious ELISA. All these four ELISA methods are utilized dependent on their need and dependent on their applications. So let us talk about Direct ELISA—where an objective protein or an objective counter acting agent is immobilized on the outside of miniaturized scale plate wells. Here, the objective proteins are immobilized in the miniaturized scale plate wells and it is hatched with a catalyst marked immune response, is the chemical named neutralizer when it is brooded together and afterward, when the substrate is included, at that point the shade of the arrangement turns yellow. So this is the procedure, we call it as immediate ELISA and yellow shading improvement demonstrates positive result of the cooperation. Next is backhanded ELISA—where the objective protein is immobilized on the outside of small scale plate wells and brooded with an immunizer to the objective protein. We call it as essential counter acting agent. At that point if the auxiliary immune response is included against the essential counter acting agent and subsequent to washing the movement of the smaller scale plate well bond compound is estimated. The protein is immobilized in the small-scale plate well. At that point, first essential antibodies are included which isn't compound connected and followed by in the subsequent stage. Furthermore, immunizer is included, which is connected with a protein and afterward when substrate is included then this catalyst connected counter acting agent with substrate cooperates and the shade of the arrangement turns yellow. Therefore, this gives again a positive sign of essence of a specific pathogen that is available in a similar example. Albeit backhanded ELISA requires more advances, at that point Direct ELISA marked optional antibodies are monetarily accessible, taking out the need to level essential immune response every single time. So it spares certain measure of cost as essential immunizer are not marked with catalyst and rather the capacity of chemical is accomplished through utilization of industrially accessible optional neutralizer. Next, is Sandwich ELISA—the neutralizer to an objective protein is, immobilized on a superficial level, small-scale plate wells and brooded first with the objective protein and afterward with another objective protein explicit counter acting agent, which is marked with a catalyst. Subsequent to washing, the movement of the smaller scale plate all around bound compound is estimated. So in sandwich ELISA most importantly, the counter acting agent is immobilized in the smaller scale plate well, rather than the antigen then antigen is included, at the second stage where essential immune response will at that point tie with the antigen followed by option of auxiliary immunizer that is connected with compound and this will tie with the antigen and afterward when substrate is included the shade of the arrangement goes to yellow. Along these lines, it shows again positive off the specific pathogens present, and this is the thing that we call it as Sandwich ELISA. Since the antigen is available in the middle of two antibodies, so this specific sort of ELISA technique is known as sandwich ELISA. The immobilized immune response, that is orange in shading and the compound marked counter acting agent which is green in shading, must perceive various epitopes of the objective protein. So both the counter acting agent's that is the yellow shading immune response just as the green shading immunizer competitive ability, of the antigens, and that indirectly tells us the story whether the amount of antigen that is present in higher quantity or lower quantity in the sample.

ELISA has become the most broadly utilized serological apparatus everywhere throughout the world. ELISA is an amazingly delicate serological method, which can identify 1 to 10 ng of infection for every ml (Clark et al., 1976). It is multiple times more delicate than the organic tests. The magnificence of ELISA is that the infection can be distinguished in unrefined concentrate of plant and bug vector with a phenomenal explicitness and affectability. For the enormous scope location of plant infections in the epidemiological examination, ELISA is the best option of every indicative instrument. Presently a-days, antisera to a wide scope of infections are economically accessible and along these lines ELISA can be performed absent a lot of issue of creation and arrangement of immune response and conjugate.

The most broadly utilized and delicate strategy is DAS ELISA is talked about in this section.

- Materials: ELISA microtitre plate (polystyrene or polyvinylchloride), micropipettes (30–300 μl), plasticwares, and crystal, rotator, hatchery, pH meter, attractive stirrer, ELISA reader, and washer.

> **Recipes for Buffers**

Coating buffer (carbonate buffer)	1.59 g Na$_2$ CO$_3$, 2.93 g NaHCO$_3$, 0.20 g NaN$_3$ made up to 1 L of distilled water adjust to pH 9.6
PBS (phosphate buffer saline)	8.0 g NaCl, 2.90 g HPO$_4$, 12 H$_2$O or 1.15 g Na$_2$ HPO$_4$, 0.20 g KH$_2$ PO$_4$, 0.20 g KCl, 0.20 g NaN$_3$ made up to 1 L of distilled water adjust to pH 7.4
Wash buffer (PBS-T/PBS-Tween)	PBS + 0.5 ml Tween 20
Conjugate buffer (PBS-TPO)	PBS-T + 20.0 g PVP (Polyvinylpyrrolidone) + 2.0 g egg albumin
Extraction buffer	2.40 g Tris (hydroxy methyl) aminomethane, 8.0 g NaCl, 20.0 g PVP, 0.20 g KC, 1 0.50 g Tween –20, 0.20 g NaN$_3$ made up to 1 L of distilled water adjust to pH 7.4, add 0.2% egg albumin just before use
Substrate buffer	97 ml Diethanolamine, 0.20 g NaN$_3$ made up to 1 L of distilled water adjust to pH 9.8
Blocking solution	5.0 g Bovine Serum Albumin (BSA)/spray dried milk (SDM + 1 L PBS-T
Stopping solution	3 M NaOH

All buffers will contain 0.02% sodium azide as a preservative.

➢ **Protocol:**

- Add 100–200 µl of 6 µg/ml purified IgG in coating buffer to each well. Incubate at room temperature for 2 h.
- Wash with PBS-T 3–4 times each for duration 3 min. Dry the plate on blotting paper.
- Add 100–200 µl of test sample (antigen) in suitable extraction buffer (use 1 part of tissue and 10–20-part extraction buffer).
- Incubate at 4–6 C overnight or 2–4 hrs at 30°C.
- Wash with PBS-T as mentioned.
- Add 100–200 µl of alkaline phosphatase enzyme linked antibody (conjugate Ab by coupling globulins with enzyme at ratio ranging between) diluted in conjugate buffer (1:1000).
- Incubate at 37°C for 2 hrs+
- Wash plate with PBS-T.
- Add 100–200 µl of PNP chromogenic substrate at 0.6 mg/ml in Di ethanol amine buffer (substrate buffer).
- Incubate at room temperature for 30 min. Take care to avoid light.
- Take observation either visually or in colorimeter (ELISA Reader) at 405 nm.

➢ **Note:**

- If you wish to stop the reaction add 50 µl of 3 M NaOH to each well.
- AP is the most widely used enzyme for plant virus assay through ELISA but horseradish oxidase and glucose oxidase are also suitable.

➢ **Flow Chart for Das Elisa:**

- Add 100–200 µl of 6 µg/ml purified IgG in coating buffer to each well and incubate for 2–3 h at 37°C-Wash 3 times with wash buffer (PBS-T).
- Add 100–200µl of test sample (1-part leaf tissue + 20 part extraction buffer) to each well and incubate for overnight at 4°C.
- Wash three times with wash buffer (PBS-T).
- Add 100–200 µl of enzyme labeled IgG diluted in conjugate buffer to each well and incubate for 2–3 h at 37°C.
- Wash 3 times with wash buffer (PBS-T).
- Add 100–200 µl of PNP substrate 0.6 mg/ml in substrate buffer to each well and incubate for 1/2 h to 2 h at room temperature in dark.
- Stop reaction with 50 µl of 3 M NaOH.
- Visual observation.
- ELISA Reader at 405 nm filter.

Dot immuno binding assay (DIBA) and tissue blotch immunoassay (TIBA), wherein the sap from the tainted plants is smudged on the nitrocellulose or nylon layers and the infection is recognized by the compound named auxiliary antibodies or chemiluminescent named tests. These systems are less work escalated than ELISA based methodology and are quick. Anyway cautious treatment of blotchs is required. Additionally, there are odds of foundation response in these techniques. Immunosorbent-EM (ISEM) was created which consolidates both serological discoveries with EM, wherein either the virions are first focused on the matrices by explicit antisera or the virions are explicitly embellished with the antisera. Parallel stream gadgets (LFD) or invulnerable strips present the promptly accessible cultivator agreeable apparatuses which can be utilized for field location of plant infections. The outcomes can be gotten in under 10 minutes. LFD use the general serological location frameworks and are normally accessible in the market. LFD comprises of permeable nitrocellulose film (NCM) which is bound to a restricted plastic strip. On this the particular monoclonal or polyclonal antibodies are immobilized and furthermore bound either to latex particles or colloidal gold, or silica. On one side of the strip test (unrefined sap) is gotten and it travels through the strip because of the hair-like powers. Presently the viral antigens tie to the antibodies in fluid stream and convey them to the immobilized antibodies catching the antigen alongside the correspondent particles. With the amassing of these bound particles a band shows up which is noticeable to the unaided eyes. Countless sidelong stream gadgets are accessible for the field identification of plant infections. LFD based discovery devices despite the fact that are easy to understand, fast, and very reasonable for the on location dynamic. Be that as it may, there are issues identified with the cost engaged with testing per test and furthermore affectability, they can't be utilized as the main strategy for infection ordering when screening the high worth and huge number of tests. Because of the low affectability Sometimes, LFD likewise gives bogus negative outcomes for the inert infections because of low affectability.

5.3.1.4.13   *Spot Immunobinding Assay (DIBA)*

DIBA is otherwise called spot ELISA. NCM or nylon layer is utilized as a strong stage rather than ELISA plate (Hibi and Saito, 1985). Test antigen separate is hovered over nylon layer and the methodology follows as ELISA. At long last, the substrate frames an insoluble hued material on the layer as a positive outcome. DIBA is exceptionally delicate and about 1.0 ng of infection can be distinguished. The system is basic, quick, and more affordable. It is valuable in routine recognition of infection in seeds or planting materials.

5.3.1.4.14   Immuno-Tissue Printing (ITP)

ITP is an adjusted form of DIBA. Rather than utilizing antigen separate, cut surface of plant organ is pushed on the nylon film. The nearness of infection in the printed part of the layer is distinguished by the method of DIBA. This strategy gives data about the dissemination of infection in the tissues.

5.3.1.4.15   Western Smudging (WB)

WB is otherwise called protein smudging. WB is named by the similarity with Southern blotching. The method includes electrophoresis, electroblotting, and strong stage immunological recognition as is accomplished for DIBA. The protein is then moved to nylon layer by electroblotting. Obstructing of the protein free territory on the layer is finished by serum egg whites. The protein bound to the layer is then recognized by the method of DIBA.

5.3.1.4.16   Immunosorbent Electron Microscopy (ISEM)

ISEM joins the immunological and electron minuscule (EM) strategies (Katz and Kohn, 1984). ISEM is a significant indicative device as it gives data of serological and morphological personality of an infection. ISEM is otherwise called serologically explicit EM or immunoelectron microscopy. ISEM is performed by catching and design of infection molecule. Catching of infection particles is done on a carbon-covered framework and was first applied by Derrick in 1973 to distinguish TMV and Potato infection Y. Derrick's method is alluded as ISEM, where the affectability of electron minute strategy could be improved altogether. ISEM system is adjusted by layering the particular immune response in abundance sum on the caught infection on the framework. These outcomes into testimony of a thick layer of neutralizer around the infection molecule that can be particularly found in EM after negative recoloring. This marvel is named as beautification by Milne and Luisoni (1977).

ISEM has numerous preferences as a serological apparatus. The test results are particular and liberated from bogus positive. ISEM is profoundly delicate as ELISA and around multiple times more impressive than the customary EM. The test technique is straightforward, fast, and a blend of infection particles in an example can be recognized by ISEM.

5.3.1.4.17   Immuno-Gold Marking (IGL)

IGL is important in situ area of infection molecule in slim segment of tainted cells. Protein is named with gold particles of around 7 to 16 nm distance across. Protein-A gold complex is then used to find neutralizer bound to infection molecule.

5.4   MOLECULAR METHODS/TECHNIQUES FOR DETECTION AND DIAGNOSIS OF PLANT VIRUS

5.4.1   POLYMERASE CHAIN REACTION (PCR)

PCR based diagnostics can be utilized for the identification of both DNA and RNA genomes of plant infections using the PCR and converse interpretation PCR (RT-PCR) individually. For PCR response groundworks explicit to viral genomic parts are required, which have an essential of earlier succession data of the viral genomes. The expanding number of incomplete and complete genome arrangements of plant infections are presently accessible in the databases, and it is conceivable to structure either explicit or deteriorate introductions for intensifying the focused on genomic locales of infections for location of the particular infection species or strain. Notwithstanding explicit discovery of the infection, the intensified items can be sequenced either legitimately or subsequent to cloning into an appropriate vector. Successions so gotten can be utilized for phylogenetic investigation. PCR based location strategies are extremely touchy, and there could be odds of sullying either or both before PCR and post-PCR prompting bogus positives. Accordingly, it is constantly prescribed to have positive and negative controls to keep away from bogus negatives or positives. In contrast with ELISA, where creation of value antibodies requires a ton of endeavors, it is a lot simpler to plan and orchestrate new groundworks. At present PCR based diagnostics are the standard diagnostics for recognition of numerous infections and infection like pathogens in various field and plant crops. A considerable lot of the infections tainting organic product trees happen at exceptionally low focus in the hosts, such infections are hard to recognize through typical PCR. Furthermore, the plant pararetro infections (family Badnavirus) have their coordinated partners in the host genome known as endogenous pararetro infections (EPRV) which have noteworthy arrangement homology to related episomal infections. The nearness of coordinated BSV like arrangements (endogenous BSV: eBSV) in the Musa genome makes the PCR based finding questionable. To determine both the issues referenced above, nucleo-and immuno-based identification devices are consolidated in immunocapture-PCR (IC-PCR) where the PCR enhancement is completed on the immuno-caught virions. Notwithstanding the capacity to intensify the episomal viral successions, IC-PCR additionally can think (total) the infection particles from the rough sap in this manner making it progressively delicate. IC-PCR is as of now considered as the best quality level for identification of pararetro infections having coordinated partners. Constant PCR Real time-PCR (rt-PCR) or qPCR is shut cylinders based PCR measures which are immediately embraced in the asset adequate labs. Ongoing PCR is an improvement over the standard PCR in which the enhancement is identified during the response procedure itself through the produced fluorescence. It includes the utilization of vague fluorescent colors (e.g., SYBR green, LUX, and so forth.) or specific probes labeled with fluorescent dyes (TaqMan, molecular beacon, etc.). The SYBR green binds to double standard DNA molecules and fluoresce. In the TaqMan-based rt-PCR sequence specific internal DNA probes labeled with dye (reporter and quencher) are used and the florescent signals are emitted during the amplification reaction. This in addition to providing good specificity and sensitivity avoids any chances of post-PCR contamination. The fluorescent signals are detected during the amplification process itself. Real time PCR has more generic application, and can be quickly standardized compared to the development of antibodies for ELISA. However, the cost involved in testing per sample is much higher than ELISA. High-throughput testing of tests, capacity to test the viral burden, less time included is explicit preferences of RT-PCR based location measures. The ELISA and ongoing PCR shares numerous advantages as the recognition apparatuses, viz. industry standard organization, vigor, high reproducibility and simple to test even by unpracticed clients:

1. **PCR and RT-PCR:** The polymerase chain response (PCR) method has created to identify exceptionally little amounts of nucleic corrosive by enhancement of a portion of the DNA arranged between two locales of realized nucleotide grouping. These sections are commonly preserved successions. Such a fragment is flanked by two oligonucleotides which fill in as introductions for a progression of responses that are catalyzed by a chemical DNA polymerase within

the sight of support and dNTPs. An enhancement cycle comprises of dissolving of the twofold abandoned format DNA at high temperature (denaturation at 94°C), hybridization of the groundworks with reciprocal base successions in the two strands at lower temperature (toughening at 54–60°C relies upon GC substance of preliminaries) and expansion of groundworks with DNA polymerase (DNA union at 72°C). Along these lines, at each pattern of high and low temperature, the succession between the groundworks will be multiplied, so that after n cycles, a 2n enhancement may have been acquired (generally 30–35 patterns of intensification are applied). To structure an introduction for specific infection data must be accessible on the nucleotide arrangement of the objective infection. The data on nucleotide succession is accessible on World Wide Web of NCBI (USA) or EMBL (Germany). Presently programming and online projects are accessible to plan the preliminaries, for example, Primer 3, GeneFisher, PCR Now and so forth. The determination of groundworks relies upon target grouping. All in all, groundworks ought to be between 18 and 25 base pair (bp) long. PCR is helpful in recognizing DNA infections. The method has been productively utilized in identifying gemini infection even from a solitary whitefly vector (Deng et al., 1994). PCR can't be applied straightforwardly to the vast majority of plant infections as their genome comprises of single strand RNA. In such cases, the RNA must be deciphered first into DNA by turn around transcriptase (RT) protein. In this manner, the strategy included is known as RT-PCR. The business units for RT-PCR are accessible which interpret the RNA into DNA and afterward combined with PCR. Diverse RT-PCR variations have been created, including immunocapture RT-PCR (IC-RT-PCR), which has been utilized with plant separates, or immobilized focuses on paper print (PC-RT-PCR) or squash catch (SC-RT-PCR), permitting the discovery of insignificant amounts of RNA focuses from plant material or bug vectors without extricate planning. In these variations, the infection particles are caught by antibodies on a surface and afterward expelled by warming with a non-ionic surfactant, for example, Triton X-100. The genome of infection is then intensified utilizing RT-PCR. This technique is helpful in concentrating infection particles from plant tests where infection titer is low or where plant mixes restraining PCR are available. Zitikaite and Staniulis (2006) recognized three infections viz., cucumber mosaic infection (CMV), tomato ring spot infection (ToRSV) and tobacco putrefaction infection (TNV) from cucumber by RT-PCR.

2. **Nested PCR (nPCR):** The technique is valuable when the infection titer is extremely low and additionally target quality is flimsy, and can't be checked by electrophoresis because of low intensification item. In this kind of PCR, two sets of preliminaries are required (Tetramers) than a solitary pair as is utilized for PCR and RT-PCR. The item from essential PCR enhancement is utilized for second PCR intensification. Be that as it may, the subsequent response can be caused to confront the danger of pollution (Lopez et al., 2008). A few infections, including Prunus necrotic ringspot infection (PNRSV), Prune overshadow infection (PDV), Plum pox infection (PPV) and Citrus tristeza infection (CTV), were distinguished by this method (Adkar-Purushothama et al., 2011; Helguera et al., 2001, 2002; Olmos et al., 1999). The settled PCR was additionally used to recognize another infection named Lettuce mosaic infection (LMV) by consolidating with Immunocapture-RT PCR (IC-RT-PCR) to expand affectability and to disentangle readiness of test (Helguera et al., 2001, 2002). This technique was applied to recognize infection even in single aphids (Moreno et al., 2007). Alone IC-RT-PCR has helped us in recognizing pepper mottle infection (PepMoV) and Potato infection Y from bean stew and potato individually (Kaur et al., 2014).

3. **Ongoing PCR:** Real time PCR is the most dynamic variation of PCR wherein the improvement of target course of action can be assessed after each PCR cycle and increase is pictographically shown through a connected screen. Ongoing PCR has picked up fame as it not just wipes out the post-PCR handling steps including gel electrophoresis, yet in addition dispenses with the introduction to cancer-causing synthetic compounds, ethidium bromide, radioactive isotopes, UV-radiation. It likewise represents a diminished danger of test pollution with amplicon and a more prominent luxury to multiplexing, synchronous testing for numerous pathogens or tests. The method can likewise be actualized in the field by utilizing convenient constant PCR machines. In spite of the fact that the continuous observing bend was raised up as the DNA exponentially enhanced, there are a few downsides to utilize ongoing PCR. Other inconvenience is that constant strategy requires exceptionally costly gear. In spite of the fact that these faults, ongoing PCR has been

progressively utilized in light of the fact that this strategy has been indicated important location for plant infections (McCartney et al., 2003). Citrus tristeza infection (CTV) in various plant tissues and TMV in soil were distinguished by continuous PCR and evaluated. Recognition of Citrus leaf smudge infection (CLBV) was likewise done by using this procedure (Ruiz-Ruiz et al., 2007, 2009; Yang et al., 2012).

4. **Multiplex PCR (M-PCR):** Multiplexing alludes to the test wherein more than one infection is distinguished in a similar response. Multiplex measures give advantage respect to cost included, effortlessness, and straightforwardness in dealing with. This procedure can identify diverse DNA or RNA targets at the same time in a solitary cylinder. In those situations where plant infections exist as blended contamination, the multiplex-PCR methods are extremely helpful. The method includes the utilization of numerous arrangements of oligonucleotides focusing on the genomic districts of various infections (inside same variety, strains of same species or infections from systematically various gatherings) in single PCR intensification. The essential of multiplexing PCR strategies is to plan the oligonucleotide groundworks so that enhancement can be completed at one toughening temperature and various preliminaries don't show any self-complementarity. The amplicon size from various viral genomes ought to appear as something else so as separate them through agarose gel electrophoresis. The multiplex-PCR has the focal points that it spares the reagents, time, is helpful and practical when more than one infection is to be recognized from a similar example. M-PCR has been normalized for various infections in various yields. The continuous based multiplex strategies can likewise be utilized, anyway as the different columnist and quencher colors should be utilized in the rt-PCR, because of the cross talk between the fluorescent signs being discharged from the correspondents, the quantity of targets is hypothetically restricted to four as it were.

5. **MT-PCR:** Multiplexed-couple PCR is an innovation stage created for exceptionally multiplexed quality articulation profiling and the quick distinguishing proof of clinically significant pathogens. The stage comprises of two rounds of enhancement. In the initial step, a M-PCR is performed at 10 to 15 cycles to permit advancement of target DNA without making rivalry between amplicons. Up to 72 diverse PCR responses can be multiplexed and performed at the same time. Fluorescence is estimated by SYBR green innovation toward

the finish of every expansion cycle, and softens bend examination gives species-explicit or quality explicit recognizable proof.

The significant confinement of all sub-atomic indicative measures is target accessibility, which is frequently directed by a push to adjust expenses and pivot time with greatest throughput. The expanded affectability of sub-atomic tests and its capacity to identify reasonable and non-suitable cells ought to likewise put overwhelming accentuation on deciphering results with other microbiological information and clinical data.

The polymerase chain response (PCR) is an *in vitro* procedure which permits the intensification of a particular deoxyribonucleic corrosive (DNA) area that lies between two areas of realized DNA arrangements. PCR is a procedure generally utilized in atomic science, microbiology, hereditary qualities, diagnostics, clinical research centers, legal science, ecological science, genetic examinations, paternity testing, and numerous different fields. This procedurc/strategy was grown first time by Kary Mullis in 1984. The name, polymerase chain response, originates from the DNA polymerase used to enhance (reproduce ordinarily) a bit of DNA by in vitro enzymatic replication. The PCR item is enhanced from the DNA layout utilizing warmth stable DNA polymerase (Taq DNA polymerase) from *Thermus aquaticus* and utilizing a computerized warm cycler (PCR machine) to get the response through at least 30 patterns of denaturing, strengthening of groundworks, and augmentation. The first particle or atoms of DNA are recreated in each cycle, in this way multiplying the quantity of DNA particles. This procedure is known as a "chain response" in which the first DNA format is expo-nentially enhanced. With PCR it is conceivable to intensify a solitary bit of DNA, or an extremely modest number of bits of DNA, over numerous cycles, producing a great many duplicates of the first DNA particle. To duplicate DNA, polymerase requires two different parts: a flexibly of the four nucleotide bases and preliminaries. DNA polymerases, regardless of whether from people, microorganisms, or infections, can't duplicate a chain of DNA without a short grouping of nucleotides to "prime" the procedure, or kick it off. Therefore, the phone has another catalyst called a primase that really makes the initial barely any nucleotides of the duplicate. This stretch of DNA is known as a preliminary. When the preliminary is made, the polymerase can assume control over creation the remainder of the new chain. All the fundamental segments required for DNA duplication in vitro are: a bit of DNA, enormous amounts of the four nucleotides, huge amounts of the groundwork arrangement, DNA polymerase, Magnesium particles, and DNA polymerase. PCR is important to analysts since it permits them to duplicate novel locales of DNA so they can be distinguished in huge genomes. Analysts in the Human Genome Project are utilizing PCR to search for markers in cloned DNA sections and to arrange DNA parts in libraries. The polymerase chain response (PCR) relies on oligonucleotide groundwork coordinated enzymatic intensification of a particular DNA arrangement of intrigue. An essential for intensifying a grouping utilizing PCR is to have known, one of kind arrangements flanking the section of DNA to be enhanced with the goal that particular oligonucleotides can be acquired. After intensification by PCR, the items are isolated by agarose gel electrophoresis and are straightforwardly imagined in the wake of recoloring with ethidium bromide. A PCR vial contains all the essential parts for DNA duplication: a bit of DNA, enormous amounts of the four nucleotides, huge amounts of the groundwork succession, and DNA polymerase. Basically, PCR is a three-advance technique viz., denaturation, annealing, and exten-sion. These three pieces of the polymerase chain response are done at various temperatures.

DNA layout that contains the DNA district (focus) to be enhanced by PCR and it can intensify an objective up to 4000 bp easily. In any case, the intensification will likewise rely upon the nucleotide organization of the format. Oligonucleotide preliminary is a short oligo nucleotide grouping integral to the furthest limit of target arrangement subsequently the one groundwork will go about as forward groundwork while other will go about as opposite preliminary. Preliminaries ought to be 18–25 nucleotides in length. Groundworks ought not to be having self-complementarities, ought not shape barrette and it ought not have crisscross at 3' end. Preliminary structuring virtual products like DNA star, Gene instruments and so forth are accessible. Deoxynucleotide triphosphates (dNTPs) are the structure obstructs from which the DNA polymerases blend another DNA strand. Reciprocal nucleotides are added to shape the new strand, DNA polymerase is chemical utilized for augmentation of DNA strand reciprocal to the layout. Various sorts of DNA polymerase are accessible like Klenow section, Taq DNA polymerase, Pfu polymerase, and so forth. Distinctive polymerase has various properties consequently choice of DNA polymerase is significant. Cushion arrangement gives an appropriate synthetic condition to ideal movement and steadiness of the DNA polymerase and blend of DNA under in vitro condition. Divalent cations that by and large contains magnesium particle (Mg^{2+}) is utilized as a fundamental co-factor required for the movement of DNA polymerase. Cycling Condition: The PCR for the most part comprises of an arrangement 30 rehashed temperature changes called cycles; each cycle ordinarily comprises of 3 discrete temperature steps. The temperatures utilized and the time spans they are applied in each cycle rely upon an assortment of boundaries. These incorporate the compound utilized for DNA combination, the grouping of divalent particles and dNTPs in the response, and the liquefying temperature (Tm) of the groundworks.

- **Initialization Denaturation Step:** This progression comprises of warming the response to a temperature of 93–96°C for 5 minutes relying up upon the multifaceted nature of genomic DNA.
- **Denaturation Step:** This progression is the principal normal cycling occasion and comprises of warming the response to 94–96C for 20–30 seconds. It causes dissolving of the DNA layout by upsetting the hydrogen securities between correlative bases, yielding single strands of DNA.
- **Annealing Step:** The response temperature is brought down to 50–65°C for 30–45 seconds permitting toughening of the preliminaries to the single-abandoned DNA layout. Normally, the toughening temperature is around 3–5 degrees Celsius underneath the Tm of temperatures for each of the primer sets must be optimized to work correctly within a single reaction, and amplicon sizes, i.e., their base pair length, should be different enough to form distinct bands when visualized by gel electrophoresis.
- **Settled PCR:** This builds the particularity of DNA intensification, by decreasing foundation due to vague enhancement of DNA. Two arrangements of preliminaries are being utilized in two progressive PCRs. In the main response, one set of groundworks is utilized to create DNA items, which other than the proposed target, may even now comprise of vaguely enhanced DNA parts. The product(s) are then utilized in a second PCR with a lot of groundworks whose coupling destinations are totally or incompletely unique in relation to and found 3' of every one of the preliminaries utilized in the main response. Settled PCR is frequently progressively fruitful in explicitly intensifying long DNA pieces than customary PCR, yet it requires increasingly point-by-point information on the objective groupings.

➢ **Applications:**

- In microbiology and sub-atomic science, for instance, PCR is utilized in research labs in DNA cloning systems, southern smearing, DNA sequencing, and recombinant DNA innovation.

- In clinical microbiology labs PCR is important for the determination of microbial contaminations and epidemiological investigations.
- PCR is additionally utilized in crime scene investigation research facilities and is particularly valuable in light of the fact that solitary a little measure of unique DNA is required, for instance, adequate DNA can be acquired from a bead of blood or a solitary hair.

5.4.2   FLUORESCENT IN SITU HYBRIDIZATION

DNA based strategy is known as FISH in short in full it is known as fluorescence *in situ* hybridization. FISH is an incredible technique for the in-situ discovery of dynamic developing living beings in natural examples. The procedure can picture the exact area of specific DNA or RNA successions in the cytoplasm, organelles, or cores of natural materials. Thus, the method can recognize metabolically dynamic creatures straightforwardly in the earth without development when RNA is available.

FISH tests frequently target groupings of ribosomal RNA or mitochondrial qualities since they are plentiful in succession databases and in various duplicates in every cell. The tests can be planned by PC helped look from target life forms for "marks." These mark locales can be explicit at various phylogenetic levels relying upon the fluctuation of the objective atom groupings. The test involves a short arrangement, going from 15 to 20 nucleotides, that are explicit for one or a few taxa at animal types, sort, or higher ordered positions. The test or oligonucleotide is then named with a fluorochrome, for instance a carboindocyanine color (CY3).

A few elements can impact the effectiveness of FISH, for instance, the sterical and electrostatical properties of rRNA, the highlights of the test, and hybridizations conditions, for example, the obsession technique, cushions, the severity of test official, and brooding time examined the distinctive openness.

This method utilized 16S or 23S ribosomal (rDNA) oligonucleotide tests named with fluorescent color in mix with fluorescence microscopy. The FISH tests of size (20–30 mers) perceived the pathogens in plant tissue cells, fixed in an infinitesimal slide and hybridized it with target quality in the pathogen in the plant tests. The test target hybridization can imagine by bright light. FISH was effectively utilized with tests to focus on the 23S rDNA to distinguish *Ralstonia solanacearum* in potato strips. So this is another DNA-based innovation where a fluorescent test is hybridized to a particular area on the microbial or pathogen genome and afterward it is pictured under fluorescent or confocal microscopy to see that whether that specific fluorescent test is noticeable or not. Obvious of fluorescent tag is corroborative for the nearness of that specific genome that is focused on. This could likewise be utilized to recognize parasites and infections and other endosymbiotic microscopic organisms that contaminate plants. The high proclivity and particularity of DNA tests give high single cells affectability in fish on the grounds that the test will tie to every one of the ribosomes in the example. In any case, the down to earth furthest reaches of discovery lie in the scope of around 103 CFU/mL. Therefore, this level or this breaking point ought to be accomplished before perception under FISH or under fluorescent magnifying instrument. Notwithstanding the location of culturable microorganisms that cause plant sicknesses FISH could likewise be utilized to identify yet to be refined alleged unculturable living beings so as to research complex microbial networks. Therefore, the test can be intended to target even those unculturable organisms which are yet to be refined. Further, FISH can likewise be utilized for discovery of pathogens in which plants indications are yet to be created. For instance, here the wool mold pathogen in impatiens plant was recognized by utilizing the fish innovation. Therefore, these are the nearness of locales of fleece buildup in the impatiens leaves and this is again a chance to distinguish pathogen before its creation of side effects. In this way, in short, we have seen that nucleic corrosive based advancements can be effectively conveyed in pathogen conclusion and recognition of even unculturable or yet to be refined or even nonculturable substances like infections could be identified effectively in nearness even before the creation of the side effects by the plants thus that it helps the ranchers or producers to take sufficient control measures at danger or fitting time before the pathogen cause serious harm to the yield plants.

5.4.3 DNA ARRAY HYBRIDIZATION

DNA cluster hybridization, otherwise called reverse dot blot hybridization (RDBH) or macroarray, is a strategy dependent on hybridization of enhanced and named genome districts important to immobilized oligonucleotides spotted on a strong help stage. It was initially evolved to distinguish changes of human qualities, is as yet a significant symptomatic instrument in clinical examination. It is presently viewed as an amazing and down to earth procedure for the location and distinguishing proof of parasites and different organisms, for example, microorganisms, from complex natural examples without the requirement for disconnection in culture. Most DNA macroarrays that have been created depend on a solitary locale for the recognition of a particular scientific classification. Among these locales, 16S ribosomal DNA has been utilized for the location of microorganisms. Different genome districts, for example, ribosomal DNA spacers (ITS), mitochondrial qualities (e.g., cytochrome oxydase c subunit 1, cox1) and some protein coding locales (β-tubulin, EF-1a, and so forth), were picked to target parasites and growth like life forms.

DNA cluster hybridization is profoundly delicate as are most PCR-based methodologies. With the boundless limit with regards to the convenience of oligonucleotides on one film and the reusability of the layers, it shows predominant multiplexing location ability at a lower cost over other PCR-based strategies. In a biodiversity study, the species profile could be uncovered by hybridizing the oligonucleotide cluster with a blended pool of DIG-named PCR items enhanced from the complete DNA of a question test of the rRNA atom for FISH tests.

5.4.4   PADLOCK PROBE TECHNOLOGY AND ROLLING CIRCLE AMPLIFICATION (RCA)

Location and portrayal of single nucleotide polymorphisms (SNPs) is getting expanding well known for pathogen ID, however was viewed as a significant test for customary continuous PCR utilizing ordinary oligonucleotides identified by fluorescent colors. So as to perceive SNPs among various genotypes, lock test methods are required. Latch tests (PLPs) are long oligonucleotides (around 100 bp) conveying a non-target-reciprocal portion flanked by the objective correlative districts at their 5' and 3' closes, which perceive nearby successions on the objective DNA).

Along these lines, on hybridization, the closures of the tests possess neighboring positions, and can be joined by enzymatic ligation. Ligation happens and the tests are circularized just when both end sections accurately perceive the objective grouping. The helical idea of twofold abandoned DNA (dsDNA) empowers the test to topologically tie to the objective strand and the test can't be dislodged.

Lock tests were at first presented for in situ DNA confinement and discovery. They were grown initially for segregation of centromeric arrangement variety in human chromosome. Be that as it may, the strategy has now been applied to location of hereditarily changed living beings.

Moving circle amplification technology (RCA) Rolling circle enhancement (RCA) is likewise an isothermal nucleic corrosive intensification which enhance the test DNA successions more than 109-crease at a solitary temperature. In RCA response, various rounds of isothermal enzymatic combination are included. Ø29 DNA polymerase expands a circle-hybridized preliminary by persistently advancing around the roundabout DNA test of a few dozen nucleotides to imitate its arrangement again and again. A significant favorable position of RCA is that dissimilar to PCR, it is unfeeling toward pollution and not at all like other isothermal advances, requires next to zero examine improvement. The limit of RCA to yield the surface bound intensification items offers noteworthy points of interest to in situ or microarray hybridization tests. In straight RCA, the result of enhancement stays fastened to the objective particle. RCA appears to be appropriate to cell and tissue-based examines related to the isothermal idea of the RCA response and the ability to restrict numerous markers all the while. Identification of Banana streak infection (BSV) secludes tainting banana utilizing RCA strategy was finished. Other than recognition just, the capacity of this technique to separate among episomal and coordinated viral genomic groupings was illustrated. For portrayal of begomo infections, this procedure is generally investigated. A couple infections which has been recognized by this strategies are Abutilon mosaic infection (AbMV), Ageratum leaf twist Cameroon infection (ALCuCMV), Chili leaf twist infection India (ChiLCuV), Tobacco wavy shoot infection [Y41] (TbCSV-[Y41]), Tomato leaf twist Kerala infection (ToLCKeV) and Watermelon chlorotic trick infection [SD] (WmCSV-SD).

5.4.5   *LOOP-MEDIATED ISOTHERMAL AMPLIFICATION (LAMP)*

Loop-mediated isothermal amplification (LAMP) is an incredible and novel nucleic corrosive enhancement technique that enhances a couple of duplicates of target DNA with high explicitness, proficiency, and quickness under isothermal conditions, utilizing a lot of four uniquely structured preliminaries and a DNA polymerase with strand dislodging action. The cycling responses can bring about the gathering of 109 to 1010-overlap duplicates of focus in under 60 minutes. Two internal and two external preliminaries are required for LAMP, in the underlying strides of the LAMP response, every one of the four groundworks are utilized, yet in the later cycling advances just the inward groundworks are utilized for strand removal DNA blend.

   The external groundworks are alluded to as F3 and B3, while the internal preliminaries are forward inward preliminary (FIB) and in reverse internal preliminary (BIP). Light depends on auto-cycling strand dislodging DNA amalgamation within the sight of Bst DNA polymerase, explicit groundworks and the objective DNA layout. The system of the LAMP intensification response incorporates three stages: creation of beginning material, cycling enhancement, and reusing.

It recognizes amplicons by means of photometry for arrangement turbidity brought about by amassing of insoluble salts from response of metal particles with the pyrophosphate particle. With SYBR Green, a shading change can be seen without gear. Light can be utilized effectively as a basic, tough screening examine and takes out the requirement for costly thermocyclers. Light measures have been produced for the whole significant classes of plant pathogenic microorganisms, including viroids, phytoplasma, microbes, and growths. This examine utilizes four exceptionally planned groundworks, a couple of external preliminaries and a couple of inward groundworks, which together perceive six unmistakable destinations flanking the enhanced DNA grouping. The RT-LAMP has been produced for basic checking of RNA infections including PVY, Tomato yellow leaf twist infection, Melon yellow spot infection and Potato leaf move infection (PLRV) (Ju, 2011; Nie, 2005) and viroids viz., Peach inert mosaic viroid, Potato shaft tuber viroid (PSTVd). Other than this, LAMP examine has been created and effectively used to identify a few infections including Tomato necrotic trick infection (ToNStV), nine rice leaf infections and tobacco infections including cucumber mosaic infection (CMV), potato infection Y (PVY), tobacco draw infection (TEV), TMV, and tobacco vein banding mosaic infection (TVBMV).

Both PCR and RT-PCR based identification strategy, requires warm cycler to give an angle of temperature to various enhancement steps. In a large number of the asset helpless research centers it is beyond the realm of imagination to expect to complete the PCR based location; in this manner, the isothermal enhancement based recognition techniques are created, which can do the intensification response at consistent temperature.

Circle intervened isothermal Amplification (LAMP) can intensify from a couple of duplicates of DNA to a size of 109 in under 60 minutes. Light uses three sets of groundworks (interior, outer, and circle introductions) for intensification, which ties at any rate at six restricting destinations. The enhancement creates an item which is having single abandoned circle areas, where the preliminaries tie without the requirement for denaturation. The intensification response is intervened by the DNA polymerase with high strand-relocation action and the entire procedure is finished in roughly one hour at around 60–65°C. Light-based identification frameworks are anything but difficult to perform and the conclusive outcomes can without much of a stretch be envisioned, subsequently simple to decipher. Because of the LAMP response, magnesium pyrophosphate (a side-effect) is created, which prompts increment the turbidity that can be seen by the unaided eyes. For the discovery of the enhancement item, the intercalating colors, for example, SYBR Green or PicoGreen can be included after intensification, and the fluorescence can be distinguished. As these colors are included post-intensification, there are odds of pollution. Be that as it may, there are reagents like calcein and $MnCl_2$ or hydroxyl naphthol blue (HNB), which can be included in the response blend itself with no restraint of the intensification. The shading changes due to calcein and $MnCl_2$ can be pictured under bright light, while due to HNB can be imagined straightforwardly. The LAMP tests are nearly as touchy as the continuous PCR. With the ongoing enhancements and by incorporation of a converse interpretation step LAMP examine can be finished in a brief period (<30 minutes) at a steady temperature of 65°C. The LAMP examine can be brought out on the DNA secluded through the typical methodology and the DNA polymerase utilized in the LAMP is open-minded to low degree of PCR inhibitors which are unavoidable during DNA extraction. The greatest bit of leeway of isothermal enhancement based analytic strategies is that these techniques can without much of a stretch be embraced by asset helpless research centers for the normal ordering and even at the purpose of dynamic that is in the plantations/fields.

5.5 OTHER MOLECULAR METHODS/TECHNIQUES FOR DETECTION AND DIAGNOSIS OF PLANT VIRUS

5.5.1 MICRO-ARRAY BASED DETECTION

With the expanding biodiversity of the plant infections, the location to a great extent depends on screening tests utilizing equal tests, which are unwieldy. In this way, testing an example for countless infections is a genuine test. Toward this path, discovery tests which are fit for distinguishing a wide scope of infections, including those not recently announced from that harvest are created. Microarrays are the hybridization-based measures which help recognize a great many infections and virioids at once. Microarray strategies depend on hybridization of named tests (viral nucleic acids) with a large number of one of a kind immobilized tests. The integral base matching between the spotted tests and the named tests (viral nucleic corrosive) is the fundamental hidden rule of the DNA microarray-based location. Each test is comprised of thousands of cDNAs or oligonucleotides which are explicit to an objective DNA or RNA arrangement. In a microarray the fluorescent color marked objective is spotted on the strong help which contains the objectives previously spotted on it. This spotting is done through automated framework. The integral response between the objective and test is perused by the age of fluorescent sign at that specific exhibit address where the hybridization has occurred. The extreme development of viral genomic groupings accessible in the open database, makes it conceivable to configuration tests dependent on the preserved genomic areas (for expansive identification) or the variable genomic district (for explicit location of an infection animal types or a strain) and recognize the infections dependent on the cross-hybridization on the clusters. Microarrays structured with the infection explicit oligo-nucleotide tests give a conventional examine wherein all the infections could be distinguished for which the grouping data is accessible and is spotted on the cluster. Also, utilization of nonexclusive tests dependent on the monitored districts for example the preserved center coat protein district from potyviruses or the saved center RT/RNase H locale of para retroviruses or the saved stem-circle area or coat protein locale of begomo infections, will have the option to identify the novel infections in that specific gathering. Microarray based location has been as of late used in the screening of field and green plants or planting materials against various infections one after another. In the further headways in the microarray-based measures the engineered oligonucleotide tests were utilized as the catch. The utilization of manufactured oligonucleotide tests permitted more prominent affectability in the location. There is an extremely high capability of microarray-based identification for plant infections in crops especially the vegetative spread ones. Microarray chips for location of various infection diseases of potato and grapevine are accessible. The microarray innovation despite the fact that has numerous preferences still the location utilizing it is expensive and expects aptitude to deal with the modern instruments. The engineered oligonucleotides offer an incredible favorable position over the PCR based tests; nonetheless, it has the impediments of framing auxiliary structures.

5.5.2   DNA MICROARRAYS

DNA microarrays are promising high throughput devices and can recognize numerous pathogens simultaneously. This is the significant preferred position of DNA microarray. Along these lines, microarray comprises of a strong framework normally a glass slide on which oligonucleotide tests or other DNA parts are put in exact areas at high thickness. So this is an average case of a DNA microarray chip where the in that microchip this is the DNA tests that are orchestrated at explicit areas at exceptionally high thickness 06:46.

The objective DNA arrangement in an example at that point hybridized to the tests and identify by fluorescence. Therefore, the objective DNA arrangement is then hybridized with the test and afterward it is distinguished with fluorescence light. The upside of microarray-based discovery is the joined incredible nucleic corrosive intensification procedures with a huge screening ability bringing about elevated level of affectability, particularity, and high throughput limit. It can identify a wide range of pathogens in a solitary examine thus, that is the reason DNA microarray is likewise being utilized to recognize different pathogens from a solitary animal categories.

5.5.3 NEXT GENERATION SEQUENCING (NGS)

Cutting edge sequencing (NGS) or profound sequencing-based strategies has been produced for the revelation of novel infections. In light of the infection contamination, plant framework delivers little RNAs (siRNAs) correlative to the viral genomic arrangements, which debases the viral RNAs, known as quieting. Profound sequencing on the siRNAs secluded from the tainted examples permits the recuperation of either full or incomplete genomic arrangements of infections. These siRNA sequencing based strategies have been successfully used to find/distinguish infection contaminations in numerous harvests. The high throughput sequencing (HTS) advancements include arrangement of the nucleic corrosive libraries, intensification of these libraries to get adequate amount of DNA followed by equal sequencing of the billions of DNA parts. These methodologies or their means are currently much progressed regarding effortlessness, cost, and running time. Expanding number of harvest genomes being sequenced, gives acceptable chance to profound sequencing based disclosure of novel plant infections, as it is conceivable in silico to allot the short groupings to be of host or non-have starting point in this way decreasing the volume of information to be examined for virome considers. Further, it is likewise critical to specify that accessibility of the genomic information for some infections is one of the pre-essentials for the delicate and dependable conclusion. NGS, notwithstanding the application in settling the etiology of viral maladies, portrayal, populace hereditary qualities, has the potential in high throughput conclusion of plant infections in both farming and agricultural yields. Of the voluminous infor-mation got from the profound sequencing venture, the most significant is to break down in silico and dispose of those got from have or other microflora and investigate just the viral determined arrangements. Either the improved viral nucleic corrosive from tainted examples or the nucleic corrosive got from infection like particles (VLP) or in part and totally refined infection particles or moving circle enhanced (RCA) advanced nucleic corrosive of the roundabout infections can be utilized for NGS based profound sequencing.

This is RNA (by and large 21–24 nt) based NGS equal sequencing of indicative and asymptomatic examples followed by all over again get together to long peruses can possibly recognize the novel uncharacterized RNA, ssDNA, switch translating dsDNA infections and viroids without the essential of earlier succession data. In the event of some infection gatherings, the moderated areas have been recognized either over the various genera or family, which empowers us to plan the conventional preliminaries focusing on the locales which can possibly distinguish various variations just as new or uncharacterized infections from that family. Little RNA (siRNA) profound sequencing innovation has end up being an effective conventional apparatus for infection ID in plants. Ongoing advancements in the sequencing innovation, high throughput test handling and mechanization has radically diminished the preparing cost per test and these advances have begun discovering place in a significant number of the testing/ordering research centers. With these progressing advancements, it is additionally expected that the NGS response cost per test will diminish further undeniably potentially making these methodology accessible for routine testing of the examples in numerous research centers. A major bit of leeway of these techniques is the expansive range capacity to identify number of infections or viroids from a solitary example just as the uncharacterized infection or infection like operators, which makes them a significantly more strong, and furthermore solid for exacting isolate testing, ordering, and accreditation programs.

5.5.4   GEL ELECTROPHORESIS

It is an expository method used to isolate DNA or RNA sections by size and reactivity. Nucleic corrosive particles which are to be broke down are set upon a thick medium, the gel, where an electric field initiates the nucleic acids (which are adversely charged because of their sugar-phosphate spine) to relocate toward the anode (which is decidedly charged on the grounds that this is an electrolytic as opposed to galvanic cell). The division of these sections is practiced by abusing the versatility's with which diverse estimated atoms can go through the gel.

Longer particles relocate all the more gradually in light of the fact that they experience more opposition inside the gel. Since the size of the particle influences its portability, littler sections end up closer to the anode than longer ones out of a given period. After some time, the voltage is expelled and the fracture inclination is broke down. For bigger partitions between comparative estimated sections, either the voltage or the run time can be expanded. Expanded stumbles into a low voltage gel yield the most exact goals. Voltage is, be that as it may, not the sole factor in deciding electrophoresis of nucleic acids. The nucleic corrosive to be isolated can be set up in a few different ways before partition by electrophoresis. On account of enormous DNA particles, the DNA is habitually cut into little pieces utilizing a DNA limitation endonuclease (or limitation protein).

5.5.5 SOUTHERN BLOT

A Southern blotch is a technique utilized in atomic science for recognition of a particular DNA succession in DNA tests. Southern smearing joins move of electrophoresis-isolated DNA sections to a channel film and resulting piece location by test hybridization. Hybridization of the test to a particular DNA section on the channel film shows that this piece contains DNA arrangement that is reciprocal to the test. The exchange venture of the DNA from the electrophoresis gel to a layer allows simple official of the marked hybridization test to the size-fractionated DNA. It additionally takes into account the obsession of the objective test half-and-halves, required for investigation via autoradiography or other identification strategies. Southern blotchs performed with limitation catalyst processed genomic DNA might be utilized to decide the quantity of arrangements (e.g., quality duplicates) in a genome.

5.5.6 NORTHERN BLOT

The northern blotch, or RNA smear, is a strategy utilized in sub-atomic science examination to consider quality articulation by discovery of RNA (or disconnected mRNA) in an example with northern smudging. It is conceivable to watch cell power over structure and capacity by deciding the specific quality articulation rates during separation and morphogenesis, just as in irregular or unhealthy conditions. Northern blotching includes the utilization of electrophoresis to isolate RNA tests by size and recognition with a hybridization test corresponding to part of or the whole objective grouping. The term 'northern smudge' really alludes explicitly to the hair like exchange of RNA from the electrophoresis gel to the smearing layer.

5.5.7   NUCLEIC ACID-BASED TECHNIQUES

The nucleic corrosive-based strategy is delicate and quite certain, and that requires that is the reason it has been utilized generally in the present setting for the discovery of plant sickness of the different root. Plant pathogens that recognize through DNA and RNA depend on defeat the specific diagnostics and pathogen scientific categorization that empowers a fast and exact discovery and measurement of plant pathogens. Important to realize that K. Mullis has gotten Nobel Prize in 1993 because of his disclosure for enhancement of nucleic corrosive arrangements utilizing the innovation known as polymerase chain response. In short famously known as PCR. In view of the constancy of DNA hybridization and replication PCR is utilized for profoundly explicit identification of parasites, microscopic organisms, infections, and phytoplasma. PCR strategy can give exceptionally high affectability and particularity because of the devotion of DNA enhancement. Achievement of PCR relies upon viability of DNA extraction and execution is influenced by inhibitors present in the example test, polymerase movement, PCR cradle, and centralization of deoxynucleoside triphosphate. What's more use of PCR for pathogen recognition requires planning of groundwork to start DNA intensification which could constrain the useful materialness of this strategy for new and obscure pathogens. On the off chance that the DNA grouping of the new or obscure pathogen isn't accessible at that point planning groundwork won't be conceivable and that is the reason this is a constraint of PCR. PCR-based strategies can be utilized for various genomes like single abandoned RNA (ssRNA) single-abandoned DNA (ssDNA) or twofold abandoned DNA (dsDNA) and PCR offers a few points of interest, for example, the ability to identify a solitary objective in complex blends this is profoundly huge and, quick, and explicit identification of different targets, and the possibility to distinguish unculturable pathogens, for example, infections, and a few microscopic organisms and phytoplasma which are not been culturable up until now. So this is the quality of the PCR that separated from the known pathogens or culturable microorganisms. It can likewise identify unculturable microorganisms or certain growths or microscopic organisms or phytoplasma alongside infections. So Genome extraction of pathogens should be possible either following manual procedures or utilizing business packs uncommonly intended to remove nucleic acids from various kinds of plant material. So DNA extraction from the plant material isn't an issue these days as we have separated from manual methods we have indicative packs or extraction units they are industrially accessible for genome extractione. All atomic discovery strategies for recognizing plant pathogens are fundamentally founded on precise structure of oligonucleotides and tests. Target groupings can be found in utilizing the Gen Bank Nucleotide Sequence Search Program that is NCBI and where the vast majority of the genomic information is put away and one can without much of a stretch discover the genomic assets from NCBI site. For looking of nucleotide bases ordinarily a device known as BLAST is utilized and the program that is BLAST and program is intended for investigation of nucleotides. Explicit nucleotide districts are chosen and introductions for explicit DNA or RNA targets can undoubtedly be structured utilizing this impact device. Preliminaries are intended to match with novel DNA locales from target living beings for DNA intensification and identification. The nearness of enhancement item affirms the nearness of the creature in the tried example. The enhanced item at that point envisioned through agarose gel electrophoresis utilizing a stain known as ethidium bromide (EtBr). Anyway these days not so much harmful but rather touchier stains like SYBR GREEN are utilized for recognition technique utilizing under UV light. For the most part, PCR can be acted in a few hours yet propelled frameworks can convey brings about couple of moments too. In this way, quickly PCR response blend is set up in a PCR tube, it made out of the layout DNA, at that point the groundworks that supplements the objective the areas of the format DNA. At that point alongside free nucleotides, at that point Taq polymerase for enhancement and afterward the blend cushion, all these are consolidated into the PCR tube at a particular extent and that PCR is run and the PCR cycles incorporate the initial step that is denaturing that is division of the two DNA resist high temperature which is at around 95°C followed by temperatures that is set for toughening of PCR groundworks which is set at around 55°C anyway it relies upon the taking care of temperature of the preliminaries. At that point, the following stage is groundwork augmentation and the temperature here set is 72°C, and this cycle is then finished for the next 32 to 35 cycles, and after the set number of cycles, it is the PCR item is checked by going through in the agarose gel electrophoresis and the nearness of a band that shows the PCR positive outcomes that affirms the nearness of the specific pathogen in that specific genome that is extricated from the plant tissues.

5.5.8   *siRNA BASED TECHNIQUES*

RNAi is an essential wonder that exist in every single living being, including plant framework, where, a remote RNA is corrupted by the acquired little nucleotide of RNAs qualities and thereby this component plant safeguard itself from certain pathogen that attacks the plant tissue framework, and this specific innovation has now been used monetarily for fundamentally create obstruction in the plants against certain plant pathogens. So what is RNAi quieting or RNA impedance—it is a cytoplasmic cell observation framework to perceive twofold abandoned, RNA, and explicitly annihilates single and twofold abandoned RNA particles, homologous to the end client utilizing little meddling RNAs regularly referred to as siRNA as a guide. Infections are the two inducers and focuses of RNAi that establishes a crucial antiviral guard system in eukaryotic life forms. It is especially significant in plants that utilization RNAi to recoup from infection maladies. The utilization of HTS of little RNAs (sRNA) from plants can effectively distinguish the infections contaminating them; including beforehand obscure infections, even in very low titer manifestations irresistible. So plants HTS helped in recognizable proof of those infections especially of obscure root on account of this HTS of the little RNA pieces that is get grouping alongside the plant RNAs. RNA hushing establishes a crucial antiviral barrier system in plants in which have compound cut viral RNA into bits of 20–24 nucleotides in size. At the point when detached and arrangement and en-mass appropriately collected, this infection infers little RNAs groupings can reconstitute genomic succession, data of the infection being focused in the plant. So once these HTS is done trailed by its gathering is done, at that point, it takes out the genomic group-ings which is contrasting from the plant beginning, and that is the reason it can now separate the genomic succession of the obscure pathogen especially the infections which taint the plants and causes illness. This methodology is autonomous of the capacity to culture or refine the infection and doesn't require a particular intensification and our improvement of viral nucleic acids. Utilizing this method known and novel DNA/RNA infections just as infection just as viroids have been distinguished at affectability levels equivalent to PCR. Thus, even nearness off a low measure of RNA or DNA genomes of the infections this strategy can be valuable to recognize of these obscure infections. So, the twofold abandoned RNA is separated or cut into piece littler bits of 20 to 24 nucleotides sizes by a plant protein known as dicer; at that point, one strand of the little twofold abandoned RNA is then consolidated into the RISC mind-boggling and afterward, alongside the RISC camp, complex this guide RNA goes for observation in the plant framework and at whatever point it finds a complimentary succession or especially mRNA of a specific quality. At that point, it proceeds to tie with that specific side and afterward, with the assistance of these, RISC complex, this mRNA is then processed and afterward further divided into little pieces. So this is the manner by which the RNAi innovation works, and in the event that it is the begun from an infection, at that point this is an innovation how this infection can be discredited in the plant framework. So here a similar instrument is portrayed where twofold abandoned RNA infection is then chopped down into little pieces and afterward one part which fills in as a guide RNA it consolidates with the RISC perplexing and afterward it proceeds to locate the new layout of the RNA which has integral sides and afterward it chops down into littler sections. So this little sections of the cut infection RNA can be exposed to HTS, trailed by gathering into contigs and scan for likeness in database utilizing the impact can assist us with identifying the infection off which these pieces have a place with or on the off chance that it doesn't have a place with any of the known arrangements in the database, it very well may be recognized as another infection. RNA hushing perceives twofold abandoned, RNA, and dispenses with RNAs homologs to the inducer RNA, by cleavage utilizing RNase III endonucleases called dicers. Plants encodes a few Dicer-like proteins that perceive and cut long twofold abandoned RNA (dsRNA) which is off 21-, 22-, up to 24-base: combines, sections, and goes about as siRNAs. siRNAs then tie to ribonuclease H-like proteins in the RNA actuated quieting complex which is in short known as (RISC) and are utilized to identify homologous single-abandoned RNA (ssRNA) particles for cleavage, delivering more siRNAs. In plants, RNAi becomes intensified when the divided RNA enlisted people and RNA-guided RNA polymerase to create all the more twofold abandoned (dsRNA), which is again separated by a dicer protein to deliver optional siRNAs, that are by and by ready to recognize and sever homologous RNA in a kind of 'degradative PCR' cycle. This prompts the gathering of a lot of siRNAs with homologous to the attacking infection. In this way, this is the fundamental component how the siRNAs are produced in the plant framework which can be even upgraded its duplicate number with the innovation utilizing RNA-coordinated RNA polymerase and afterward further aggregation of siRNAs which we call it as auxiliary siRNAs and this amass siRNAs are as a rule presently sequenced for recognizable proof of the new infection.

Use of next generation sequencing (NGS) for siRNA recognition: NGS stages, for example, illumina (Solexa) and SOLid creates huge sums but instead short peruses of nucleotide successions. With them a lot of bioinformatics apparatuses are utilized to once more get together of such short peruses and profound sequencing of siRNAs could in this way lead to identification and conclusion of plant infections. Thus, in this specific case for recognition of viral creature; which is respectable in nature. The NGS stages illumina and SOLiD are being end up being useful. NGS stages have been utilized to recognize reverse transcribing viruses, known viruses, new viruses, and even blended diseases or Defective RNA/DNAs just as contamination. Along these lines, NGS stages have been utilized for different purposes for effectively recognizing and distinguishing the causal specialist of plant side effects. Thus, with this we have reached a conclusion of the siRNAs based analytic innovation. To sum things up, it manages especially the NGS innovations are utilized for sequencing of siRNAs and this nucleotide sequenced siRNAs can be gathered and afterward we can get a thought of the whole genome of infection that is causing the plant manifestation and in the event that it is exposed to comparable successions in the NCBI database we can recognize the current infection that is being contaminating plants and distinctive plant species. On the other hand if the groupings are altogether not the same as the arrangements information being saved in NCBI database, at that point it ordered to be another infection that is causing the plant sickness.

5.5.9   GENOMICS BASED DIAGNOSIS

Genomic-based sequencing is otherwise called NGS or cutting edge sequencing innovation. It is essentially pyrosequencing or high-throughput sequencing, and it is changing the field of pathogen discovery in an assortment of plant tests. In this way, this is the best bit of leeway of this specific genomics helped determination program, where we don't require any earlier information on the microorganism that is really making the sickness the plant. Along these lines, this methodology is definitely not a quick test yet significant for recognizable proof of obscure infections and viroids, has no past grouping information of the life form is required. Revelation of new infections/viroids and new has increment quickly after the presentation of the NGS. Along these lines, use of NGS in finding of plant pathogens let us go for a contextual investigation how it is helping new pathogens to be recognized. PCR strategy recognizes designs like organisms, infections, and microbes by duplicating explicit pieces of their DNA. The disservice of this is the inquiry is exceptionally specifically one base his/her appraisal of which pathogen is probably going to be available on specific manifestations and afterward one adjusts the investigation likewise. So if there should arise an occurrence of PCR one needs to assume that most likely this pathogen is contaminating the plant and one will structure groundworks as needs be to recognize that specific pathogen though in the event of NGS no such supposition that is required as its succession the whole genome whether from the pathogenic parasites microbes or infection or from the plant and afterward it separate the genome or gatherings that isn't of plant starting point or that isn't of the host plant so it can then are contrasted and the NCBI database and afterward observe that obscure arrangements coordinate with an another known pathogen and afterward it tends to be effectively distinguished based on this NCBI search. However, in the event that it is search gives no outcome, on the other hand one can reason that the succession that is gotten from the conceivable pathogen-related with the host plant is of another or obscure pathogen beginning. So billions of building squares for every example is produced along these lines, not at all like old sequencing strategies, NGS maps billions of hereditary arrangements for a whole plant test. One can check billions of nucleotides, the structure squares of DNA, and the request for this nucleotide decides to which type they have a place. So again NGS creates billions of database and this billion structure squares can be then collected and afterward they can be looked on NCBI database to come out to a resolution which is very not quite the same as the other nucleic corrosive based innovation like PCR innovation which scan for a distinct length site explicit enhancement of the groundworks and afterward it intensifies just a specific quality or fragment of a quality for recognizable proof of the pathogen root. This uncovers plant's own arrangements and permits us to see which different groupings are available in the example. One can lead his/her investigation utilizing propelled programming that have 'took care of' with hereditary data on plants and pathogens which empower us to rapidly give a dependable and authoritative answer. So relying upon the nature of the DNA test that answer can be even given inside a day. Therefore, this is the manner by which this NGS stages are making a difference.

5.5.10 ADVANCED AND MOLECULAR DIAGNOSTIC TECHNIQUES

DNA-based propelled purpose of care symptomatic procedures. Therefore, purpose of care strategies is picking up ubiquity since they don't require advanced types of gear. They can settle on quick choices and they are extremely simple to deal with. Therefore, purpose of care indicative papers performed were a less expensive rate too and that is the reason they are likewise in extremely appeal. We realize that PCR-based strategies, in spite of the fact that they have a few focal points however certain restrictions of PCR-based innovations is that this innovation can't be taken legitimately into the fields and it requires power and other supporting gadgets. So it's genuinely restricts its ampleness for purpose of care applications. As another option, isothermal DNA enhancement techniques are undeniably fit to conquer this impediment of PCR. For example, Isothermal enhancement joined with horizontal stream stripes and compact fluorometers have been effectively utilized for purpose of care, location of pathogen DNA. In plant tissues DNA extraction strategy expects capacity to productively evacuate in number of synthetic substances that can restrain the DNA intensification response. Therefore, this is significant, extremely important that to have an effective PCR run, we have to have a DNA separated appropriately from the plant tests and expel all the synthetic compounds that can repress its enhancement. So parallel stream gadget to put it plainly, we call it as (LFD) DNA extraction technique has been accounted for as fast and effective purpose of care testing and has been effectively utilized in plant pathogen recognition. Therefore, these LFDs are by and large helpful in purpose of care recognition of plant pathogens since it requires extremely less measure of substrate also, less, less an ideal opportunity to get the DNA. This strategy includes test disturbance in extraction cradle utilizing metal rollers before moving the lysate into the discharge cushion of a LFD nitrocellulose layer. The innovation is exceptionally straightforward. Take the example, put it into the example support in the cylinder that contains two metal balls, that is metallic root and afterward it pounds the tissues in a manner and the tissue liquid is then moved to the parallel stream gadget, at that point NCM. At that point little bit of film is extracted and included into the DNA intensification response, for example, PCR or other isothermal enhancement techniques. So the layer ties the DNA that is available in the tissue and this little bit of the film is still moved to the PCR tubes or isothermal enhancement cylinders and this fills in as a DNA that is available in the example. The segregated DNA is entirely steady on the film at room temperature, which permits the extraction to be acted in the field condition. So that is the reason the purpose of care DNA extraction techniques utilizing nitrocellulose layer has been mainstream to such an extent that it ties the DNA, it doesn't require a particular temperatures to be kept up and that is the means by which it is being utilized for purpose of care identification. Another, strategy is utilization of basic cellulose-based dipstick that permits plants test handling as meager as 30 seconds. Plant tissues are here essentially macerated by shaking in a cylinder containing extraction support and 1 to 2 metal rollers for 8 to 10 seconds. At that point cellulose dipstick is embedded in the cylinder containing the example before washing it multiple times in a subsequent cylinder containing wash cradle, lastly into a cylinder containing the intensification blend. So this is again a straightforward technique, where does the cellulose based dipsticks are inundated, embedded into the plant tissues. In addition, afterward it is macerated by the assistance of the extraction support and the metal rollers. It is simply accomplished for 8 to 10 seconds and afterward the cellulose dipstick when embedded into the cylinder, at that point it followed the example and into it and afterward subsequent to washing it in a second cylinder then that dipstick can in a roundabout way move to the PCR or intensification blender. So the innovation works productively in numerous developed species including rice, wheat, tomato, and sorghum just as some hard to seclude plant species like leaves from develop trees of mandarin, lime, and lemon. Therefore, it tends to be utilized to distinguish pathogen DNA just as RNA from contaminated tissues and works with different intensification strategies, for example, PCR, LAMP, and RPA. Therefore, this is another extremely simple purpose of care, DNA extraction technique utilizing cellulose-based dipstick. So utilization of nucleic corrosive, isothermal enhancement strategies in plant sickness location. So despite the fact that PCR is exceptionally delicate and hearty, it is compelled by various specialized restrictions. For example, explicitness is profoundly reliant on the groundworks utilized, and its characteristic affectability makes it inclined to bogus positives because of test cross sullying. So PCR, despite the fact that it's a valuable and touchy pack which can identify pathogen, which is available in an extremely low sum, however, it has certain restrictions what's more PCR likewise requires electrically fueled gear to play out the warm cycling, which constrains its utilization of purpose of care diagnostics. Various option isothermal methods are currently accessible that can forestall the need of a warm cycler. So one of those isothermal procedure is Loop-intervened isothermal enhancement, to put it plainly, it is known as LAMP. So it has broadly been utilized to its high efficiencies, particularity, straightforwardness, and snappiness. Light requires two long external groundworks and two short internal preliminaries that perceive six explicit arrangements in the objective DNA. The primary internal preliminary containing sense and antisense successions in the DNA will hybridize the objective grouping and start DNA combination. In the following stage, the external preliminary completes the abandoned uprooting DNA combination and produces a solitary abandoned DNA which functions as a layout for the second internal and external preliminaries creating a DNA particle with a circle structure. The unremitting cycling, response collects items with rehashed groupings of target DNA of various sizes. The reactant tube is brooded at 63°C to 65°C in a customary research center water shower or warmth hinder that helps in keeping up a steady temperature. Therefore, this is probably the best-preferred position of this specific strategy. We hear that we don't require distinctive temperature systems all things considered on account of PCR where we require temperature for denaturation of DNA, at that point, we required temperature for toughening of preliminary, and afterward we required an alternate temperature for expansion of groundwork. Yet, here we require just a single temperature that is around 63 to 65°C, which can be kept up in a research facility water shower or warmth square. The enhanced item can be distinguished by unaided eyes as a white accelerate or a yellow green shading arrangement after expansion of SYBR green to the response tube. So here we don't require even the gel electrophoresis or ETM bromide recoloring for recognition of the PCR intensification item. Here the item can be envisioned as white accelerate on the cylinder by even unaided eyes or in the event that we include SYBR green, at that point response blender turns yellow, green in shading. So this is an affirmation shading change is an affirmation of the inspiration of the response with the preliminaries. There are three significant points of interest of this specific strategy. To start with, it tends to be done at a steady temperature with a short response time. This quick isothermal procedure makes it perfect for purpose of care location of plant pathogens in the field. Since it doesn't require any electrical help for running the response just as we don't require any gel electrophoresis framework to envision the result response. What's more, that is the reason it is an excellent apparatus for utilization of purpose of care, location of plant pathogens. Furthermore, it is high intensification, productivity, and affectability as it creates huge measure of PCR item with low measures of info DNA. Lastly, this strategy is moderately financially savvy. Besides, there have been reports that LAMP produces amplicons with a few transformed rehashes, which could be possibly used to expand the affectability in hybridization measures, for example, LAMP-ELISA hybridization, and LAMP joined with colorimetric gold nanoparticle hybridization tests. So development of the circles is likewise been viewed as a positive result of LAMP responses in light of the fact that in strategies like LAMP-ELISA hybridization and LAMP joined with colorimetric gold nanoparticle hybridization process, it helps a ton. So in a word we can see the response method of the LAMP strategy where this is the format DNA and this is the first forward inside preliminary that divides and in the complimentary site. In addition, afterward it reaches out to this bearing. Thereby shaping this specific strand and this strand is presently utilized by second groundwork that is in reverse inward preliminary. At that point, it stretches out to these bearings and leads this blend of this new quality. Lastly, since this segment is complimentary to this, this will come and structure a circle and this is complimentary to this area so this will come and structure a circle. In addition, this is the second circle it creates. Furthermore, presently this second introduction for the subsequent preliminary, this is uncovered and afterward-external groundwork is then job becomes possibly the most important factor. At that point, it begins intensifying starting here to the next point and it strands the specifics in new strain. In addition, afterward here the second external preliminary ties and afterward it enhances through this segment till this and it creates a new quality. So this is the way the intensification procedure happens in the event of LAMP strategy where two interior preliminaries and two external groundworks, they perceives six destinations first, second, at that point third, fourth, fifth, and sixth. So every one of these destinations are perceived. What's more, this LAMP method is helpful for location of plant pathogens under field conditions. The joining of LAMP with electro concoction sensors offered a hearty stage for pathogen identification as it was exceptionally touchy, identifying as low as 10 duplicates of the pathogen genomic DNA. Therefore, this is once more, we can see that it is profoundly touchy as just 10 duplicates of the pathogen.

5.6 RECENT ADVANCES IN DIAGNOSTIC TECHNIQUES

5.6.1 BIOSENSORS BASED DIAGNOSIS

The assortment of biosensors have been grown as of late and it is been popularized for different applications and they are helping the diagnosticians to identify plant pathogens in a proficient and exact. Thus, contingent upon the working rule of the sensor the analytes could be distinguished that utilizing a sensor dependent on electrical, chemical, electrochemical, and optical, attractive or even vibrational signs. So dependent on these techniques of biosensing these biosensors are created for identification of plant pathogens. The restriction of location could be upgraded by the utilization of nanomaterial grids as transducers and the explicitness could be improved by the utilization of bio acknowledgment components, for example, DNA, immune response, proteins, etc. So organic tests can be of different birthplaces like antibodies, nucleic acids, and numerous different components like bacteriophages, etc. All tests can distinguish the pathogen target and their sign is then transduce through a fitting transduction stage and afterward the sign is enhanced for identification signal in the yield gadgets can push us to really recognize the pathogen that is related with a plant illness. So biosensor stages dependent on nanomaterials, nanomaterials show electronic and optical properties and can be integrated utilizing various kinds of materials for hardware and detecting applications. The ubiquity of nanomaterials for sensor improvement could be ascribed to high surface territory, high electronic conductivity, and plasmonics properties of nanomaterials that upgrade the restriction of identification. Therefore, that is the reason nanomaterials is picking up ubiquity for being utilized as biosensors for discovery of plant pathology. The immobilization of the bio acknowledgment components, for example, DNA, neutralizer in compound can be accomplished utilizing different methodologies including biomolecule adsorption, covalent connection and capsulation or a complex blend of these techniques. The nanomaterials utilized for wire sensors development incorporate metal and metal oxide nanoparticles, quantum dabs, and carbon nanomaterials just as carbon nanotubes and graphene just as polymeric nanomaterials. Therefore, this incorporates by advancement of development of nanomaterial biosensors. Gold nanoparticles based optical immuno sensors have been produced for identification of pathogens. Notwithstanding single test sensors nano chips had likewise been made of miniaturized scale clusters which contained fluorescent oligo tests that were additionally created for identifying single nucleotide change in microorganisms and infections with high affectability and particularity dependent on DNA hybridization. Along these lines, assortments of utilized with of nanoparticles whether it has been utilized alone or in mix with different chemicals and biomolecules, its application is being white in plant pathology for identification of plant pathogens. Second gathering is Affinity Biosensors so contrasted with the vague nanoparticle-based biosensors consideration of a bio acknowledgment component can extraordinarily expand the particularity of the sensor. Thusly, different kinds of biosensors have been created and among them partiality biosensors are mainstream. The proclivity biosensors the detecting is accomplished dependent on response on the bio acknowledgment component and the objective analyte. So it's not the fundamentally the molecule; however, it is a molecule related with the biosensor its partiality for distinguishing the objective analyte, it is significant and this bio-recognition essentially help the biosensors to identify and afterward transduce the sign and followed by intensification of the sign. Therefore, partiality biosensors can be created utilizing immune response and DNA as acknowledgment components. So counter acting agent based biosensor-the immune response based biosensors give a few preferences, for example, quick discovery, improved affectability, constant examination, and potential for measurement. The biosensors empower the pathogen location in air, water, and seeds with various stages for nurseries on field even post-reap stockpiles of processors and wholesalers of yields and organic products. The standard of setting up counter-acting agent-based immunosensors lies in the coupling of explicit immunizer with a transducer, which converts the coupling component to a sign that can be broken down, for example, polyacetylene, polypyrrole, or polyaniline. So counter-acting agent joined with a nanoparticle that really perceived the analyte and it is being utilized generally these days, for identification of certain plant pathogens. In this way, in fact the biosensor is signal far off transfused and afterward it went for information investigation lastly we get the result of the outcome. Regardless of whether the biosensor might be immunizer based or it might be founded on DNA. The different procedure continues as before and the last yield is dissected dependent on the information it creates from the biosensor. So immune response based biosensors could distinguish plant pathogens, for example, cowpea mosaic infection, TMV, lettuce mosaic infection. So it has been broadly used to recognize different viral pathogens from various plant species. Immune response based biosensors innovation has gained colossal ground upon usage on nanotechnology. Gold nanorods (AuNRs) functionalized by antibodies have been utilized to recognize cymbidium mosaic infection (CymMV), odontoglossum ringspot infection (ORSV) for fast finding of viral diseases. Therefore, nanoparticle 3D shape-based biosensors has been additionally improved when it is labeled with a counter-acting agent. The quality of the nanoparticles has even gone past its unique breaking point with the utilization of this biosensor, and this consolidated biosensor of nanoparticles and counteracting agent gives high explicit acknowledgment of the pathogen analyte and it gives a legitimate recognizable proof of the pathogen that is related with the plant malady. Next is DNA/RNA-based partiality biosensor. Because of plausibility of confinement at a sub-atomic level the DNA-based biosensors empowers early location of maladies before any visual side effects show up. Therefore, this is a significant technique for breaking down plant pathogen. There is RNA based proclivity biosensor and on the grounds that it can recognize pathogens before the pathogen duplicates by and large and produce a clear side effect on the host plants.

Correlative RNA was additionally misused for location of plant infections, for example, cymbidium mosaic infection just as odontoglossum ring spot infection. Therefore, it is reciprocal RNA based hybridization. That is, inno-vation is likewise being utilized for recognition of certain plant infections.

5.6.2   *VOLATILE ORGANIC COMPOUND (VOC) BASED DIAGNOSIS*

Unpredictable natural mixes are discharged by plants and the example of or the profile of unstable natural mixes gets changed when the plant associates with its encompassing, or its air in an agreeable circumstance or with a pressure condition. Along these lines, that is the guideline behind the example of various unpredictable natural mixes being discharged or radiated by the plants into its encompassing. Along these lines, unstable natural mixes radiated from leaf surfaces are terminal metabolites of the host plant and can show its physiological wellbeing status. Unstable natural profiling may portray 'plant-to-plant' and 'plant-to-bother correspondence and in this manner it's picking up significance. VOC markers like Hexenols, hexenals, hexanyl esters and classes of terpenoids and indoles may help in quick segregation of parasitic contamination and creepy crawly vector taking care of. Since the example of outflow of this specific compound get changed and it follows an unmistakable example when it is tainted by a contagious pathogen and when it is being feeded by an herbivory Insect. So dependent on this whether the plant is being tested by creepy crawly or certain parasitic pathogens at that point arrival of all VOCs gets fluctuated. So utilizations of VOC profiling and there are sure examples of overcoming adversity. Likewise, cucumber mosaic infection (CMV) in developed squash plants. The CMB tainted plants indicated a general net increment in amounts of VOCs like Hexenal, methyl, hepten thus numerous others in all the plants yet no major subjective distinction in VOC profile could be recognized in contaminated plants. Both bug vectors, *Aphis gossypii* and *Myzus persicae*, were specially pulled in to CMV-tainted plants, along these lines with beet armyworm (BAW) to mildew covered nut plants. Notwithstanding the littler size and substandard nature of CMV-contaminated plants. Along these lines, this is another model were *Aphis gossypii* and *Myzus persicae* these creepy crawlies were pulled in towards the Cucumber mosaic infection (CMV) plants when it was contaminated with the cucumber mosaic infection (CMV). This exhibit the plant is instigating modified VOC profile because of viral disease and this sort of instrument is famously known as very typical upgrade. Therefore, this is too typical upgrade in light of the fact that the earlier pathogen is contaminating the host plants and altering the VOC profiling and this adjusted VOC profiling is drawing in specific creepy crawlies and which is anything but an ordinary wonder and that is the reason it is 06:54 known as excessively ordinary boost. The unpredictable mark of plants could be investigated utilizing a gas-chromatography strategy to break down the nearness of explicit VOC that is demonstrative of a specific infection. To upgrade the exhibition of mixes partition and examination, the gas chromatography is regularly joined with mass spectrometry to distinguish obscure mixes in the unpredictable example. Therefore, this is ordinarily sent for the enlightenment of known mixes; however, when it is joined with mass spectrometry at that point, even the obscure mixes can be recognized in the VOC profile. GC/GC-MS can give progressively exact data about the plant ailment because of its high explicitness. It likewise permits the identification of sicknesses at various stages dependent on quantitative data gathered from the VOC test. So gas chromatography combined with mass spectrometry is an excellent framework to precisely profile the VOC of sound and contaminated plants and dependent on these examples one can recognize the wellbeing status of the plant whether it is tainted or it isn't contaminated by a specific pathogen and afterward dependent on this mark atom one can begin securing the plants through other restorative measures or defensive measures if the profile shows the nearness of a pathogen. Therefore, with this, we have seen that how volatile organic compounds (VOC) can help in the determination and location of certain plant pathogens by implication in such a case that these pathogens are related, then unquestionably the VOC profile will be of a specific kind in contrast with a solid one. With the goal that causes us to decide whether the seed parcel or the plants are whether contaminated by any pathogen or not. What's more, in the event that it is contaminated, we can take either suitable control measures for ensuring the seed part or the plants from the speculated pathogen.

5.6.3 MICRONEEDLE (MN)-BASED DIAGNOSIS

Nucleic corrosive based advances which we realize that it is unmistakable and significant; however, in field determination nucleic corrosive based advances has difficulties and to beat these difficulties there is steady improvement of this specific innovation where it is being rearranged and adjusted and one among this innovation is utilization of small scale needle for DNA extraction for heading of plant pathogens from tainted plants under field condition. Therefore, in field sub-atomic conclusion of plant ailments by means of nucleic corrosive intensification (NAA) is as of now restricted by unwieldy conventions for removing and segregating pathogen DNA from plant tests. To address this test a quick plant extraction strategy has been created utilizing a dispensable polymeric Microneedle (MN) fix. By applying MN fix on plant leaves intensification examine prepared DNA can be extricated inside a moment from various plant species. MN removed DNA has been utilized for direct polymerase chain response intensification of plant plastic DNA without filtration. So the smaller scale needle based extraction strategy is tantamount with generally normal and most favored DNA extraction technique that is CTAB technique and that is the reason

MN based DNA extraction strategy is having huge possibility to be utilized under field conditions in coming years. This basic cell lysis free and purging free DNA extraction technique could be transformative way to deal with encourage quick example groundwork for atomic analysis of different plant illnesses straightforwardly in the field. The MN fix is postage stamp-sized it can straightforwardly be put in a leaf and squeezed on the off chance that we see the MN fix, at that point it includes little cone like projections which is fundamentally extremely cheap polymers and this postage size stamp of MN fix is helpful to utilize and this cone like structures of polymers are not really 8 millimeters in length. Therefore, when this fix is fixed or squeezed against a leaf, it at that point infiltrates the leaf epidermis and the DNA is get clung to the polymer from the solid plant tissues of Mesophyll cells just as contaminated tissues of Mesophyll cells including pathogen DNA too. Therefore, the DNA gets tie with the polymer of the MN fix and this fix would now be able to be taken for PCR intensification. That MN technique has focal points over CTAB strategy since DNA extraction by CTAB strategy requires a ton of hardware and it takes a few hours. Further, CTAB extraction is a multi-step process including tissue pounding to natural solvents and axes. By differentiate the new smaller scale needle-based DNA extraction procedure includes just scarcely any means and require just a MN fix and a fluid support arrangement and it can separate DNA inside couple of moments or minutes. Therefore, this is a concise blueprint of the CTAB strategy which includes a few procedures and that is the reason it is very tedious and ordinarily it takes three to four hours for extraction of DNA. In contrast with the CTAB technique MN fix can separate DNA inside a moment. So for a rancher or a specialist can straightforwardly apply the MN fix to the plant they think to be infected, at that point hold the fix set up for a couple of moments and afterward strip it off. The fix is then flushed with the cushion arrangement washing hereditary material of the smaller scale needles and into a sterile holder. The whole procedure takes pretty much a moment. The miniaturized scale needle procedures immaculateness levels were tantamount to other approved lab strategies for DNA extraction, we have just discussed it and above all the slight distinction in virtue levels between the smaller scale needle and CTAB tests didn't meddle with the capacity to precisely test the examples by a PCR or LAMP examine. Therefore, despite the fact that the nature of DNA might be undermined somewhat yet it isn't having any impact on PCR enhancement or LAMP examine process and that is the reason MN-based DNA extraction strategy is especially suggested these days for extraction of DNA legitimately from field samples. Since DNA extraction has been considered as a critical obstacle to the improvement of on location testing instruments, the smaller scale needle-based innovation has now given an answer for the issue. It is coordinated ease, field versatile gadget that can play out each progression of the procedure, from taking an example to distinguishing the pathogen and revealing the aftereffects of a measure. Further, a lateral flow device (LFD) could be utilized to precisely identify the pathogen DNA extricated through the miniaturized scale needle innovation. So both miniaturized scale needle innovation alongside the little stream gadget can be utilized under field conditions so the issue in regards to extraction of DNA is illuminated through small scale needle-based innovation and the PCR based innovation is supplanted with little stream gadgets and that is the manner by which in-field the pathogen can be legitimately recognized by utilizing these novel applications like smaller scale needle-based and fix just as LFD. In this way, we have seen the utility of smaller scale needle-based innovation for extraction of plant DNA from the field tests and it is profoundly practically identical to the best quality level CTAB technique and it gives almost 100% results from the extracted DNA that was utilized through MN fix.

5.6.4 ARTIFICIAL INTELLIGENCE (AI)-BASED DIAGNOSIS

Plant disease diagnosis has been traditionally been done by visual observation where involvement of experts is an essential criterion and it is always a expensive matter because one has to travel to the field and at certain times finding an important expert is also a difficult for growers in certain regions. So bringing an expert from a far of place it again a time-consuming and expensive process so artificial intelligence (AI) is coming up in a way that it is helping to solve certain issues like those. Due to which consulting experts is not a mandatory incurrent context because AI is trying to replace some of the human activities through use of certain machine and machine learning processes. The use of technology to replace human activities and guarantee efficiency is known as AI. However, the question arises can AI help improve agricultural productivity? Therefore, we will just see into it what are the technologies in this field has been gaining and how it is helping the growers to identify and detect certain pathogens. So artificial application in agriculture is broadly being used for in terms of: (a) agricultural robotics; (b) soil and crop monitoring; and (c) predictive analytics. Farmers are increasingly using sensors and soil sampling to gather data and this data is stored in farm management systems that allow better processing and analysis. The availability of this data and other related data is paving away to deploy AI in agriculture. As a result, in number of tech companies is investing in algorithms that are being useful in agriculture. For example, image recognition is used in potatoes by AgVoice developed by a Georgia based startup for using natural language toolkit for field notes and yield prediction algorithm based on satellite imagery. Various resources developed several artificial intelligent devices that can identify diseases in plants. For example, Tensor Flow is a technique known for transfer learning to teach the AI to recognize crop diseases and pests damage. It uses Google's open source library to build a library of around two thousand 2,756 AI images of cassava leaves from plants in Tanzania. The success was that the AI was able to identify the disease with 98% accuracy. Therefore, this shows the effectiveness of AI to being deployed in agriculture for detection and diagnosis of plant pathogens. Another technique is using of image segmentation and soft computing technique to detect plant diseases. In image segmentation, the process basically is of separating our grouping an image into different parts. These parts normally correspond to something that humans can easily separate and view as individual objects. The segmentation process is based on various features found in the image. This might be color information, boundaries, or segments of an image. Therefore, this is the process how the image segmentation technique offers first-image acquisition takes place then, image pre-processing is required then, image segmentation is done, then features extraction in image is done, and finally, detection and classification of plant diseases are obtained. Therefore, these are some of the examples of how this technology works. Therefore, this is an input, and this is the output after image segmentation and based on this input and output images, the AI detects the particular disease that is occurring on the plant leaf. iPathology: it is another AI-based application that is robotic applications and management of plant and plant diseases in agriculture. Therefore, robots application can help in precision plant protection technologies. Therefore, intelligence technologies using machine vision or learning have been developed for plant disease detection and identification. A recognition method based on visible spectrum image processing to detect symptoms of religious like citrus greening which is also named as Huanglongbing (HLB) caused by *Candidatus liberobacter* species on citrus leaves. The experimental results showed that detection accuracy is as high as 91.93%. This is again a significant improvement where AI-based diagnosis is able to detect more than nine times out of ten times accurately. The huanglongbing detection system can detect the pathogen at the pre-symptomatic stage; this is again a very significant application of AI-based technologies because the pre-symptomatic stage is normally not able to trace by human eyes. Then there are certain mobile apps like for example Plantix, which is being widely used in several countries of the world and this is a simple app where even laymen can use for detection and diagnosis of a particular plant disease problem and then and get the information regarding how to manage this problem.

Plantix is a free mobile application which offers farmers and gardeners the possibility to receive decision support directly on their smartphone. Due to image recognition the app is able to identify plant type as well as appearance of a possible disease, pest, or nutrient deficiency in the plant. The app can be used very easily by the users where the users have to simply take the image of the infected plant leaf and upload it to the app and then the app recognizes the leaf damage pattern and based on it, it gives an output in terms of information about the probable disease-causing agent and we also provide recommendations for taking adequate control measures for management of that particular problem associated with the plant. Therefore, it's a very simple tool based on AI being developed and it can be used by all growers and gardeners which may not have very scientific understanding of the disease problem. Therefore, this is how AI is coming in a big way in agriculture system. Particularly it is helping plant disease diagnosis and helping us to understand the causal agent of the plant diseases.

5.6.5   PROXIMAL SENSING OF PLANT DISEASES

The common method for diagnosis of plant disease includes visual examination, microscopic evaluation, as well as molecular, serological, and microbiological techniques. However, the new sensor-based methods assess the optical properties of plants within different regions of the electromagnetic spectrum and are able to utilize information beyond the visible range. They enable the detection of early changes in plant physiology due to biotic stresses, which is a limiting factor for human eye and that is why we normally have to wait till the symptom production stage for management of the particular pathogen. Currently, the most promising techniques are sensors that measure reflectance, temperature, or fluorescence of the leave. Remote sensing is a method used to obtain information from plants or crops without direct contact or invasive manipulation. The concept has been recently enlarged by proximal, close-range or small-scale sensing of plant material. Proximal sensors may be hand-held, machine-mounted or attached to suitable unmanned aerial vehicles (UAVs). Proximal sensing vs. remote sensing. So, this comparative between proximal and remote sensing tools. In case of proximal sensing normally autonomous systems are that can go very close the plants are used which are ground based as well as certain machine mountain cameras and another tools being used then sometimes even unmanned vehicles can also be used for proximal sensing but mostly these are used as remote sensing tools. Further, aircraft or satellite is used for remote sensing detection of plant diseases. Various sensors, like sensors like RGB, multispectral, hyperspectral, thermal, chlorophyll fluorescence and 3D sensors are used for detection of plants symptom at various level at cell, leaf, plant, plot, field, or even at the ecosystem level. So depending on the uses of the equipment or depending on the requirement this different sensors are used to detect and diagnose plant diseases. Systems for proximal disease sensing—first one is thermography. The thermal imaging is a non-contact technique to determine the temperature distribution of any object in a short period of time. Infrared radiation emitted from plant surfaces may be recorded by detectors sensitive to radiation in the infrared region from 8 to 12 micrometers. Each pixel of image is related to a temperature value of the objects surface and may be illustrated in false-color image. The technology can be used from microscope applications to ground-based equipment covering a range from leaf tissues to crop canopies. It is used to detect pathogens like tobacco plants infected with tobacco mosaic virus (TMV).

5.6.6   REMOTE SENSING OF PLANT DISEASES

Remote sensing is a method used to obtain information from plants or crops without direct contact or invasive manipulation. To obtain information on an object by measuring the electromagnetic energy reflected/backscattered or emitted by the surface of the earth. The measurements are possessed and analyzed to retrieve information on the object observed, for example, plant health in this case. Therefore, remote sensing is an indirect assessment technique which is able to monitor vegetation conditions from a distance and evaluate the spatial extent and patterns of vegetation characteristics and plant health, in this application. So a plant which is in stressed conditions, that is (induced by a disease), reacts with potassium mechanisms that lead to suboptimal growth which show up as changes in variables such as leaf area index (LAI), chlorophyll content, or surface temperature and thus producing a spectral signature different from the signature of healthy or unstressed vegetation. Therefore, this is the basic principle how remote sensing works.

The remote sensing community defines plant disease monitoring as: detection that is (deviation from the healthy), identification that is (diagnosis of specific symptoms among others and differentiation of previous diseases), and quantification that is measurement of disease severity that is percent leaf area affected and so on. Different sensors and techniques are required for detecting plant response to various diseases and disease severity. Therefore, with this, we have seen the remote sensing devices, how it works, and how it can be used for collecting data at large scale and getting appropriate information very accurately and correctly at a very short time interval. Therefore, this is also a technology that can be used for other agricultural purposes for measurement of abiotic stresses or pest infestations, but it is widely being used in diagnosis and detection of plant pathogens causing plant diseases.

5.6.7   ON-SITE TESTING MOVING DECISION MAKING FROM THE LAB TO THE FIELD

On-location testing fundamentally is a term that is regularly used to portray two particular exercises, right off the bat discovery that is the underlying situating of the bug or pathogen tainted example which is in many occasions is performed outwardly. The second is movement is distinguishing proof, and it is typically accomplished by sending speculated tests to a research facility. In this way, giving innovative answers for empower progressively fast dynamic is an absolute necessity these days. It isn't really just examination administrations who profit by these methods; they can be sent all through the ranch to fork to confine misfortunes brought about by pathogens. Conveying disentangled recognition and recognizable proof techniques remotely assists with accelerating assessment just as encourages exchange. In this way, that is the reason nearby testing is picking up fame and gadgets have been created to settle on dynamic procedure in the field itself. Performing diagnostics is a piece of a dynamic procedure to forestall or constrain spread of pathogens. The quicker the choices are made progressively successful the activity might be. Generally, once a potential infection has been found, examples are sent to a research center for testing that causes delay in the dynamic procedure. Moreover, if pathogens go unnoticed at pre-indicative disease organizes; this disappointment of visual perception can prompt its spreading unchecked until it has developed to such a level, that it very well may be seen. Along these lines, dynamic at field level is exceptionally critical. These issues have driven the drive to create mechanical arrangements that would satisfy two corresponding jobs. Initially, placing devices under the control of those on the cutting edge to empower quick distinguishing proof of pathogens and it would forestall delays. Also, creating discovery devices that direct those on the cutting edge to the site of the issue, at the pre-indicative disease. Therefore, these instruments are fundamental for advancement for mechanical arrangements at the on-location testing. These devices, when connected together empower an increasingly productive discovery and finding process empowering quicker sending of control measures. Certainly, quicker arrangement of control measures would prompt least misfortunes by the producers. Strategies dependent on latex agglutination have been performed for plant maladies since in the mid-1980s. For instance, Ani Biotech has built up a potato infection test packs by Ani Biotech based on latex agglutination technique. From that point forward progressively refined techniques have been created to empower quick ID. Early test packs dependent on latex agglutination on glass or plastic slides, required: an enormous number of temperature labile reagents, had numerous means in which reagents were included successively, and the translation of the outcome was regularly abstract, requiring a considerable measure of preparing and experience to the duplicate. A portion of these subsequent age units (e.g., ready packs by Neogen) likewise fused compound substrates, successfully reproducing research center ELISA procedure, yet performed quickly on a strong help. This gave focal points regarding both convenience and understanding of results which were not; at this point abstract and effectively deciphered by non-pros even in the field circumstance. Thus, this is a mechanical headway that is the manner by which it is helping on the on location testing or dynamic at the field level. The hugest advancement came in the late 1990s with the utilization of homogeneous test unit groups created and misused. The LFD position was abused at first in the phyto-diagnostics field for the recognition of potato infections for use in seed affirmation frameworks and end up being an extensive improvement over past configuration. The hidden science in a LFD is viably equivalent to a latex agglutination unit; the amassing of immune response covered latex or might be colloidal gold particles brought about by the nearness of the immunizer target. The key distinction anyway is that the coupling happens during the slender progression of test and reagents along a layer, yet not in the arrangement as it was in the previous cases. Testing dependent on LFD innovation remains the least complex and most quick alternative for field use where explicit restricting reagents for the objectives of intrigue are accessible. The main huge downsides to LFD ways to deal with field identification are the accessibility of reagents with explicitness suitable for the application and the characteristic absence of intensification that limits affectability. Along these lines, LFD approaches have certain restrictions, as we have seen that.

It is useful for certain basic pathogens like organisms, microscopic organisms and infections, yet in the event that it is an unpredictable malady, at that point unquestionably, these LFD's are not adequate assistance for the diagnosticians under field condition. Along these lines, in-field identification where utilizations of LFD's advancements are utilized rapidly they can gather the plant tissues and they can get the tissue macerated and the tissue concentrate can be then exposed to the LFD gadget and inside couple of moments the nearness or nonattendance of the particular groups can assist us with identifying that whether the plant is influenced by the creature being tried is there or not. Additionally, there are different gadgets which require no power rather a versatile or convenient water shower or warmth square can be utilized to go for intensification test and afterward subject to its examination and this again help us to take choice in the field itself yet without going to the research center. For more prominent affectability and particularity, sub-atomic science strategies, for example, PCR is utilized to enhance target nucleic acids. Be that as it may, usage of these techniques on location has been a test. Various organizations have created fieldable continuous PCR gear. While versatile continuous PCR has been assessed widely there are a few noteworthy downsides to it execution. Right off the bat, in PCR techniques, extraction of nucleic corrosive by and large requires sensibly expounds extraction strategies to keep away from co-refinement of mixes which repress the chemicals. Furthermore, while rough, convenient, and sometimes battery controlled gear is accessible it stays costly, to a great extent because of the requirement for cautious temperature guideline and delicate identifiers that are required for its identification. To take care of both of these issues ensuing examination has been centered around assessment of isothermal intensification sciences, which could dispose of both the issues related with the ordinary PCR responses. Isothermal enhancements are strategies in which the intensification response is hatched at a solitary temperature. This gives preferences as far as effortlessness over PCR, since the responses don't should be cycled precisely between temperatures, accordingly water-showers, dry-squares or hatcheries can be utilized to brood responses. Plant pathogen examines by the loop intervened intensification (LAMP) technique is the most broadly received strategy to date under field circumstance. Discovering pathogens-to finding a pathogen require a feeling of having a plant tainted by a pathogen which can be illuminated as: Sniffing Pathogen Infection. At that point, one must see the disease from a separation and one must utilize Surveillance Tools for identification and determination of the conceivable event of plant pathogens.

Therefore, these little supplies that are coming up that are supplanting some high delicate types of gear like PCR and these are helping the agents to take a choice at the on-location level without reclaiming the examples to the lab for its check. It is just in specific situations where on-location testing isn't corroborative at exactly that point the examples are then taken back to the research center for additional investigation. Along these lines, this is the manner by which one can see that how dynamic procedures has been currently moved from the lab to field. In light of improvement of these advancements like LFDs and isothermal enhancements forms. With this we have reached a conclusion of the subject in the vicinity testing and dynamic from lab to field. In the following, we will discuss virtual diagnostics systems and we will perceive how it is helping any examiners or producers to take a choice for plant illness the executives.

5.6.8   VIRTUAL DIAGNOSTICS NETWORK

Plant is consistently in danger as a result of a few reasons one of the hugest explanation is plant pathogen. Also, some of them be begun in a similar level, or same territory or though, some of them has been recognized to be originating from a far-off spot or from a fresher, another nation. Along these lines, it is consistently critical to have an information base on most likely dangers to edit plants in such a case that the pathogen isn't existing prior in a specific zone individuals may not know about the pathogen which perhaps of isolate importance from of that region without appropriate database or without a legitimate instrument to go for ID. Along these lines, in this setting virtual demonstrative systems become possibly the most important factor a major job and it helps the producers and leaders into an extraordinary degree.

Plant Health Risk: The control and moderation of outlandish plant pathogens are reliant upon early location and precise findings additionally in a time span to empower successful reaction. Increment reliance on universal exchange to address worldwide monetary and food security challenges pathogen invasion will likewise proceed. Be that as it may, an inquiry emerges is that do we have a reasonable plant biosecurity procedure and satisfactory plant biosecurity foundation to shield plant frameworks from the pathogens that undermine the plant wellbeing? The capacity of enigmatic satellites to adjust the host scope of Gemini infections may introduce new difficulties to plant wellbeing and jumble our capacity to analyze and relieve these new infection illnesses or whether it is a satellite variated change in infections. These every single imaginable event of new strains of infections could assist us with identifying dependent on the current database that is accessible that can control someone to help and analyze the issue. Therefore, there is requirement for Robust Plant Diagnostic Systems. Regarding exchange, the simple nearness of a pathogen can stop shipments whether the illness was showed.

5.7   CONCLUSION

Economic loss has been estimated more than several billion dollars per year worldwide due to plant viral diseases and there is no commercialized chemical to manage those. Plant diseases caused by viruses can be effectively controlled when means of manage are applied at the initial step of viral disease development or by planting virus-free crops. This is reason why accurate diagnosis is important. Symptomatic diagnosis is still useful but often has erroneous results, because of confusion associated with high variable symptoms by interactions between host and virus or by abiotic stresses. Therefore, reliable diagnostic platforms which can be accepted officially are required. The methods based on serological principle and molecular biology has been used for virus diagnosis. ELISA associated with serology was initiated and adapted as a diagnostic tool worldwide since it is easy to use with durability. After PCR invented, PCR based diagnostics have been adapted as a diagnostic system comparable with ELISA even becoming predominant method. There are several reasons for this change. PCR is standardized in industrial level as ELISA resulting in worldwide use in diagnostic facilities. Furthermore, PCR-based assay has better sensitivity over ELISA and even faster. Isothermal nucleic acid amplification methods including LAMP are under development for virus detection because it is faster and has greater sensitivity over conventional PCR. Since there is no an ideal detection method to fulfill all requirement to detect, it is very important to develop appropriate and effective techniques which can be applied for management of viral diseases in worldwide level. By doing that, sustainable agriculture will be achieved. The ideal future diagnostic tool, therefore, should be multiplexed and gather information from the genome, the proteome, and the metabolome, and also provide this information quickly and easily at an affordable cost. Many hurdles must be overcome before these techniques can be transformed from research tools into routine agricultural practices.

KEYWORDS

- **agar double diffusion test**
- **artificial intelligence**
- **citrus leaf smudge infection**
- **electron microscopy**
- **fluorescence *in-situ* hybridization**
- **gas chromatography-mass spectrometry**

REFERENCES

Abu, S. H. S., Murant, A. F., & Daft, M. J., (1968). The use of antibody sensitized latex particles to detect plant viruses. *J. Gen. Virol., 3*, 299.

Ball, E. M., (1973). Solid phase radioimmunoassay for plant viruses. *Virology, 55*, 515–520.

Baranwal, V. K., & Sharma, S. K., (2015). *Diagnostics of Plant Viruses: Recent Advances* (pp. RC-63–70). ICAR-Sponsored winter school on recent advances in integrated pest management.

Bercks, R., & Querfurth, G., (1971). The use of the latex test for the detection of distant serological relationships among plant viruses. *J. Gen. Virol., 12*, 25.

Biswas, K. K., (2017). *Plant Viruses, Diseases and Their Management*. I.K. International Pvt. Ltd.

Dijkstra, J., & De Jager, C. P., (1998). *Practical Plant Virology, Protocols and Exercises* (p. 459). Springer, New York.

Emmons, R. W., & Riggs, J. L., (1977). Application of immune fluorescence to diagnosis of viral infections. In: Maramorosch, K., & Koprowski, H., (eds.), *Methods in Virology* (Vol. VI, pp. 1–28). Academic Press, New York.

Hampton, R., Ball, E., & De Boer, S., (1990). *Serological Methods for Detection and Identification of Viral and Bacterial Plant Pathogen, A Laboratory Manual*. APS Press, St. Paul, MN.

Havranek, P., (1978). Use of quantitative immunoelectrophoresis in cucumber mosaic virus assay, II Strain and isolate-specific modifications of the antigenic complex. *Phytopathol. Z., 93*, 5.

Hibi, I., & Suito, Y. A., (1985). Dot immunobinding assay for the detection of tobacco mosaic virus in infected tissues. *J. Gen. Virol., 66*, 1191–1194.

Hull, R., (2002). *Matthew's Plant Virology*. Academic Press.

Katz, D., & Kohn, A., (1984). Immunosorbent electron microscopy for detection of viruses. *Advances in Virus Research., 29*, 169–194.

Lau, H. Y., & Botella, J. R., (2017). Advanced DNA-based point-of-care diagnostic methods for plant diseases detection. *Front. Plant Sci., 8*, 2016. doi: 10.3389/fpls.2017.02016.

Maree, H. J., Fox, A., Al Rwahnih, M., Boonham, N., & Candresse, T., (2018). Application of HTS for routine plant virus diagnostics: State of the art and challenges. *Front. Plant Sci., 9*, 1082. doi: 10.3389/fpls.2018.01082.

Matthews, R. E. F., (1967). Serological techniques for plant viruses. In: Maramorosch, K., & Koprowski, H., (eds.), *Methods in Virology* (Vol. III, pp. 199–241). Academic Press, New York.

National Agricultural Higher Education Project (NAHEP-CAAST), (2019). *Short Term Training Program on Genome Assisted Diagnosis of Plant Viruses, Viroids and Phytoplasmas*. ICAR-Indian Agricultural Research Institute, New Delhi. http://nahep-caast.iari. res.in (accessed on 9 February 2021).

Pallás, V., Sánchez-Navarro, J. A., & James, D., (2018). Recent advances on the multiplex molecular detection of plant viruses and viroids. *Front. Microbiol., 9*, 2087. doi: 10.3389/ fmicb.2018.02087.

Prabha, K., Baranwal, V. K., & Jain, R. K., (2013). Applications of next generation high throughput sequencing technologies in characterization, discovery and molecular interaction of plant viruses. *Indian J. Virol., 24*, 157–165.

Rajasulochana, P., & Jayalakshmi, T., (2018). *International Journal of Pure and Applied Mathematics, 119*, 2969–2981.

Rochow, W. F., & Duffus, J. E., (1978). Relationship between barley yellow dwarf and beet western yellows viruses. *Phytopathology, 68,* 51.

Sharma, S. K., Kumar, P. V., & Baranwal, V. K., (2014). Immuno-diagnosis of episomal banana streak MY virus using polyclonal antibodies to an expressed putative coat protein. *J. Virol. Methods, 207,* 86–94.

Sharma, S. K., Vignesh, K. P., Poswal, R., Rai, R., Geetanjali, A. S., Prabha, K., Jain, R. K., & Baranwal, V. K., (2014). Occurrence and distribution of banana streak disease and standardization of a reliable detection procedure for routine indexing of banana streak viruses in India. *Sci. Hortic., 179*, 277–283.

Sheriff, S., Silverton, E. W., Padlam, E. A., Cohen, G. H., Smith-Gill, S. J., Finzel, B. S., Van, L. J. W. M., & Verduin, B. J. M., (1986). Detection of viral protein and particles in thin sections of infected plant tissue using immune gold labeling. In: Jones, R. A. C., & Torrance, L., (eds.), *Developments and Applications in Virus Testing* (pp. 193–211). Association of Appl. Biologists. Wellesbourne.

Van, R. M. H. V., (1988). Molecular dissection of protein antigens and the prediction of epitopes. In: Budon, R. N., & Van, K. P. H., (eds.), *Laboratory Techniques in Biochemistry and Molecular Biology* (pp. 1–40, 227). Elsevier Science Publ. Co. Inc. New York.

CHAPTER 6

Development of Recombinant Coat Protein for Detection of Banana Viruses

VIMI LOUIS,[1] K. C. DARSANA DILIP,[2] and ALAN C. ANTONY[2]

[1]*Banana Research Station, Kerala Agricultural University, Kannara, Thrissur, Kerala–680652, India, E-mail: vimilouis@gmail.com*

[2]*College of Horticulture, Kerala Agricultural University, Thrissur, Kerala–680656, India*

ABSTRACT

One of the major constraints in banana production is viral diseases caused by *Banana bunchy top virus* (BBTV), *Banana bract mosaic virus* (BBrMV), *Cucumber mosaic virus* (CMV), and *Banana streak virus* (BSV). Apart from incurring yield loss of about 40–100%, it affects germplasm multiplication and international exchange as viruses are primarily transmitted through planting materials other than being vector borne. Consequently, procuring quality disease free planting materials has been the foremost measure that should be considered with utmost importance, considering the latent period for viral disease symptoms to be showcased, especially in tissue cultured planting materials. Thus, it is very evident as to how important virus indexing is for banana producers and export-based enterprises. Although PCR-based detection methods are more sensitive and efficient, serological method-enzyme-linked immunosorbent assay is always preferred over it as it is both economical and reliable. Research always has been about perfecting the method to obtain quality antigen to raise antiserum whether it be partial purification of the virus from the infected sample or producing recombinant coat protein *in vivo* which now has taken up to protein production *in vitro*. Acknowledging both advantages and disadvantages of all the available methods, this chapter will walk you through the advances in research for developing viral antigen for raising antiserum so as to utilize them for serological based detection of banana. However, the fact that recombinant coat protein production has limitless applications other than in serological assay development cannot be ignored.

6.1   INTRODUCTION

Banana, belonging to the genus *Musa* in *Musaceae* family of the order *Zingiberales* evolved in South East Asia with India as one of the centers of origin. It is regarded as 'the most affordable nutritious gold mine' as it is a rich source of carbohydrate, vitamin B, potassium, phosphorus, calcium, and magnesium. Cholesterol content being negligible reduces risk of heart diseases when used regularly and is recommended for patients suffering from high blood pressure, arthritis, ulcer, gastroenteritis, and kidney disorders. Banana is the most consumed and produced fruit in India occupying 20% of the total area under cultivation and 37% of total fruit production in India. In a developing country like India, it is important to utilize the available land and maximize the yield in minimum time period. After the advent and commercialization of tissue culture in banana, the farmers are able to produce export quality fruits with lesser investment. As a water and fertilizer intensive crop, management of diseases and pest infestation during any stage of the crop proves to be exorbitant to the farmers. Diseases like Panama wilt of banana caused by *Fusarium oxysporum* f. sp. *cubense*, bacterial wilt caused by *Ralstonia solanacearum* and *Pectobacterium carotovorum*, leaf spot caused by *Mycosphaerella sp.*, *Deigtoniella sp.*, Banana bunchy top disease caused by *Banana bunchy top virus* (BBTV), banana bract mosaic disease (BBrMD) caused by *Banana bract mosaic virus*, infectious chlorosis caused by *Cucumber mosaic virus* (CMV), and Banana streak disease causing *Banana streak virus* apart from post-harvest diseases like anthracnose and cigar end rot pose a major threat to banana cultivators. Viral diseases showcase symptoms at a later stage is however most economically calamitous among other biotic and abiotic factors. Virus indexing at an early stage is the most economical way to manage viral diseases.

6.2   BANANA BUNCHY TOP DISEASE

Banana bunchy top disease accounts for a yield loss of 50–100%in infected plants. BBTV is type member of genus *Babuvirus* in the family *Nanoviridae*. The infected plants show dwarfing and stunted growth. The disease is named after the most conspicuous symptom, i.e., small and brittle leaves exhibits rosetting. Marginal necrosis sets in after few days of infection. The virion is isometric and measures 18–20 nm in diameter. It is persistently transmitted by banana black aphid, *Pentalonia nigronervosa* (Hu et al., 1996).

Based on the nucleotide sequence analysis of BBTV, it is broadly classified into two phylogenetic groups, *viz.,* Pacific Indian Oceans (PIO) and South-East Asian (SEA) (Karan et al., 1994). PIO comprises of the isolates from Myanmar, Pakistan, Sri Lanka, India, Australia, Hawaii, Egypt, and Tonga whereas, SEA includes the isolates from Indonesia, China, Philippines, Japan, Vietnam, and Taiwan. All the Indian isolates reported so far belong to the PIO group (Banerjee et al., 2014).

6.2.1   GENOME ORGANIZATION AND ARCHITECTURE OF BANANA BUNCHY TOP BABUVIRUS

BBTV is multipartite with six circular single-stranded DNA segments of approximately 1 Kb, which are encapsidated separately by the icosahedral viral coat protein of 18–20 nm diameter (Burns et al., 1995). DNA-R,-S, -M, -C, and -N segments codes for master replication initiator protein (Rep), coat protein, cell to cell movement protein (MP), cell cycle link protein and nuclear shuttle protein, respectively (Beetham et al., 1997; Wanitchakorn et al., 1997, 2000a, b). However, the function of protein encoded by DNA-U is not characterized hitherto (Beetham et al., 1999; Tian and Zhuang, 2005).

6.3   BANANA BRACT MOSAIC DISEASE (BBrMD)

Banana bract mosaic disease (BBrMD) was first reported in 1979 in the Philippines, and it subsequently became widespread throughout the Philippines, India, Sri Lanka, Vietnam, and Western Samoa. This disease is regarded as one of the limiting factors of banana and plantain production in India (Selvarajan et al., 2009). The typical symptom of BBrMD is dark reddish-brown linear streaks on the bracts of the inflorescence of infected banana, from where the name of the virus originates. This distinguishes it from all other known viral diseases of bananas. Initial symptoms are observed on the pseudostem, petioles, and midrib, which are green or reddish-brown (depending on cultivar) streaks or spindle-shaped lesions. Leaf arrangement resembles that of a traveler's palm. Distinct dark-colored linear streaks are detected on the pseudostem on removing the dead leaf sheaths. At severe conditions, the infected leaf sheaths are seen separated from the pseudostem by itself. Chlorotic streaks may occur on the bunch stalks, and high disease intensity is associated with distorted fingers and reduction in bunch weight. This reduced the economic value of the bunch considerably.

BBrMV is nonpersistently transmitted by banana aphid, *P. nigronervosa* and other aphid species *viz.*, *Rhopalosiphum maidis*, and *Aphis gossypii* (Magnaye and Espino, 1990). Mechanical transmission of BBrMV is not proven till date. Apart from *Musa*, BBrMV is also reported to infect small cardamom (*Elettaria cardamomum*) in India and also infects flowering ginger, *Alpinia purpurata* (Vieill.) K. Schum in Hawaii (Wang et al., 2010; Siljo et al., 2012).

6.3.1 GENOME ORGANIZATION AND ARCHITECTURE OF BANANA BRACT MOSAIC POTYVIRUS

BBrMV, the causal agent of bract mosaic disease (BBrMD), belongs to the genus *Potyvirus* in the family *Potyviridae*. It is a monopartite single-stranded RNA virus which is encapsidated in a flexuous filamentous rod of approximately 725 nm long. ssRNA genome of 9,711 bp encodes for a polyprotein which is post translationally modified into at least ten mature proteins by viral proteinases (Ha et al., 2008).

The difficulty in assessing infection in vegetative phase of the banana is the major concern as symptoms of leaf petiole and sometimes even on the pseudostem is inconspicuous. Repeated reports of a latent phase for symptom expression and detection of virus from symptomless plants is alarming.

6.4 INFECTIOUS CHLOROSIS OF BANANA

CMV infection in banana was first reported in Australia by Magee (1930). The disease was named as infectious chlorosis, heart rot, virus sheath rot, cucumber mosaic, and banana mosaic, according to the external symptoms (Stover, 1972). Mosaic patterns or discontinuous linear streaking bands, extending from leaf margin to midrib are the characteristic symptoms of infectious chlorosis. Advanced infection is characterized by curling of leaves, rosette leaf arrangement and dying suckers. Even at low titer of virus, the whole leaf produces parallel chlorotic streaks due to breakdown of chloroplasts and decline in chlorophyll production (Dheepa and Paranjothi, 2010) which later become distorted with irregular wavy margin along with necrosis (KAU, 2016). The symptoms often are sporadic and hence majority of the leaves appear healthy. The expression of symptoms is influenced by virus strain and temperature (Hitchborn, 1956). Among all the strains of CMV infecting banana, heart rot strain causes significant losses, due to rotting of inner leaves leading to death of the plant (Lockhart, 2000). It has been reported in many cases that the symptoms of infection are masked and such plants act as virus reservoirs. Nevertheless, the typical symptoms of infectious chlorosis mimics calcium and boron deficiency in the field and vice versa. Hence, early detection of the virus is highly relevant.

CMV is geographically wide spread and has abroad host range; however, it was first reported to cause mosaic disease on cucumber and other cucurbits. It is now known to occur worldwide in both temperate and tropical climates, affecting many agricultural and horticultural crops (Sokhandan-Bashir et al., 2012). Estelitta et al. (1996) reported CMV to be a major threat to banana farmers in Kerala, especially where cucurbitaceous vegetables are cultivated as intercrops in banana. In Kerala, infection of CMV was reported and well characterized in banana varieties such as *Karpooravally*, *Nendran*, *Palayankodan*, *Peykunnan*, *Kosthabontha*, *Mottapoovan*, *Bhimkhel*, *Dhakh-insagar*, *Madhuraga*, *Rasthali*, and *Musa ornate* (KAU, 2016).

6.4.1 GENOME ORGANIZATION AND ARCHITECTURE OF CUCUMBER MOSAIC VIRUS (CMV)

According to ICTV (2012), CMV causing infectious chlorosis in banana belongs to the family *Bromoviridae*. Members of *Bromoviridae* family exhibit remarkable diversity in their capsid architectures. Members of this family are spherical or quasi-spherical, having T = 3 icosahedral symmetry, and a diameter of 26–35 nm. Total genome length of CMV is approximately 8 kb. Genomes consist of three linear, positive sense ssRNAs with 5'terminal cap. The 3'termini are not polyadenylated, but generally are highly conserved within a species or isolate. They are either tRNA-like and can be amino acylated.

CMV is the type species of the genus *Cucumovirus* in the family *Bromoviridae*. It contains three spherical particles, each approximately 28 nm in diameter. A third non-structural protein P2b is expressed, as subgenomic RNA (sgRNA), from RNA2 and functions in cell-to-cell movement and post-transcriptional gene silencing (PTGS). RNA 3 represents non-structural MP (P3a, cell-to-cell MP) and the structural capsid protein (CP) or coat protein (P3b, CP) that is expressed via subgenomic RNA (i.e., RNA4) (Hull, 2009).

6.5   BANANA STREAK DISEASE

Banana streak disease is caused by *Banana streak virus* (BSV) of genus *Badnavirus*, family *Caulimoviride*. Different species of BSV causing streak disease in banana are *Banana streak Mysore virus* (BSMyV), *Banana streak OL virus* (BSOLV) and *Banana streak GF virus* (BSGFV) (ICTV, 2012). Characteristic symptoms of BSV infection in banana plants are continuous or discontinuous chlorotic or necrotic streaks on the leaves, necrosis of pseudostem, and stunting of plant. Infection occasionally leads to plant death. Aberrant bunch emergence, fruit peel splitting and necrotic fruit spots are other symptoms (Jones, 2000). A variety of factors including virus isolate, host genotype, level of crop management apart from environmental conditions influence symptom expression (Lockhart, 1986; Harper et al., 2004). Although, vegetative propagation is the principal method of virus spread, mealy bugs (*Planococcus citri* and *Sacchariococcus sacchari*) colonizing banana transmit the virus in a semi persistent manner (Lockhart and Olszewski, 1994).

BSV was traditionally not considered a serious problem until the discovery of endogenous BSV (eBSV) sequences in the nuclear genome of *M. balbisiana* (Harper et al., 1999; Gayral et al., 2008). During mass multiplication by micropropagation and hybridization, recombination events occur in the endogenous sequences allowing the reconstituted viral genome to be activated and thereby resulting in episomal infection.

6.5.1   GENOME ORGANIZATION AND ARCHITECTURE OF BANANA STREAK BADNA VIRUS

Badna viruses are non-enveloped bacilliform virus particles of 30 x 130–150 nm with double stranded circular DNA genome of 7.2–8 kb which is noncovalently closed which typically encode three open reading frames. Function of ORF 1 is unknown and that of ORF2 is virus associated. ORF3 encode a multifunction polyprotein which is cleaved post translationally to MP, coat protein, aspartic protease, reverse transcriptase (RT) and ribonulease H (King et al., 2012).

6.6   DETECTION OF BANANA VIRUSES

Resistance breeding in banana and plantain is hard due to the polyploid nature of most of the cultivated varieties. The absence of effective control measures for viral diseases narrows down the options to early detection and roguing of infected planting materials for the management of diseases caused by viruses. This is most relevant in the case of tissue culture plants, which are produced and cultivated with a lot of added investment. As discussed earlier in the chapter, diagnosing viral infection based on symptomatology is not favored as it is sometimes confused with nutrient deficiencies. Moreover, the latent phase of viral infection creates an impression of the infected plant being healthy.

Detection of banana viruses is primarily done using serological techniques like enzyme-linked immunosorbent assay (ELISA), direct immunobinding assay (DIBA), lateral flow assay (LFA), or by nucleic acid-based methods like polymerase chain reaction (PCR), loop-mediated isothermal amplification (LAMPS), rolling circle amplification (RCA), nucleic acid spot hybridization (NASH), or by combination of both like immunocapture-polymerase chain reaction (IC-PCR). It can also be done using electron microscopy (EM) and combination of serology and EM *viz.,* Immunosorbent electron microscopy (ISEM). Immunocapture PCR combined with ISEM is currently used as standard method for BSV indexing worldwide (Sharma et al., 2014). Basis of any serological assay is the antiserum used for the same and the sensitivity and reliability of any assay depends on the quality of antiserum used for the assay. In all the serological assays employed for detection of banana viruses, antisera are developed against the surface epitopes present on coat protein of the virus/virion.

Iskra et al. and Wu and Su reported the purification of virus-like particles (VLPs) of BBTV from BBTD-affected banana plants in 1989. Monoclonal antibodies (MAbs) raised against Taiwanese BBTV purified from infected sample by Wu and Su were used to develop ELISA and immunofluorescence (IF) assays to detect BBTV and provide for the first time a diagnostic tool for the disease in 1990. A variant of ELISA, plate-trapped antigen (PTA)-ELISA was also developed using the monoclonal antibody produced against the viral coat protein. Purification of an Australian isolate of BBTV was reported by Thomas et al. in 1991. They used a sucrose density ultracentrifugation for partial purification of the virions which was later subjected to equilibrium centrifugation in Cs_2SO_4. The antigen to raise the antisera was obtained after a laborious process that continued for approximately 5 days. Antiserum to BBTV was raised in a New Zealand White rabbit following two intramuscular injections of thus purified virions which was used for the development of ELISA.

Recently, many more rapid nucleic acid-based detection methods have been developed for detection of viruses. Although these methods are more rapid and sensitive than serological assays, the requirement of skilled personnel and sophisticated infrastructure for the same makes it less preferable for diagnosis. Nevertheless, WHO has designated a perfect diagnostic technique to be "ASSURED," i.e., affordable, sensitive, specific, user-friendly, rapid and robust, equipment-free, and deliverable. Molecular-based techniques are indeed specific, sensitive, and rapid but it definitely cannot be user friendly as it needs sophisticated equipment. Serological based assays can be perfected to fit in all the parameters of good diagnostic assay should have provided; it is made more specific and sensitive. In case of banana viruses, when antiserum is raised against partially purified virus, phenols, and secondary metabolites present in the host give rise to a lot of interference. Low concentration of *Banana streak virus* is another problem encountered in the virus purification from infected plants. This notably challenges the specificity of the antisera as well as considerably reduces sensitivity of the assay. Development of recombinant coat protein and its use in immunizing the animal greatly solves the above problems.

6.7   DEVELOPMENT OF RECOMBINANT COAT PROTEIN

Attempts to produce recombinant coat protein of BBTV was made by Wanitchakorn et al. (1997); Abdelkader et al. (2004); Shilpa et al. (2016); and Arumugan et al. (2017). They were successful in expressing BBTV coat protein in pMAL-c2, pQE-30, and pET28a (+). Wanitchakorn et al. raised polyclonal antiserum against the fusion protein with maltose binding domain (MBD) which gave some interference in western blot analysis. They have discussed the importance of raising polyclonal antisera against BBTV-CP without the tag, which is being attempted in our lab (unpublished).

The recombinant coat protein production of BBrMV has been discussed in 1997 and 1999 by Wanitchakorn et al. and Rodoni et al., respectively. The coat protein gene amplified using gene specific primers was puri-fied and cloned into pMAL-c2 and transformed in to *E. coli* (DE3) BL21 strain by Wanitchakorn et al. Whereas, Rodoni et al. cloned the entire coat protein-coding region of BBrMV into pProEX-1 expression vector (Life Technologies). *E. coli* M15 cells containing the pREP4 repressor plasmid was the expression host used in the experiment. The purified fusion protein was used to raise antiserum in New Zealand white rabbit.

CMV coat protein has been expressed in a series of pET vectors like pET21a, pET21-d, apart from pQE30. Rosetta and BL21 (DE3) strains of *E. coli* are the commonly used expression hosts (Rostami et al., 2014; Kim et al., 2016; Koolivand et al., 2017). Currently, *E. coli* BL21 (DE3) pLysS is the most commonly used expression host. DE3 is an arrangement of T7 RNA Polymerase gene, under the control of LacUV 5 promotor on a phage genome whereas; pLysS is a plasmid that encodes T7 lysozyme that minimizes leaky expression of the protein. Expression of CMV CP was also attempted in pRSET-C/BL 21 (DE3) pLysS system in our lab. Interference of host 25 KDa protein was observed in this case (unpublished). Thus, it is very clear that vector and host used for *in vivo* expression plays a major role in recombinant protein production apart from the protein itself.

The recombinant coat protein production of BSMyV was carried out by Sharma et al. (2014). They cloned two putative coat protein-coding regions (p37 and p48) of BSMyV in expression vectors pMAL-c2X and pET28a. Although, they succeeded in overexpressing p37 as a fusion protein of size 79KDa (consisting of 42KDa MBP and 37KDa p37) in pMAL-c2X, was unsuccessful in pET28a. Overexpression of p48 was also unproductive. The specificity and antigenicity of recombinant protein was confirmed by western blot. Cleaving of MBP37 fusion protein using specific protease factor Xa resulted in partial cleavage. The mixture of partially cleaved and uncleaved MBP37 protein antigen was used in raising polyclonal antiserum in rabbits. The developed antiserum was evaluated in ISEM, ACP-ELISA, and IC-PCR assays based on the indexing of BSMyV in field and tissue culture-derived plants.

Although *in vivo* expression of viral coat proteins are preferred than partial purification of viruses from the infected hosts for antibody production, it has to be admitted that cell-based systems are slow and cumbersome. *In vitro* cell-free protein expression utilizing cell lysate of *Escherichia coli,* wheat germ, or rabbit reticulocytes is gaining momentum among the scientific society as it is simple, fast, and can overcome toxicity and solubility issues sometimes experienced in the traditional *E. coli*-based expression systems. As opined by Rosenblum and Cooperman (2013) cell-free protein expression offers a convenient method for generating small amounts of protein for a wide array of applications. The significance of cell-free translation systems is that biologically active proteins can be synthesized. However, primarily issues apart from affordability are protein folding and post-translational modifications which have to be addressed. A clear advantage that this method has over *in vivo* protein synthesis is that the environmental conditions can be adjusted easily. Nevertheless, there is always scope for improvement in both the techniques with the advent of more technological advancements.

6.8   APPLICATIONS AND FUTURE PROSPECTS OF RECOMBINANT PROTEIN

The role of recombinant coat protein does not end in developing quality antiserum alone. Viruses like particles (VLP) that are assembled during purification of viral coat protein in heterologous systems have immense applications. These multimeric nanostructures are lately in limelight of investigations in vaccinology, drug delivery, and gene therapy. Plant VLP's without the infectious nucleic acids have the proficiency to replace nanocarriers for presenting immunogenic peptides to the immune system. Marusic et al. demonstrated in 2001 that PVX-derived chimeric viral particles with glycoprotein of HIV in its N terminal administered to mice via different routes elicited high levels of HIV-1-specific immunoglobulin G (IgG) and IgA antibodies. Here, the coat protein gene of PVX was modified such that when expressed the capsomeres with the glycoprotein will form a chimeric VLP. Further, researches have been successfully carried out demonstrating their utility for targeted drug delivery and their role in delivering nucleic acids forming the basis of gene therapy (Pan et al., 2012). Even though recombinant coat proteins of banana viruses have significant role in antiserum production and virus detection, there is undoubtedly vast opportunities that lies ahead. On the other hand, fact that applications of recombinant coat protein/viral-like protein in unraveling many mysteries of virus-host relationship in future, cannot be ignored.

6.9   CONCLUSION

Contamination of plant protein besides presence of polyphenols interfering the process of purification and high cost of *in vitro* synthesis makes recombinant coat protein production all the more superior. It has been demonstrated with viruses like *Citrus psorosis virus* (Salem et al., 2018) that antisera raised against recombinant coat protein is more sensitive and specific than antisera raised against viral coat protein. In a perennial crop with a complex genome like banana that makes resistance breeding problematic, diagnosis of the devastating diseases at an early stage is the ideal way to save the stakeholders depending this crop for income. As we are moving forward to fulfill sustainable developmental goals of the United Nations, what better than providing quality-planting materials to the farmers at an affordable rate for inculcating "responsible production." This will definitely be a step forward in reducing losses incurred due to virus infestation and a leap towards food security and zero hunger.

ACKNOWLEDGMENT

We thank Kerala Agricultural University, India, for financial support. Darsana Dilip K. C. acknowledges the Department of Science and Technology, India, for INSPIRE fellowship.

CONFLICT OF INTEREST STATEMENT

The authors declare that there is no conflict of interest with the contents of this article.

KEYWORDS

- **banana viruses**
- **direct immunobinding assay**
- **enzyme-linked immunosorbent assay**
- **immunosorbent electron microscopy**
- **lateral flow assay**
- **recombinant coat protein**

REFERENCES

Abdelkader, H. S., Abdel-Sala, A. M., El Saghir, S. M., & Hussein, M. H., (2004). Molecular cloning and expression of recombinant coat protein gene of banana bunchy top virus in *E. coli* and its use in the production of diagnostic antibodies. *Arab J. Biotechnol., 7*(2), 173–188.

Arumugan, C., Kalaimughilan, K., & Kathithachalam, A., (2017). Banana bunchy top viral coat protein (CP) gene expression studies at molecular level in hill banana cv. sirumalai (AAB). *Int. J. Curr. Microbiol. App. Sci., 6*(6), 398–411.

Banerjee, A., Roy, S., Behere, G. T., Roy, S. S., Dutta, S. K., & Ngachan, S. V., (2014). Identification and characterization of a distinct banana bunchy top virus isolate of Pacific-Indian Oceans group from North-East India. *Virus Res., 183*, 41–49.

Beetham, P. R., Hafner, G. J., Harding, R. M., & Dale, J. L., (1997). Two mRNAs are transcribed from banana bunchy top virus DNA-1. *J. Gen. Virol., 78*, 229–236.

Beetham, P. R., Harding, R. M., & Dale, J. L., (1999). Banana bunchy top virus DNA-2-6 are monocistronic. *Arch. Virol., 144*, 89–105.

Burns, T. M., Harding, R. M., & Dale, J. L., (1995). The genome organization of banana bunchy top virus: Analysis of six ssDNA components. *J. Gen. Virol., 76,* 1471–1482.

Dheepa, R., & Paranjothi, S., (2010). Transmission of cucumber mosaic virus infecting banana by aphid and mechanical methods. *Emirates J. Food Agric., 28*(8), 117–129.

Estelitta, S., Radhakrishnan, T. C., & Paul, T. S., (1996). Infectious chlorosis disease of banana in Kerala. *Infomusa., 5*(2), 25–26.

Gayral, P., Noa-Carrazana, J. C., Lescot, M., Lheureux, F., Lockhart, B. E., & Matsumoto, T., (2008). A single banana streak virus integration event in the banana genome as the origin of infectious endogenous pararetro virus. *J. Virol., 82*(13), 6697–6710.

Ha, C., Coombs, S., Revil, P. A., Harding, R. M., Vu, M., & Dale, J. L., (2008). Design and application of two novel regenerate primer pairs for the detection and complete genomic characterization of potyviruses. *Arch. Virol., 153,* 25–36.

Harper, G., Hart, D., Moult, S., & Hull, R., (2004). Banana streak virus is very diverse in Uganda. *Virus Res., 100,* 51–56.

Harper, G., Osujii, J. O., Heslo-Harrison, J. S., & Hull, R., (1999). Integration of banana streak badna virus into the Musa genome: Molecular and cytogenetic evidence. *Virology, 255,* 207–213.

Hitchborn, J. H., (1956). The effect of temperature on infection with strains of *cucumber mosaic virus. Ann. Appl. Biol., 44*(4), 590–598.

Hu, J. S., Wange, M., Sether, D., Xie, W., & Leonhardt, K. W., (1996). Use of polymerase chain reaction (PCR) to study transmission of banana bunchy top virus by the banana aphid (*Pentalonia nigronervosa*). *Ann. Appl. Biol., 128,* 55–64.

Hull, R., (2009). *Comparative Plant Virology* (2ⁿᵈ edn.). Academic Press: China.

ICTV (International Committee on Taxonomy of Viruses), (2012). *In International Committee on Taxonomy of Viruses* (p. 976). Ninth Report on Bromoviridae.

Iskra, M., Garnier, M., & Bove, J. M., (1989). Purification of *banana bunchy top virus* (BBTV). *Fruits, 44,* 63–66.

Jones, D. R., (2000). *Diseases of Banana, Abaca and Ensete.* Oxon: CABI.

Karan, M., Harding, R. M., & Dale, J. L., (1994). Evidence for two groups of banana bunchy top virus isolates. *J. Gen. Virol., 75,* 3541–3546.

KAU (Kerala Agricultural University), (2016). *Package of Practices Recommendations: Crops* (15ᵗʰ edn.). Kerala Agricultural University: Thrissur.

Kim, H. S., Kim, D. W., & Jung, Y. T., (2016). Development of serological procedures for sensitive, rapid detection of cucumber mosaic virus in lilium. *Hortic. Environ. Biotechnol., 57*(6), 633–639.

King, A. M. Q., Adams, M. J., Lefkowitz, E. J., & Carstens, E. B., (2012). *Virus Taxonomy: Ninth Report of the International Committee on Taxonomy of Viruses.* Academic Press.

Koolivand, D., Sokhandan, B. N., & Rostami, A., (2017). Preparation of polyclonal antibody against recombinant coat protein of cucumber mosaic virus isolate B13. *J. Crop Protect, 6*(1), 25–34.

Lockhart, B. E. L., & Olszewski, N. E., (1994). Serological and genomic heterogeneity of banana streak badna virus: Implications for virus detection in Musa germplasm. In: Ganry, J., (ed.), *Breeding Banana and Plantain for Resistance to Diseases and Pests* (pp. 105–113). CIRAD/INIBAP, Montpellier, France.

Lockhart, B. E. L., (1986). Purification and serology of a bacilliform virus associated with banana streak disease. *Phytopathology, 76,* 1995–1999.

Lockhart, B. E. L., (2000). Virus diseases of Musa in Africa: Epidemiology, detection and control. In: Craenen, K., Ortiz, R., Karamura, E. B., & Vuylsteke, D. L., (eds.), *International Conference on Banana: International Conference on Banana and Plantain for Africa* (p. 359). Acta Horticulturea, Uganda.

Magee, C. J., (1930). A new virus disease of bananas. *Agric. Gaz. New South Wales, 41*(12), 140–156.

Magnaye, L. V., & Espino, R. R. C., (1990). Banana bract mosaic, a new disease of banana-symptomatology. *Philipp. Agric., 73*, 55–59.

Marusic, C., Rizza, P., Lattanzi, L., Mancini, C., Spada, M., Belardelli, F., Benvenuto, E., & Capone, I., (2001). Chimeric plant virus particles as immunogens for inducing murine and human immune responses against human immunodeficiency virus type-1. *J. Virol.,* 8434–8439.

Pan, Y., Jia, T., Zhang, Y., Zhang, K., Zhang, R., Li, J., & Wang, L., (2012). MS2 VLP-based delivery of microRNA146a inhibits autoantibody production in lupus-prone mice. *Int. J. Nanomed., 7*, 5957–5967.

Rodoni, B. C., Dale, J. L., & Harding, R. M., (1999). Characterization and expression of coat protein coding region of banana bract mosaic potyvirus, development of diagnostics assays and detection of the virus in banana plats from five countries in southeast Asia. *Arch. Virol., 144*, 1725–1737.

Rosenblum, G., & Cooperman, B., (2014). *Engine out of the Chassis: Cell-Free Protein Synthesis and its Uses* (Vol. 588, pp. 261–268). Elsevier.

Rostami, A., Bashir, N. S., Pirniakan, P., & Masoudi, N., (2014). Expression of cucumber mosaic virus coat protein and its assembly into virus like particles. *Biotechnol. Health Sci., 1*(3), 180.

Salem, R., Arif, I. A., Salama, M., & Osman, G. E. H., (2018). Polyclonal antibodies against the recombinantly expressed coat protein of the citrus psorosis virus. *Saudi J. Biol. Sci., 25*(4), 733–738.

Selvarajan, R., Balasubramanian, V., Jeyabaskaran, K. J., Pandey, S. D., & Mustaffa, M. M., (2009). Management of banana bract mosaic disease through higher dose of fertilizer application in banana cv. neypoovan (AB). *Indian. J. Hortic., 66*(3), 301–305.

Sharma, S. K., Vignesh, K. P., & Baranwal, V. K., (2014). Immunodiagnosis of episcopal banana streak MY virus using polyclonal antibodies to an expressed coat protein. *J. Virol. Methods, 207*, 86–94.

Shilpa, B. S., Moger, N., Vani, A., Rabinal, C., Hedge, R. V., & Byadagi, A. S., (2016). Cloning and expression of banana bunchy top virus (BBTV) coat protein gene in *E. coli. Elixir Agriculture, 94*, 40000–40004.

Siljo, A., Bhat, A. I., Biju, C. N., & Venugopal, M. N., (2012). Occurrence of Banana bract mosaic virus on cardamom. *Phytoparasitica, 40*, 77–85.

Sokhandan, B. N., Nematollahi, S., Rakhshandehroo, F., & Zamanizadeh, H. R., (2012). Phylogenetic analysis of new isolates of Cucumber mosaic virus from Iran on the basis of different genomic regions. *Plant Pathol. J., 28*(4), 381–389.

Stover, R. H., (1972). *Banana, Plantain and Abaca Diseases*. Commonwealth Mycological Institute, Kew, UK.

Thomas, J. E., & Ralf, G. D., (1991). Purification, characterization and serological detection of virus-like particles associated with banana bunchy top disease in Australia. *J. Gen. Virol., 72*, 217–224.

Tian, E., & Zhuang, L., (2005). Cloning and sequencing of DNA components of banana bunchy top virus Hainan isolate. *Chinese J. Agric. Biotechnol., 2*(2), 91–97.

Wang, I. C., Sether, D. M., Melzer, M. J., Borth, W. B., & Hu, J. S., (2010). First report of banana bract mosaic virus in flowering ginger in Hawaii. *Plant Dis., 94*(7), 921.

Wanitchakorn, R., Hafner, G. J., Harding, R. M., & Dale, J. L., (2000b). Functional analysis of proteins encoded by banana bunchy top virus DNA-4 to-6. *J. Gen. Virol., 81,* 299–306.

Wanitchakorn, R., Harding, R. M., & Dale, J. L., (1997). Banana bunchy top virus DNA-3encodes the viral coat protein. *Arch. Virol., 142*(8), 1673–1680.

Wanitchakorn, R., Harding, R. M., & Dale, J. L., (2000a). Sequence variability in the coat protein gene of two groups of banana bunchy top isolates. *Arch. Virol., 145*, 593–602.

Wu, R. Y., & Su, H. J., (1990). Production of monoclonal antibodies against banana bunchy top virus and their use in enzyme-linked immunosorbent assay. *J. Phytopathol., 128*, 203–208.

CHAPTER 7

Management of Plant Viruses through Host RNAi Defense Mechanism

NANDLAL CHOUDHARY

Amity Institute of Virology and Immunology, Amity University
Uttar Pradesh, Noida–201313, Uttar Pradesh, India,
E-mail: nchoudhary@amity.edu

ABSTRACT

RNA silencing is a form of nucleic acid-based plant immunity against the plant pathogens including viruses. It has been demonstrated first in plant systems which protect crops from pathogens and eventually enhance the quality production. RNAi is a highly structured template dependent control mechanism evolved in eukaryotes for genome protection from alien genomes as well as altering the levels of gene expression. The potato crops were the first RNAi based transgenic generated resistant to potato virus Y (PVY) targeting the viral proteinase gene. Afterwards, many transgenic crops were developed for other economical viruses such as begomovirus, sugarcane mosaic virus (SCMV), tobacco mosaic virus (TMV), cucumber mosaic virus (CMV), tomato spotted wilt virus (TSWV), rice tungro bacilliform virus (RTBV) which led to improved yield. The presence of dsRNA is very crucial to trigger to activate the gene silencing mechanism followed by diced or sliced into siRNA molecules by cytoplasm dicer enzyme. These siRNA molecules then associate with argonaute (AGO) protein of RNA induced silencing complex (RISC). RISC controls the gene expression through either cleavage of target gene sequence or inhibits translation process. RNAi mechanism is an efficient defense process against viral pathogens plant.

7.1   INTRODUCTION

The agriculture sector is very important for the Indian economy as it contributes a major sum in GDP. The potential productivity of agriculture crops has not yet been achieved due to many factors such as phytopathogens, insect pests, environmental damages, lack of quality seed, etc. Phytopathogens like viruses, bacteria, fungi are considered as important biotic stress-causing significant quality yield loss. The integrated disease management (IDM) has been adopted from last 3–4 decades to minimize the losses from phytopathogens which are safe to environment, human, and animals; sustainable and inexpensive compared to chemical controls (Mandal et al., 2012). Among phytopathogens, viruses found most devastating due to their characteristic's association with host and ability to cause systemic infections. The primary infection of virus responsible to decreases the crop productivity and then reduces seed quality as well if infection persist. At present, approx.: 1200 plant viruses reported from agriculture crops, of which approx.: 250 are characterized severe which are potential threat to agriculture industry (Briddon and Markham, 2000). Plant viruses are composed of nucleic acids or genome packed in a thick layer capsid or coat protein. Naturally, viruses evolved as an obligate parasite that is dependent on infecting host cell resources for replication and multiplication. Host plant does not contain any receptor as if animal cells therefore plant viruses enter plant cells only through physical injury like wound/cut or via insect vectors like aphid. The plant viruses containing ssRNA genome are found most pathogenic like sugarcane mosaic virus (SCMV), potato virus Y (PVY), etc., as it accumulates in higher quantity. The viruses from infected cells move to the healthy neighboring cells via plasmodesmata (PD) and reach to the phloem to move eventually to the distal or systemic site to cause significant yield loss. Some crop variety tolerant to virus has been developed by plant breeding methods which seem to be least expensive and promising. However, it has limitation because of continuous evolution of virus strain which can re-start infecting the resistant crops. The biotechnological methods using the disease resistant gene are found promising to develop transgenic resistant plants following regulatory guidelines (Mann et al., 2008; Sanghera et al., 2011).

The understanding of gene silencing mechanism based on RNA interference (RNAi) has improved in last 20 years, which is a natural defense process evolved in plant system against viral pathogens. In this process, the dsRNA suspected as potent trigger molecules to induce RNAi mechanism to degrade the targets homologous mRNAs and thereby inhibit transcription/ or translation. This process popularly known as post-transcriptional gene silencing (PTGS) (De Bakker et al., 2002; Almeida and Allshire, 2005). The induction of resistance against viral pathogen via RNAi mediated is considered a powerful and eco-friendly molecular tool for crop improvement (Baulcombe, 2004). The target gene of the viral pathogen can be silenced partially or completely by transgenic plants generation: Example-PVY resistant potato and SCMV resistant sugarcane crops.

7.2   HISTORY OF RNAi

The evolutionary story of gene silencing was started in the early 1990s in the laboratory of Professor Richard Jorgensen, and his wife and collaborator, Carolyn Napoli, a plant scientist at the University of Arizona, Tucson. They exogenously introduced a chalcone synthase gene (CHS) into petunias crops to increase the CHS gene product that is an enzyme involved in the production of pigments in petunia flower. However, unexpectedly, flowers were de-pigmented showing variegated color or no color. This indicates that exogenously introduced transgene becomes inactive and suppress the expression of endogenous CHS gene too (Napoli et al., 1990). Napoli, named the variegated flower a "Cossack dancer" as it looks like a man in voluminous costume with arms and legs outstretched. Professor Jorgensen called this phenomenon "cosuppression" because it appeared both sets of CHS genes interfere the expression of each other. The similar phenomenon was also found in fungus *Neurospora crassa* and the unknown event named as gene quelling (Romano et al., 1992). The mechanism of unexpected findings was not explained that time and remained unsolved for next many years (Jorgensen et al., 1996; Cogoni and Macino, 2000).

In the year 1998, the mystery mechanism was solved and described by Dr. Andrew Fire, Johns Hopkins University, Baltimore, and Dr. Craig Mello, University of Massachusetts, Boston, when they were studying the switching on/off gene by injecting exogenously antisense gene into nematode worm but found both endogenous and exogenous gene off. They correlate his finding with similar report earlier of co-suppression in petunias and gene quelling in yeast. Dr. Fire thought his control experiment not working properly because of some contaminant as it repetitively producing similar results as actual experiment. They investigate further and find out that sense and antisense strand of RNA forms a dsRNA molecule and that was the contaminant involved in silencing of gene expression at post-transcriptional level (PTGS). Drs. Fire and Mello named their discovery RNAi and published results in Nature Journal on 19 February 1998 (Fire et al., 1998). Both jointly received the Nobel Prize in Physiology or Medicine in the year 2006 for their discovery of RNAi-gene silencing by double-stranded RNA (dsRNAs).

7.3   MECHANISM OF RNA INTERFERENCE (RNAi)

RNAi mechanism is a multistep process where host cellular pathway degrades the target mRNA sequences by small RNA on homology basis. RNAi is a natural defense pathway evolved in plant system which triggered by long dsRNA of variable source of origin like viral RNA replication intermediates, transcription of inverted repeats, genomic containment of retrotransposons, and post-transcriptional regulation of gene expression, stress-induced overlapping antisense transcripts and RNA-Directed RNA Polymerase (RDR) transcription of aberrant transcripts (Hamilton and Baulcombe, 1999; Mette et al., 1999; Borsani et al., 2005; Luo and Chen, 2007). Two different RNAi molecules; microRNA (miRNA) and small interfering RNA (siRNA), have been characterized in plant system. Upon virus infection, viral RNAs form a long dsRNA replication intermediate in host cells which act as substrate for RNase III endonuclease Dicer-like (DCL) enzyme in cytoplasm to produce short interfering dsRNAs of 21–24nt (Zou et al., 2005; Shi, 2003). This mechanism was studied in model plant Arabidopsis (*Arabidopsis thaliana*) which contains four DCL; DCL2, DCL3 and DCL4 to generate the 22, 24 and 21nt siRNAs, respectively, whereas DCL1 recognizes host genome-encoded imperfect hairpin RNAs for biogenesis of 21/or 22nt micro RNAs (miRNAs) (Bologna and Voinnet, 2014; Borges and Martienssen, 2015). The siRNA generated of each size has unique function and participate in host defense. One strand of siRNA which is complementary to target mRNA is called guide RNA and another strand known as passenger strand. Both siRNA and miRNA exhibit 2-nt overhangs at their 3' site which are prone for nuclease degradation and this protected by methylation process catalyzed by hue enhancer-1 (HEN1) methyltransferase enzyme (Yang et al., 2006).

The 21nt siRNAs class is usually the most dominant siRNAs followed by 22nt size class. The 21 nt size dsRNA recruited by multifunctional molecules-induced silencing complex (RISC), which comprises accessory proteins argonaute (AGO) for catalytic endonucleases function, helicase, RNA binding proteins (RBP), and some trans-acting RNA-binding proteins (TRBP), etc., (Liu et al., 2009). A helicase component in RISC unwinds the duplex siRNA and passenger strand degraded by AGO. Guide RNA-RISC-AGO complex then scan to search complementary target mRNA for degradation by PTGS if 100% homology found and in case of imperfect homology, the complex only interferes in translation of target mRNA (Kim, 2008). The target mRNA degradation starts from 10nt upstream of siRNA-target mRNA duplex by endonuclease RNase (or slicer) function possessed in RISC (Elbashir et al., 2001).

The 22 ntdsRNA will load onto AGO1 protein however, it alters the conformation of AGO1 to not allow for PTGS of target mRNA to occur. In this condition, the guides RNA recruited by RNA-dependent RNA polymerase (RDR6) at 3' end and induce to form target mRNA into dsRNA. This dsRNA will be cleaved as described above that lead to generate additional secondary siRNAs. The secondary siRNA accumulation also provides protection to the host plant if same virus infects again by the mechanism known as transitive silencing (Chen et al., 2010). Some report suggests that RDR6 recruitment not necessarily mediated by 22-nt sRNAs, bubby any size of sRNA containing an asymmetric bulge in its duplex structure (Manavella et al., 2012). The siRNA quantity should surpass the certain threshold for onset of systemic silencing. The intron less genes compared to intron containing gene are found more susceptible to RDR6 processivity and thus transitivity and systemic silencing (Dadami et al., 2014).

The 24nt sRNAs are incorporated on AGO4 which recognizes cognate viral DNA to induce methylation of cytosine residues of both DNA strands by methyltransferase enzyme, this process called RNA-directed DNA methylation (RdDM). RdDM pathway is unique to plants which mediates de novo DNA methylation for transcriptional gene silencing (TGS) which requires homology sequence to promoter region where PTGS require homology with coding region (Dubrovina et al., 2019). Upon DNA virus infection, DNA transcribes to RNA transcript by plant pol II in nucleus which processed by DCL2/4 and DCL3 to produce virus specific siRNA molecules of 21, 22, and 24 nt siRNAs like endogenous tasiRNA pathway (Wang et al., 2010). The 24-nt siRNA as like others siRNA gets methylated at 3'-OH by HEN1, load onto AGO4-RISC involving both nuclear and cytoplasmic steps and then direct RdDM for DNA methylation for TGS process. It is proposed that DNA methylation interfere the transcription factors for its interaction with promoter region, however it is not yet known whether only DNA methylation alone enough for silencing or not. The virus infected localized cells produce RNAi molecules which moves to 10–15 neighboring healthy cells through PD and then reach to vascular system to move to distant parts of plant parts to provide the systemic resistance against viruses. This phenomenon is called systemic silencing (Voinnet and Baulcombe, 1997).

7.4   VIRAL SUPPRESSOR OF RNAi SILENCING (VSR)

Both PTGS and TGS pathways are involved in plant defense against plant viruses, however viruses have evolved to suppress or resist the host RNAi silencing mediated RNA degradation as counter defense by encoding strong multi-functional proteins called viral suppressors of RNA silencing (VSRs). More than 40 types of VSR are characterized in plant viruses to disrupt the host RNAi pathways majorly by binding to siRNA/AGO1/AGO4/miR168/HEN1/DICER, or downregulate RDR6, disrupt SA/JA systemic signaling pathway or also escape from recognition from host RNAi mechanism by making stable secondary structures and/or specific subcellular localization by RNA molecules (Ruiz and Voinnet, 2007). Some specific VSR like p69 encoded by Turnip yellow mosaic virus prevent the RDR-dependent secondary dsRNA synthesis (Chen et al., 2004), P14 encoded by aureus viruses sequestering both long dsRNA and siRNA without size specificity (Merai et al., 2005), p20 and CP of Citrus tristeza virus suppress silencing signal, and p23 inhibit intracellular silencing, Pns10 of Rice dwarf virus targets an upstream step of dsRNA formation.

7.5   TYPES OF RNAi MOLECULES TO INDUCE SILENCING

The antiviral silencing mechanism for crop improvement can be activated majorly either by full-length or partial or mutated viral sense gene induce post-transcriptional gene silencing (S-PTGS), hairpin RNA induced PTGS (hp-PTGS), artificial miRNA induced PTGS (AMIR), and trans-acting siRNA (TAS) induced PTGS (TAS).

7.5.1   S-PTGS

S-PTGS mechanism activated by sense gene construct in host plant and silencing state can achieve before or after viral infection. In both cases, plant RDR transcribes the ssRNA sequence of virus into dsRNA which processed into siRNA for gene silencing (Ding and Voinnet, 2007). The construct of same viral gene may act differently and accumulate varying level of siRNA in different crops therefore provide different resistance level against viral pathogens. The antisense gene was also used for construct preparation, which observed providing similar gene silencing response as S-PTGS. This mechanism termed as AS-PTGS (Prins et al., 1997). Both S-PTGS and AS-PTGS have been used for crop improvement against plant viruses before the RNAi mechanism explained.

7.5.2 hp-PTGS

hp-PTGS technology: the silencing vector was constructed with sense and antisense of inverted repeats which separated by a non-complementary spacer region under control of plant promoters and terminators. The sliceable intron as a replacement of spacer region can also use which also give better gene silencing response in plants. The construct when transformed in plant they transcribed RNA sequence complementary to each other which easily form dsRNA. Therefore, this method does not need RDRs to generate dsRNA. This dsRNA then processed into siRNAs following RNAi mechanism for gene silencing or knockdown. First hpPTGS mediated virus resistance against PVY was showed in tobacco crops in 1998 (Waterhouse et al., 1998) and later extended to many crops.

The hpRNA acted primarily by DCL4 to generate 21-nt siRNAs, but DCL2 and DCL3 also involve producing low levels of 22 and 24-nt siRNAs. Interestingly, hpRNA expressed by RNA polymerase III promoter or containing an intron are processed in 24-nt siRNAs indicates that hpRNA specifically targeted by nucleolar localized DCL3. It also possible that target mRNA or unprocessed hpRNA may serve template for RDRs to generate secondary 24-nt siRNAs to strengthen hpRNA-induced silencing. The 24-nt siRNA induce for methylation at coding region but not promoter region. An hpRNA transgene found inducing effective gene silencing in RdDM mutant plants than in wild type which suggest that hpRNA subject to RdDM caused transcriptional self-silencing. The cloning vectors like pHannibal and pHellsgate vectors use to prepare the hpRNA constructs. The 3 or 4-copy of inverted direct repeats can induce almost 100% efficiency against viruses which showed in arabidopsis and maize, however, such construct do not produce dsRNA directly, but how they induce effective gene silencing remains unclear. The exact size of siRNA has not determined in this case, but the accumulation of siRNAs suggested for PTGS. The intrinsic direct repeats can induce transitive silencing suggests that siRNAs derived from the direct repeats induce secondary siRNA production from downstream sequence (Guo et al., 2016).

7.5.3   ARTIFICIAL microRNA (AMIR)

In plant system, the primary miRNA transcript (pri-miRNA) synthesize naturally from intronic region in nucleus are processed by specific ribonuclease Drosha enzyme to produce a hairpin intermediate of 70 nt. The pre-miRNA then transported by exportin-5/Ran-GTP into cytoplasm to be cleave by Dicer into miRNA duplexes. This miRNA duplex as like siRNA forma RISC-AGO complex to degrade the target mRNAs or inhibits translation. In AMIR approach, complementary RNA sequence target to virus gene were replaced in natural pri-miRNA transcript to create an artificial mature miRNA. The engineered construct when transformed into plants they processed into mature miRNA comprises desired sequences to confer specific virus resistance. This strategy has advantages over conventional RNAi strategies: (a) require only short sequence to find easy target sequence and biosafe compared to long dsRNA; (b) tissue-or cell-specific knockdowns of genes of interest if specific promoter uses (Guo et al., 2016).

7.5.4   TRANS-ACTING siRNA (TAS)

The tasi-RNAs is a class of 21-nt sRNAs like miRNAs derived from transcript of non-coding genetic loci transcribed by RNA polymerase II known as ta-siRNA-generating loci (TAS genes). The tasi-siRNA biogenesis induced specifically by 22nt miRNAs and not by other size of miRNA to convert the precursor transcript into long dsRNA by RDR6. This dsRNA then processed by DCL1 into 21-nt siRNAs in the nucleus which like other siRNA methylated by HEN1 and interact with either AGO1 or AGO7 to direct the degradation of target mRNAs. The large number of tasiRNA generated from coding and non coding gene which collectively known as phased siRNAs or phasiRNAs. However, the functions of phasiRNAs have not yet fully understood (Guo et al., 2016).

7.6   TRANSFORMATION METHODS OF dsRNA MOLECULES

In RNAi research field, the success with dsRNA transgenes has not been achieved as expected due to challenges of effective delivery of dsRNA in plants. RNAi mechanism can be induced against plant viruses by ds RNA of various lengths ranging from 50–150 bp of shorter size or up to 2.5 kb of larger size (Pooggin, 2017). The efficiency of gene silencing varies depending on transformation methods because all has different efficiency. Since discovery of RNAi mechanism, the different methods have used for expression of dsRNA in plants to induce gene silencing against plant viruses. Some are as below:

1. **Agroinfiltration:** Agro inoculation or agroinfiltration is the powerful methods where agrobacterium carrying DNA constructs were injected into intracellular spaces of leaves for triggering gene-silencing mechanism. In plants, cytoplasmic RNAi can be induced efficiently by agroinfiltration, as delivery of T-DNA vectors constructs by *Agrobacterium tumefaciens* for transient expression. The infiltration of hairpin constructs is especially effective, because their dsRNA can be processed directly into siRNAs, while constructs expressing long dsRNA also useful to induce gene silencing of target virus (Sharma et al., 2013).

2. **Micro-Bombardment:** In this method, particles coated with dsRNA, siRNA or DNA encoding hpRNA, sense, or antisense RNA is delivered using biolistic pressure by micro-bombardment to the nucleus of cells for gene silencing. The silencing effect due to RNAi mechanism has been observed in a day which may continue for 3–4 days post bombardment. Systemic spread of silencing molecules also observed after 15 days indicates that siRNAs induced for de novo formation of siRNAs, which accumulate to cause gene silencing (Sharma et al., 2013). The chemically synthesized 22-nt sRNAs when applied to leaves of *Nicotiana benthamiana* they also trigger local and systemic RNAi mechanisms (Dalakouras et al., 2018). Confocal microscopy study revealed that when hpRNA applied through petiole absorption and/or trunk injection in grapevine (*Vitis vinifera*) and apple (*Malus domestica*) trees it transported through the xylem but restricted to the apoplast and no RNAi observed. The microcompartment method at least ensures for efficient sRNA delivery into plant cells for local RNAi activation for virus resistance.

7.7 MANAGEMENT OF PLANT VIRUSES BY RNAi MOLECULES

Upon virus infection, host plant defend naturally through hypersensitive response (HR) and extreme resistance response (ER) which induces the production of secondary metabolites, elevate the level of ethylene, jasmonic acid (JA), salicylic acid (SA), nitric oxide (NO), pathogenesis related protein

(PR protein) to block entry of virus/or to eliminate by increased rate of ion flux. The acquired virus resistance mechanisms in plants are of two types: (a) gene silencing independent virus resistance; and (b) gene silencing dependable virus resistance via TGS and PTGS. Gene silencing independent virus resistance approach is like cross protection in which the susceptible plant inoculated with a milder strain for accumulation of defense molecules before onset of severe target virus. The major drawback is the potential cause of serious diseases on other varieties growing nearby. To compensate this drawback, pathogen derived resistance (PDR) based strategies was used by expressing coat protein of PVY in potato, movement protein (MP) of potato leaf roll virus (PLRV) and replicase protein gene to achieve virus resistance. This strategy provides promising resistance, but potential threats are also associated like environmental safety concerns due to constitutive expression of gene, interaction of viral gene product which may alter biological properties of virus to create new virus strain with novel pathogenic properties, host range and transmission specificity. Antisense RNA method is another powerful tool which is small and untranslatable RNA molecules complementary to target viral RNA. It prevents all potential interaction of viral gene with other nucleic acids or protein factors to provide virus resistance. The small piece of viral genome can also transfer into susceptible crops for viral resistance but there is potential chance of its integration into host genome or, it can activate the numerous proteins including ribonucleases that cleave all type of RNAs. However, the viral gene once integrate also in host genome will not cause any disease because entire viral genome required for disease development.

7.7.1   TRANSGENIC DEVELOPMENT

Transgenic development with dsRNAs/siRNAs molecules is a well-established antiviral crop protection strategy before virus infection. The identification of hotspot region of target viral sequence is critical to design dsRNA molecules to achieve resistance against plant viruses. For example, 3' UTR sequence of CMV can be a hotspot region to design dsRNA molecules because it is essential for replication and found conserved among CMV strains. In other example, mutation study suggests a hotspot region in RISC to design a miRNA construct which confer broad-spectrum resistance to CMV compared to only targeting VSR. Transgenic potato and sugarcane plants were developed using shRNA sequence synthesized of non-viral origin targeting CP gene of PVY and SCMV for degradation of viral mRNA by silencing mechanism. The non-coding intergenic region (IR) of mung bean yellow mosaic India virus

(MYMIV) was expressed as hairpin construct under control of 35S promoter for black gram crops, shows complete recovery. The P69 and HC-Pro VSR of Turnip yellow mosaic virus and turnip mosaic virus (TuMV), respectively are also disrupted by using miR159 of 273 bp of pre-miRNA transcript of Arabidopsis. The amiRNA vectors target to 2b viral suppressor used against cucumber mosaic virus (CMV) by blocking the slicer activity of AGO1 (Khalid et al., 2017; Pooggin, 2017).

The efficacy of sRNA construct widely tested in model crops tobacco first and then applies to major crops for transgenic development. At present sRNA based many virus resistant transgenic crops has been generated like cereals, vegetables, fruits, ornamentals, and some cash crop. In 1998, transgenic Papaya was the first crops successfully released in Hawaii using PDR approach expressing CP transgene against papaya ring spot virus (PRSV) (Gonsalves, 2006). Some viral resistant transgenic fruits crops are developed such as papaya targeting HcPro gene of PRSV, banana targeting rep gene of banana bunchy top virus (BBTV), and plum against CP gene of plum poxvirus with S-PTGS mechanism. Citrus with single VSR molecules employing S-PTGS mechanisms did not work effectively against therefore target multiple VSRs of CTV with hp-PTGS mechanism to achieve resistance. Banana and citrus transgenic lines are not yet approved to release commercially. Some virus resistant transgenic vegetable crops are developed such as squash against CP of SqMV, cucumber against rep gene of CFMMV, watermelon against CP of multiple viruses by S-PTGS mechanism and cantaloupe against PRSV by hp-PTGS mechanism, PVX-resistant potato, CMV-resistant tomato, and pepper with dual resistance to ToMV and CMV by S-PTGS. Transgenic potato against CP of PLRV and PVX, and helper-component proteinase (Hc-Pro) of PVY simultaneously also developed using hp-PTGS mechanism (Waterhouse et al., 1998). Transgenic tomato against rep gene was also developed using both S-PTGS and hp-PTGS mechanism.

Some virus resistant transgenic cereals crops are developed such as maize against CP and P1 gene of Maize streak virus (MSV), maize against rep gene of maize dwarf mosaic virus (MDMV) with hp-PTGS mechanism, wheat against wheat streak mosaic virus (WSMV), rice against RBSDV, RSV, RTBV, and RTSV using hp-PTGS mechanism, cassava against AC1 gene of ACMV and SLCMV using S-PTGS, peanuts against PStV-and TSV with S-PTGS mechanism, soybean against CP/Hc-Pro of soybean mosaic virus (SMV), common bean against CP using S-PTGS and AC1 using hp-PTGS of BGMV. BGMV-resistant common bean was approved for commercial release in Brazil (Khalid et al., 2017).

7.7.2   NONTRANSGENIC STRATEGY

Some of countries has strict guidelines and policy for prepared for transgenic crop development and also do not allow to be imported any transgenic foods. In addition, some crops even found recalcitrant to transformation or regeneration in tissue culture. Therefore, alternative non-transgenic approach is developed to provide virus resistance to crop plantlike exogenous application of dsRNA molecules. The in vitro transcribed 997 bp, 1483 bp and 1124 bp dsRNAs of pepper mild mottle virus (PMMoV), tobacco etch virus (TEV) and alfalfa mosaic virus (AMV), respectively were mechanically inoculated in leaves of *N. benthamiana* to demonstrate resistance (Tenllado and Diaz-Ruiz, 2001). The dsRNA can be produced in bacterial expression systems an easy and cost-effective approach and spray on crops in fields. The management of viral pathogens in field roughly required 10 g dsRNA per hectare that vary based on sensitivity and systemic RNAi induction (Zotti et al., 2018). Such huge dsRNA production cannot meet by *in vitro* production but now 'RNAagri' proprietary fermenter technology offers to produce as much need with the help of engineered bacteria and yeast (Tenllado et al., 2004). This methods has been used to generate dsRNA of CP for SCMV CP; 279 bp dsRNA CP for papaya ringspot virus (PRSV) CP; 500 bp dsRNA for pea seed-borne mosaic virus (PSbMV); dsRNA CP for cymbidium mosaic virus (CymMV); 588 bp dsRNA of HcPro targeting the zucchini yellow mosaic virus (ZYMV), 480 bp dsRNA of CP or MP for TMV (Dalakouras et al., 2020). An alternative *Pseudomonas syringae* harboring components of bacteriophage phi6 encoded RdRP used to generate dsRNA by RNA transcription (Niehl et al., 2018). It has advantage since the RdRP of phage phi6 converts ssRNA templates into dsRNA without primer and de novo. The major drawback associated with dsRNA is that it degrades after 5–10 days so this approach needs frequent spraying to prolong antiviral protection.

Recently, a nanoparticle-based approach has been developed where dsRNA was attached to the BioClay, double hydroxide clay nanosheets of an average particle size of 80 to 300 nm and spray to crops plant. This dsRNA/ BioClay complex on leaf surface was sticky, stable, and protected from nuclease and even did not get washed with rinsing. The dsRNA/BioClay complex gradually broken down due to CO_2 and moisture and released dsRNA got entry in cells by diffusion or active transport (Mitter et al., 2017). This method successfully demonstrated with 461 bp dsRNA/BioClay target to CP of bean common mosaic virus (BCMV) on *N. benthamiana* and cowpea

(Worrall et al., 2019; Dubrovina and Kiselev, 2019). This method found effective up to 25–30 days in single dose means do not frequent spraying.

7.8 CONCLUSION

RNAi is a unique and powerful defense mechanism evolved in plant system against plant viruses. Upon infection by RNA virus, RNA genome undergoes to forms a long dsRNA which activate the RNAi silencing mechanism of plant system. This long dsRNA gets cleave into small dsRNA of 21–24nt size by the action of host encoded DICERs enzymes. The primary dsRNA of 21nt size load onto RNA Induced Silencing Complex (RISC), a multifunctional protein to play role of helicase, RNase, and other important function to cleaves viral genome into smaller fragment by process called PTGS. However, host defense often fails because plant viruses encode virus suppressor RNAi silencing (VSR) to counter the host RNAi mechanism. Plant viruses cause the significant loss to agricultural crop production, and like other pathogens, they cannot be managed by spraying chemicals. Therefore, many transgenics crops exploiting the RNAi mechanism has been developed and commercially released over the past years by using the viral coding and suppressor gene. Due to strict guidelines imposed on transgenic development, the application of RNAi in non-transgenic way also demonstrated like exogenous application of bacterially expressed dsRNA to induce viral gene silencing. Recently, dsRNA molecules combined with BioClay nanoparticles and sprayed on leaves that enhance the performance of molecules and stability. RNAi research has provided a lot of information over the last two decades, which proves to apply for the management of plant viruses for agricultural crop development.

KEYWORDS

- **alfalfa mosaic virus**
- **bean common mosaic virus**
- **chalcone synthase gene**
- **cucumber mosaic virus**
- **maize dwarf mosaic virus**
- **pathogen-derived resistance**

REFERENCES

Almeida, R., & Allshire, R. C., (2005). RNA silencing and genome regulation. *Trends in Cell Biol., 15,* 251–258.

Baulcombe, D., (2004). RNA silencing in plants. *Nature, 431,* 356–363.

Bologna, N. G., & Voinnet, O., (2014). The diversity, biogenesis, and activities of endogenous silencing small RNAs in Arabidopsis. *Annu. Rev. Plant Biol., 65,* 473–503.

Borges, F., & Martienssen, R. A., (2015). The expanding world of small RNAs in plants. *Nat. Rev. Mol. Cell Biol., 16*(12), 727–741.

Borsani, O., Zhu, J., Verslues, P. E., Sunkar, R., & Zhu, J. K., (2005). Endogenous siRNAs derived from a pair of natural cis-antisense transcripts regulate salt tolerance in *Arabidopsis. Cell, 123,* 1279–1291.

Briddon, R. W., & Markham, P. J., (2000). Cotton leaf curl virus disease. *Virus Res., 71,* 151–159.

Chen, H. M., Chen, L. T., Patel, K., Li, Y. H., Baulcombe, D. C., & Wu, S. H., (2010). 22-nucleotide RNAs trigger secondary siRNA biogenesis in plants. *PNAS USA, 107*(34), 15269–15274.

Chen, J., Li, W. X., Xie, D., Peng, J. R., & Ding, S. W., (2004). Viral virulence protein suppresses RNA silencing-mediated defense but up regulates the role of microRNA in host gene regulation. *Plant Cell, 16,* 1302–1313.

Cogoni, C., & Macino, G., (2000). Homology dependent gene silencing in plants and fungi: A number of variations on the same theme. *Curr. Opinion. Microbiol., 2,* 657–662.

Dadami, E., Dalakouras, A., Zwiebel, M., Krczal, G., & Wassenegger, M., (2014). An endogene-resembling transgene is resistant to DNA methylation and systemic silencing. *RNA Biol., 11,* 934–941.

Dalakouras, A., Jarausch, W., Buchholz, G., Bassler, A., Braun, M., Manthey, T., Krczal, G., & Wassenegger, M., (2018). Delivery of hairpin RNAs and small RNAs into woody and herbaceous plants by trunk injection and petiole absorption. *Front Plant Sci., 9,* 1253.

Dalakouras, A., Wassenegger, M., Dadami, E., Ganopoulos, I., Pappas, M. L., & Papadopoulou, K., (2020). Genetically modified organism-free RNA interference: Exogenous application of RNA molecules in plants. *Plant Physiol., 182,* 38–50.

DeBakker, M. D., Raponi, M., & Arndt, G. M., (2002). RNA-mediated gene silencing in non-pathogenic and pathogenic fungi. *Curr. Opinion. Microbiol., 5,* 323–329.

Ding, S. W., & Vionnet, O., (2007). Antiviral immunity directed by small RNAs. *Cell, 130,* 413–426.

Dubrovina, A. S., & Kiselev, K. V., (2019). Exogenous RNAs for gene regulation and plant resistance. *Int. J. Mol. Sci., 20,* 2282.

Dubrovina, A. S., Aleynova, O. A., Kalachev, A. V., Suprun, A. R., Ogneva, Z. V., & Kiselev, K. V., (2019). Induction of transgene suppression in plants via external application of synthetic dsRNA. *Int. J. Mol. Sci., 20,* 1585.

Elbashir, S. M., Harborth, J., Lendeckel, W., Yalcin, A., Weber, K., & Tuschl, T., (2001). Duplexes of 21 nucleotide RNAs mediate RNA interference in cultured mammalian cells. *Nature, 411,* 494–498.

Fire, A., Xu, S., Montgomery, M. K., Kostas, S. A., Driver, S. E., & Mello, C. C., (1998). Potent and specific genetic interference by double-stranded RNA in Caenorhabditis elegans. *Nature, 391,* 806–811.

Gonsalves, D., (2006). Transgenic papaya: Development, release, impact and challenges. *Adv. Virus Res., 67,* 317–354.

Guo, Q., Liu, Q., Smith, N. A., Liang, G., & Wang, M. B., (2016). RNA Silencing in plants: Mechanisms, technologies and applications in horticultural crops. *Current Genomics, 17,* 476–489.

Hamilton, A. J., & Baulcoumbe, D. C., (1999). A species of small antisense RNA in posttranscriptional gene silencing in plants. *Science, 286*(5441), 950–952.

Jorgensen, R. A., Cluster, P. D., English, J., Que, Q., & Napoli, C. A., (1996). Chalcone synthase co suppression phenotypes in petunia flowers: Comparison of sense vs. antisense constructs and single-copy vs. complex T-DNA sequences. *Plant Mol. Biol., 31,* 957–973.

Khalid, A., Zhang, Q., Yasir, M., & Li, F., (2017). Small RNA based genetic engineering for plant viral resistance: Application in crop protection. *Front Microbiol., 8,* 43.

Kim, Y. S., Lee, Y. H., Kim, H. S., Kim, M. S., Hahn, K. W., Ko, J. H., Joung, H., & Jeon, J. H., (2008). Development of patatin knockdown potato tubers using RNA interference (RNAi) technology, for the production of human-therapeutic glycoproteins. *BMC Biotechnol., 8,* 36.

Liu, Y., Ye, X., Jiang, F., Liang, C., Chen, D., Peng, J., Kinch, L. N., Grishin, N. V., & Liu, Q., (2009). C3PO, an endoribonuclease that promotes RNAi by facilitating RISC activation. *Science, 325,* 750–753.

Luo, Z., & Chen, Z., (2007). Improperly terminated, un polyadenylated mRNA of sense transgenes is targeted by RDR6-mediated RNA silencing in *Arabidopsis. Plant Cell, 19,* 943–958.

Manavella, P. A., Koenig, D., & Weigel, D., (2012). Plant secondary siRNA production determined by microRNA-duplex structure. *PNAS USA, 109*(7), 2461–2466.

Mandal, A. K., Kashyap, P. L., Gurjar, M. S., Sanghera, G. S., Kumar, S., & Dubey, S. C., (2012). Recent biotechnological achievements in plant disease management. In: *Curr. Concepts in Crop Protect.*

Mann, S. K., Kashyap, P. L., Sanghera, G. S., Singh, G., & Singh, S., (2008). RNA interference: An eco-friendly tool for plant disease management. *Transgen. Plant J., 2*(2), 110–126.

Merai, Z., Kerenyi, Z., Molnar, A., Barta, E., Valcozi, A., Bistray, G., Havelda, Z., Burgyan, J., & Silhavy, D., (2005). Aureus virus P14 is an efficient RNA silencing suppressor that binds double stranded RNAs without size specificity. *J. Virol., 79,* 7217–7226.

Mette, M. F., Van, D., Winden, J., Matzke, M. A., & Matzke, A. J., (1999). Production of aberrant promoter transcripts contributes to methylation and silencing of unlinked homologous promoters in trans. *EMBO J., 18,* 241–248.

Mitter, N., Worrall, E. A., Robinson, K. E., Li, P., Jain, R. G., Taochy, C., Fletcher, S. J., Carroll, B. J., Lu, G. Q., & Xu, Z. P., (2017). Clay nanosheets for topical delivery of RNAi for sustained protection against plant viruses. *Nat. Plants, 3,* 16207.

Napoli, C., Lemieux, C., & Jorgensen, R., (1990). Introduction of a chimeric chalcone synthase gene into petunia results in reversible co-suppression of homologous genes in trans. *The Plant Cell, 2,* 279–289.

Niehl, A., Soininen, M., Poranen, M. M., & Heinlein, M., (2018). Synthetic biology approach for plant protection using dsRNA. *Plant Biotechnol. J., 16,* 1679–1687.

Pooggin, M. M., (2017). RNAi-mediated resistance to viruses: A critical assessment of methodologies. *Curr. Opin. Virol., 26,* 28–35.

Prins, M., Laimer, M., Noris, E., Schubert, J., Wassenegger, M., & Tepfer, M., (2008). Strategies for antiviral resistance in transgenic plants. *Molecular Plant Pathology, 9*(1), 73–83.

Romano, N., & Macino, G., (1992). Quelling: Transient inactivation of gene expression in Neurospora crassa by transformation with homologous sequences. *Mol. Microbiol., 6,* 3343–3353.

Ruiz, F. V., & Voinnet, O., (2007). Roles of plant small RNAs in biotic stress responses. *Annu. Rev. Plant Biol., 60*, 485–510.

Sanghera, G. S., Kashyap, P. L., Singh, G., & Teixeira, D. S. J. A., (2011). Transgenics: Fast track to plant stress amelioration. *Transgen. Plant J., 5*(1), 1–26.

Sharma, V. K., Sanghera, G. S., Kashyap, P. L., Sharma, B. B., & Chandel, C., (2013). RNA interference: A novel tool for plant disease management. *African Journal of Biotechnology, 12*, 2303–2312.

Voinnet, O., & Baulcombe, D. C., (1997). Systemic signaling in gene silencing. *Nature, 389*, 553.

Wang, X. B., Wu, Q., Ito, T., Cillo, F., Li, W. X., Chen, X., Yu, J. L., & Ding, S. W., (2010). RNAi-mediated viral immunity requires amplification of virus-derived siRNAs in *Arabidopsis thaliana. PNAS USA, 107*, 484–489.

Waterhouse, P. M., Graham, M. W., & Wang, M. B., (1998). Virus resistance and gene silencing in plants can be induced by simultaneous expression of sense and antisense RNA. *PNAS USA, 95*, 13959–13964.

Worrall, E. A., Bravo, C. A., Nilon, A. T., Fletcher, S. J., Robinson, K. E., Carr, J. P., & Mitter, N., (2019). Exogenous application of RNAi-inducing double-stranded RNA inhibits aphid-mediated transmission of a plant virus. *Front Plant Sci., 10*, 265.

Yang, Z., Ebright, Y. W., Yu, B., & Chen, X., (2006). HEN1 recognizes 21-24 nt small RNA duplexes and deposits a methyl group onto the 2′ OH of the 3′ terminal nucleotide. *Nucleic Acids Res., 34*, 667–675.

Zotti, M., Dos, S. E. A., Cagliari, D., Christiaens, O., Taning, C. N. T., & Smagghe, G., (2018). RNA interference technology in crop protection against arthropod pests, pathogens and nematodes. *Pest Manag. Sci., 74*, 1239–1250.

CHAPTER 8

Diagnosis and Management of Viruses Infecting Ornamental Plants

NITIKA GUPTA,[1] PRITAM JADHAV,[1] and RICHA RAI[2]

[1]*ICAR-Directorate of Floricultural Research, Pune–411005, Maharashtra, India, E-mail: nitika.iari@gmail.com (N. Gupta)*

[2]*Division of Plant Pathology, ICAR-Indian Agricultural Research Institute, New Delhi–110012, India*

ABSTRACT

Floriculture has flourished as an important commercial sector of agriculture in the last few decades. Ornamental plants are an important part of the global horticulture industry. They are cherished for their esthetic value and used around the world for enhancing the beauty and the impact of a landscape. The demand of high-quality propagating material and agrotechniques for specific objectives has made the role of diagnosis of viral diseases in ornamental plants a conspicuous activity. The ornamental plants are more conducive to the multiplication and spread of viruses. Mostly ornamental plants are vegetatively propagated, and the viruses and virus-like pathogens are transmitted mechanically. A small number of infected plants in a production area can antagonistically influence the quality of the net product. The problem is heightened by the fact that most field-grown plants are virus-infected and also that symptom may be masked or vanished due to environmental conditions. Many a times, lack of knowledge and absence of information on proper identification of the pests and diseases often results in more input costs and losses as the right management is not opted. Identification and diagnosis of pests and diseases involve many processes and diagnostic tests. In most of the cases, failure to identify the right cause results in major crop loss and heavy chemical inputs in agriculture in general and floriculture in particular. Many of the morphological manifestations of pests and diseases are often misunderstood as novel variations in the case of ornamental crops, and they are multiplied and distributed across the country. Thus, the right identification of a pest or disease attack or a nutrient deficiency at the right time is vitally important for undertaking the right decisions for saving the crop from damages. Plant viruses are widespread and economically important plant pathogens, cause significant economic losses in ornamentals. Vegetative propagation, changed cultural practices, worldwide exchange, and movement of planting materials to newer zones with different climatic conditions lead to the spread of pathogens and disease outbreaks. To manage viral diseases, it is essential to diagnose the viruses efficiently and effectively. Recent advances in molecular techniques have gained importance leading to reliable detection of these viral pathogens. The methodologies for detection of viruses in ornamentals have become indispensable to the ornamental industry as no practical remedy exists against viruses in the field. In the present chapter, we have tried our best to describe the diagnosis and management of different viruses occurring in important ornamental plants.

8.1   INTRODUCTION

Ornamental plants are very popular and economically important worldwide. The international market of ornamental plants is constantly increasing. India has a tremendous potential to become a major contributor to the world floriculture trade. The diverse types of climatic conditions in different regions of India furnish it for cultivation of almost all the popular ornamental species of the world, either from temperate, tropical, or subtropical climates (Zaidi et al., 2011). There is always demand for new attractive flowers in the international market. With novelty, the economics of the floricultural trade depends on the quality and quantity of ornamentals. However, viruses, viroid, and phytoplasma are persistent threats in the production of ornamental plants. Viruses are easily transmitted through the import of virus-infected tubers, bulbs, or cuttings and can lead to infestation in entire cultivation. Unfortunately, due to vegetative propagation, viruses are propagated from the mother stock to the next generation as well. The introduction of new virus vectors (e.g., *Thrips palmi, Frankliniella occidentalis, Bemisia tabaci*) and their establishment has increased the risk of virus infections in ornamentals (Engelmann and Hamacher, 2008). Entry of novel virus species into production areas; introduction of novel exotic genera or species to widen the range of ornamentals; worldwide trade with ornamentals; and production of plant material in countries representing different standards of production and other virus pressure are the other important aspects of concern. As the number of ornamental plant species and varieties are very high and their phylogenetic as well as geographic origins vary considerably, the number of infecting agents, especially viruses, differs consequently, giving rise to an immense range of virus-plant pathosystems. Viruses infecting ornamentals have extremely wide host ranges, and many of them such as *Tomato spotted wilt virus* (TSWV), *Impatiens necrotic spot virus* (INSV), or *Cucumber mosaic virus* (CMV) are omnipresent. Some of them however, are very specific and infect only certain species or genera of plants, such as, for example, *Pelargonium flower break virus* (PFBV) or *Angelonia flower mottle virus* (AFMoV) (Engelmann and Hamacher, 2008). Chemical treatment against virus is ineffective but the viral diseases can be managed by preventive measures. Phytosanitary measures like use of virus free panting material, rouging of infected individuals, elimination of alternative hosts and vectors and quarantine practices are currently being followed to prevent virus infestation. *In-vitro* techniques have been standardized to produce virus free ornamental crops. An accredited lab under the National certification system for tissue culture raised plants (NCS-TCP) of the Indian Department of Biotechnology; keeps check on the tissue culture raised plants for fidelity and viruses, and thereafter certifies the material for the growers. In this way, the planting materials are generated and provided to the growers.

Apart from precautionary measures, early detection and accurate diagnosis of the viral outbreak helps in preventing spread of viral diseases efficiently. Some viral diseases can be diagnosed quickly by visual examination of symptoms, while many do not show any symptoms or show similar symptoms, such condition demands the use of serodiagnostics and molecular tests for their proper identification. There are several techniques available for plant virus diagnosis. Advances in molecular techniques have evolved immensely leading to reliable detection of the viruses (Aini et al., 2008; Fang and Ramaswamy, 2015; Lau et al., 2017). Techniques like electron microscopy (EM), immunosorbent electron microscopy (ISEM), ELISA, PCR/RT-PCR/IC-RT-PCR and nucleic acid hybridization have been used in floricultural crops for the detection of viruses (Webster et al., 2004).

8.1.1   SERODIAGNOSIS

Plant pathogens like viruses cannot be cultivated artificially, to address this issue serological assays were developed, which are used to detect other plant pathogens too (Caruso et al., 2002). Serological detection of diseases involves identification of pathogens using antibodies supported by color change in the assay. Antibodies are the immunoglobulin (Ig) proteins produced in the body of an animal in response to the presence of antigens which are foreign proteins, polynucleotides, complex carbohydrates, or lipopolysaccharides. Each antibody is specific and binds to a particular antigen. ELISA (Enzyme-Linked Immunosorbent Assay) is the most commonly used serological method, developed for the detection of plant viruses (Clark, 1981). It was modified according to the specific needs of the detection assay viz., Direct ELISA (conjugated antibody-enzyme complex directly binds to antigen) and Indirect ELISA (antigen does not directly bind to antibody-enzyme conjugate) (Khurana, 2006). Presently, ELISA is the most favored method and used tremendously for the detection of plant viruses using monoclonal and polyclonal antisera.

8.1.2 MOLECULAR METHODS

Polymerase chain reaction (PCR) is *in-vitro* reaction driven by enzyme and initiated by primer, capable of exponential amplification of DNA. Virus detection techniques based on PCR, are rapid, sensitive, specific, efficient, versatile, and relatively economical, require less time and less resources and the results obtained are highly accurate (Henson and French, 1993). Hence, are the apt means for the detection of viruses. Methods involving nucleic acid largely involve DNA as principle nucleic acid, such as fluorescence in-situ hybridization (FISH) and many PCR variants like nested PCR (nPCR), multiplex PCR (M-PCR), real-time PCR (RT-PCR), cooperative PCR (Co-PCR), and DNA fingerprinting. However, RNA-based methods like reverse transcriptase PCR, nucleic acid sequence-based amplification (NASBA) and AmpliDet RNA are also useful (Balodi et al., 2017). All of each mentioned methods are capable of rapid and accurate detection and finally quantification of plant viruses.

In the present chapter, we have tried our best to describe the diagnosis and management strategies utilized for different viruses occurring in important ornamental plants.

8.2 GLADIOLUS

The gladiolus crop is attacked by different viruses out of which *Bean yellow mosaic virus* (BYMV) and *Cucumber mosaic virus* (CMV) occur more frequently. The other viruses seen infecting gladiolus are *Arabis mosaic virus* (ArMV), *Broad bean wilt virus* (BBWV), *Tobacco mosaic virus* (TMV), *Tobacco necrosis virus* (TobNV), *Tobacco black ring virus* (TobBRV), *Tobacco ring spot virus* (ToRSV), *Tomato ring spot virus* (TRSV), *Tomato spotted wilt virus* (TSWV), *Soybean mosaic virus* (SMV), and *Strawberry latent ring spot virus* (Katoch et al., 2003). Double antibody sandwich (DAS) DAS-ELISA, direct tissue blotting immunoassay (DTBIA), immunosorbant electron microscopy (ISEM) and RT-PCR are the techniques successfully utilized to detect the various viruses infecting gladiolus. In which RT-PCR followed by Southern hybridization test is the most sensitive and reliable method for detection of CMV in Gladiolus (Raj et al., 2002).

8.3   ROSE

Rose is one of the most important commercial crops known for its beauty and essential oil. However, quality of bud and essential oil is severely affected by different viral diseases (Lockhart et al., 2011). Rose virus gets easily transmitted from seeds, pollens, vectors, and grafts, which makes it very difficult to control the spread of virus. *Arabis mosaic virus* (ArMV), *Prunus necrotic ringspot virus* (PNRSV), and S*trawberry latent ringspot virus* (SLRV) either alone or in combination frequently infect commercially important field grown roses (Thomas, 1980). In recent years many effective molecular and serological diagnostic methods were developed to prevent their spread. ELISA is used successfully for the detection of these viruses in rose. ELISA test in rose is available for PNRSV, *Arabis mosaic virus* (ArMV), *Apple mosaic virus* (ApMV), and *Prune dwarf virus* (PDV). Recently, polyclonal and monoclonal antibodies (MAbs) were developed against *Rose rosette virus* (RRV) using bacterially-expressed RRV 316 a nucleocapsid (NP) and used successfully for RRV detection in rose (Jordan et al., 2018).

RT-PCR-based diagnostic methods can be used for the detection of RRV. A more sensitive and cost-effective probe-based method was developed using the nucleoprotein (NP) gene of the RRV called isothermal reverse transcription-recombinase polymerase amplification (RT-exoRPA) assay. The primers and probe sequences for the RT-exoRPA assay are designed based on conserved regions of RRV genomic RNA segment 3. RT-exoRPA assay is performed using RT-exoRPA primers and probe. This method can detect as less as 1 fg/μl *in-vitro* transcript of virus in just 25 minutes from various parts like leaves, stems, petals, pollens, roots, and from different varieties of rose with less sophisticated equipment. It is very precise and gives no false positives for other viruses infecting roses belonging to same or distinct genus. This assay can be useful for rapid detection of RRV in large-scale cultivations of commercial nurseries and landscapes (Babu et al., 2017).

Recently, *Rose cryptic virus-1* (RoCV1) was identified using high throughput sequencing (HTS) of the nucleic acid extracted from infected samples. The infected samples were collected and sequenced. The sequence obtained was inspected using MEGAN community edition for the presence of RoCV1 sequences in the samples. Further, the primers and probes for the conserved regions of RoCV1 were designed from the sequences obtained. Using those designed primers and probes, samples could be successfully tested using an RT-PCR for the presence of RoCV1 (Vazquez-Iglesias et al., 2019).

8.4   CHRYSANTHEMUM

Chrysanthemum, one of the most popular ornamentals worldwide, is vastly affected by viral and viroid infections which cause about 30% losses in production (Zhao et al., 2015). About twenty viruses have been identified on chrysanthemum so far. The most damaging among them are the RNA viruses such as CMV and *Tomato aspermy virus* (TAV) from the genus *Cucumovirus* (family *Bromoviridae*), *Chrysanthemum virus B* (CVB) (genus *Carlavirus,* family *Betaflexiviridae*), *Tobacco mosaic virus* (TMV, genus *Tobamovirus*, family *Virgaviridae*) and *Potato virus Y* (PVY, genus *Potyvirus*, family *Potyviridae*) (Verma et al., 2003; Song et al., 2012; Choi et al., 2015; Zhao et al., 2015). Other known chrysanthemum viruses belong to families *Potyviridae* (*Turnip mosaic virus*) (TuMV), *Zucchini yellow mosaic virus* (ZYMV), *Chrysanthemum spot virus* (CSV), SMV from the genus *Potyvirus)* (Mitrofanova et al., 2018) and *Bunyaviridae* (*Chrysanthemum stem necrosis virus* (CSNV), *Tomato spotted wilt virus* (TSWV), *Impatience necrotic spot virus* (INSV) from the genus *Tospovirus*) (Dullemans et al., 2015; Wu et al., 2015; Mitrofanova et al., 2018). The poty and bunyaviruses are highly pathogenic viruses, having wide range of hosts and affecting many economically important crops.

ELISA has proved to be a sensitive and reliable method for detecting viruses in Chrysanthemum. The polyclonal rabbit antibodies against most of the viruses infecting Chrysanthemum are commercially available. *Chrysanthemum carlavirus B* (CVB) and CSNV can be detected accurately using a polyclonal antibodies in DAS-ELISA (Verhoeven et al., 1996; Verma et al., 2003; Matsuura et al., 2007).

A specific qRT-PCR based method was developed for the detection of CSNV. The 70 bp long region of N (nucleo capsid) viral gene was targeted for detection. The method is found to be highly specific and can distinguish accurately the closely related tospoviruses i.e, TSWV and INSV (Boben et al., 2007).

A multiplex reverse transcription loop-mediated isothermal amplification (mRT-LAMP) assay was developed for the simultaneous detection of *Chrysanthemum stunt viroid* (CSVd) and CVB (Xing-Liang et al., 2014). The assay is done using the set of primers designed using the coat protein gene of CVB and the complete nucleotide sequence of CSVd. The mRT-LAMP could distinguish between CVB and CSVd in a single reaction more accurately than classical PCR. This method is highly reliable and sensitive (Xing-Liang et al., 2014).

The CVB is the most damage causing viruses infecting Chrysanthemum. For sensitive detection of CVB, the coat protein gene specific, nPCR-based method was developed. It could detect the virus up to 10^{-9} dilution of the cDNA (Zhiyong et al., 2017).

8.5   CARNATION

Carnation, an important cut-flower crop, is susceptible to infection by several viruses, out of which six viruses viz., *Carnation necrotic fleck virus* (CarNFV), *Carnation ringspot virus* (CRSV), *Carnation etched ring virus* (CERV), *Carnation mottle virus* (CarMV) and *Carnation vein mottle virus* (CVMoV) are reported to cause significant losses (Raikhy et al., 2006a). Most of these viruses could be detected serologically with DAS-ELISA using poly-clonal IgG (Singh et al., 2005). RT-PCR and IC-RT-PCR were used for sensitive detection of *Carnation mottle virus* (CarMV), CRSV and CVMoV (Raikhy et al., 2006b; Singh et al., 2005).

For the development of a sensitive ELISA-based diagnostic kit for CERV and CarMV, the coat protein and movement protein (MP) genes of respective viruses were amplified using the multiplex RT-PCR (mRT-PCR). The gene specific primers for the mRT-PCR were designed from the available sequence of an Indian isolate of CERV and CarMV. Amplified CP CERV and CP CarMV was cloned in expression vectors and the *in-vitro* expressed protein was purified for the production of antisera against the viruses for their effective diagnosis using ELISA (Raikhy et al., 2006b, 2007; Singh et al., 2005).

CarMV, CERV, CVMoV, CRSV, *Carnation Italian ring spot virus* and *Carnation latent virus* are the common viruses infecting carnation plants. For simultaneous detection of these viruses, the non-isotopic molecular hybridization technique was developed (Sanchez-Navarro et al., 1999). Specific digoxigenin (DIG)-labeled RNA transcripts were developed for hybridization with the corresponding virus. The extracts are applied onto the nylon membranes and it is hybridized with a mixture of the riboprobes of viruses. In a single assay, this technique could detect the mixture of viruses with five fold dilutions of extracts of carnation plants infected with the different viruses.

CRSV causes serious losses in carnation. CRSV can be detected by RT-PCR. However, to overcome the limitations of RT-PCR, a faster and less labor-intensive technique was developed, called reverse transcription-loop mediated isothermal amplification (RT-LAMP) assay (Yusuke et al., 2018).

8.6   ORCHID

Most orchid plants are susceptible to infection with *Odontoglossum ring spot virus* (ORSV), *Cymbidium mosaic virus* (CymMV) and *Cymbidium ring spot virus* (CyRSV) (Sharma et al., 2005a). CymMV and ORSV are the most predominant viruses infecting orchid (Sherpa et al., 2004). However, symptoms like leaf necrosis, flower color break, and necrotic spotting shown by CymMV were also displayed by some other viruses. RT-PCR can be used effectively to diagnose infection of CymMV from other viruses (Sherpa et al., 2003). For more accuracy and ease of detection of virus infection, the diagnostic methods based on coat protein gene were developed (Sherpa et al., 2006). The CP gene was targeted for detection of different isolates of a virus-infecting orchid because on aligning the sequences of nine different isolates of CymMV and ORSV, it was found that the CP gene was highly conserved. Targeting CP region of viruses prevents the occurrence of false negative results in detection process.

The coat protein gene is found to be highly conserved in viruses. Hence, viruses can also be accurately detected using the molecular mass of the coat protein gene of viruses. With liquid chromatography/mass spectrometry (LC/MS) and Matrix-Assisted Laser Desorption-Ionization (MALDI) Mass Spectrometry, the molecular weight of the coat protein can be obtained easily with high speed, accuracy, and sensitivity. This technique can be applied to all viruses with known coat protein molecular weights. The two most prevalent orchid viruses, namely CymMV and ORSV were detected simultaneously and rapidly using the LC/MS and MALDI (Stella et al., 2000). Automated analysis of LC/MS and MALDI allows rapid virus detection in large number of samples with simple sample preparation steps.

A fiber optic particle plasmon resonance (FOPPR) immunosensor has been developed for quick and sensitive detection of virus affected orchid plants. This technique uses the gold nanorod (AuNR) as labeling material for antibodies instead of gold nanospheres (AuNSs). Hence, it could overcome the problem of color interference of sample matrix due to use of AuNSs. The gold nanorods (AuNRs) are immobilized on the unclad fiber core surface coated with virus antibodies. This technique is rapid and sensitive (48 and 42 pg/mL) in detecting infection of the CymMV and ORSV. It can identify healthy and infected orchids in 10 min and can quantify the infection level also (Hsing-Ying et al., 2014).

8.7 IRIS

Bulbous iris is severely affected by *Iris mild mosaic virus* (IMMV) and *Iris severe mosaic virus* (ISMV) while rhizomatous irises are susceptible to infection of *Iris fulva mosaic virus* (IFMV). These viruses belong to the potyvirus group and are non-persistently transmitted by aphids. *Bean yellow mosaic virus* (BYMV) and *Narcissus latent virus* (NLV) have also been found on iris. CMV, BBWV, and *Tobacco ringspot virus* (TRSV) are less commonly seen.

Iris mild mosaic potyvirus (IMMV), *Iris severe mosaic potyvirus* (ISMV), and *Narcissus latent carlavirus* (NLV) can be tested in leaf and bulb using ELISA. However, with bulb material only the IMMV antiserum gave good results. Hence, detection through ELISA has some limitations (Hammond et al., 1985; Derks et al., 1986; Van Schadewijk et al., 1988). Therefore, to overcome these limitations the methods like RT-PCR and nPCR has been developed. The gene sequences of the RNA segment and the reference strain (TSWV, INSV, *Peanut yellow spot virus*, *Tomato chlorotic spot virus*), are used to design primers for diagnosis using RT-PCR. For nPCR, primers are designed based on the location of RT-PCR primer sets and amplification obtained on 100-fold dilution of the RT-PCR product (Yong-Gil and Jae-Young, 2014).

8.8 LILY

Lilies are the most important flower bulb crop in the world and their production and quality is severely affected by infection of *Lily symptomless virus* (LSV), *Lily mottle virus* (LMoV) and CMV. These viruses are transmitted in non-persistent manner by aphids and by vegetative propagation (Chinestra et al., 2010). These viruses can be indexed using ELISA and reverse transcription polymerase chain reaction (RT-PCR) at early and flowering stages. The outer and inner scales of bulbs and the leaves are used for indexing (Sharma et al., 2005b).

8.9   BULBOUS ORNAMENTALS

Potyviruses are commonly and widely seen infecting ornamentals such as *Anemone, Galtonia, Muscari, Ornithogalum, Allium, Stenomesson, and Veltheimia.* Differentiation and accurate identification of various potyviruses is needed for better management. Hence, molecular cloning techniques were developed (Pham et al., 2011). The coat protein amplicons of potyviruses infecting ornamentals were amplified using degenerate primers and sequenced. Using the sequence information, the specific primers were developed to diagnose these viruses using RT-PCR (Pham et al., 2011).

8.10   MANAGEMENT

Plant viruses are obligate intracellular parasites that replicate within a cellular environment but are inert entities outside of cells. They are transmitted from infected plants to the healthy ones by insect vectors or from one generation to next generation through vegetative propagation (Hull, 2002). To prevent the use of infected plant material, *in-vitro* techniques have been standardized to test and produce virus free ornamental crops. Farmers and NGO' are being trained for multiplication and maintenance of virus-free quality planting material. Thus, the virus free planting materials are produced and provided to the growers. Researchers are making efforts to understand the host-pathogen interaction and the role of viral suppressor molecules in disease management. Recently various approaches like transcriptomics, development of cDNA chips, biosensors, etc., are being studied to achieve rapid and sensitive diagnosis.

Nowadays, insect pests became common problems in greenhouse and field production systems associated with floricultural crops. Many insect pests cause direct harm to crops and a few of them cause indirect damage by transmitting pathogens, for example, fungi, and viruses. The primary insect vectors of floricultural crops are the western flower thrips (*Frankliniella occidentalis*), fungus gnats (*Bradysia* spp.), shore flies (*Scatella* spp.), green peach aphid (*Myzus persicae*), and sweet potato whitefly (*Bemisia tabaci*)

(Cloyd, 2016). Intensive pest management strategies need to be implemented against insect vectors, especially in regards to the western flower thrips, due to the direct transmission of INSV (Cloyd, 2009). The common strategy followed is extensive application of insecticides which may cause development of resistance in insect towards insecticide. Therefore, in addition to the use of insecticides, cultural, and sanitation practices should be considered and biological control agents (natural enemies) should be utilized. Integrated management strategies must be effective in sustaining populations at very low levels and reduce the potential disease transmission to floricultural crops. Because of the truly global nature of the ornamental industry we need to be vigilant with our virus detection and improve elimination techniques.

Many of the common landscape ornamentals are vegetatively propagated. Vegetative propagation allows maintenance of the systemic infections of viruses and phytoplasmas for further multiplication and transmission. Viruses and phytoplasmas are microscopic entities, not visible with the unaided eye, are capable of moving with plant sap and difficult to eliminate from planting materials once they get infected. Some viruses are highly seed-transmitted so the seeds from infected plants should be avoided. Therefore, viral diseases should be managed properly by eliminating the source and spread of the virus when possible and the elimination of the source is achieved by using virus-free planting materials.

The best management approach for ornamental nursery and floricultural crops is the use of stock propagated from virus-free or certified sources. The *in-vitro* regenerated and multiplied virus free clones of the original plants obtained from meristem tip, is another important way of getting planting material free from viruses, particularly from herbaceous plants.

Several kinds of therapies are being developed like use of micro shoot tip culture, meristem culture, embryogenic culture, and micro grafting to eliminate viruses and viroids from a plant. These treatments may also be combined with heat, cryogenic or chemotherapy (Milosevic et al., 2012). Micro shoot tip therapy is found to be the most reliable therapy and has been successfully utilized on a wide range of ornamentals and crop plants. The meristematic tissue has the potential to eliminate virus because the virus cannot keep up the pace with the fast dividing cells of the meristem tip and hence, gets eliminated in the time run. In addition, the microshoot tips are not directly connected to the vessels in the plant where many viruses are located, although many recent studies have shown that the mechanism of gene silencing is, in fact, the main reason behind this (Foster et al., 2002). The virus-free Dahlia, freesia, geranium, lily, etc., have been grown using meristem culture (Ram et al., 2005). According to Zaayen (1992), meristem culture applied to *in-vitro* cultures of lily, Alstroemeria, and Delphinium yielded virus free plantlets than the usual procedure in plants grown in the glasshouse. An increasing number of ornamental crops are being multiplied *in-vitro*. Zaayen (1992) also concluded that the application of meristem culture to make plants free from viruses is useful and cost effective.

8.11 CONCLUSION

Many viruses have extremely wide host range; they can spread quickly to various neighboring plantings of different crops through vectors. Vector transmission causes rapid and easy viral distribution that leads to the formation of stable natural foci of viral infection. The studies on genetic diversity, geographical distribution, and biological properties of ornamental crop viruses can contribute immensely to the development of effective molecular and serological diagnostics to prevent their further spread. The recent advancements in the field of plant pathology along with molecular biology, biotechnology, and bioinformatics have opened new pathways for the development of specific and sensitive procedures of diagnosis (Balodi et al., 2017).

   Molecular methods like PCR, quantitative PCR along with serological methods have high sensitivity and specificity and are less time consuming. However, at the same time, it is required that these methods should be standardized firstly, to make them specific and sensitive and secondly, for global acceptance of the method as standard protocol for quarantine purposes. The diagnostic tests provide information for epidemiological purposes, which ultimately can help to control the infection and to develop disease-free stock of flower. With continuous advances in molecular biology and immunology, scientists, and farmers will get the opportunity to improve the plant disease diagnosis. Efforts are constantly being made to improve the existing methods and produce better diagnostic kits to detect viruses in ornamental plants. For instance, diagnostic kits are being developed by the Department of Biotechnology of India's Ministry of Science and Technology to detect viruses in fruits, ornamentals, spices, and plantation crops. Diagnostic kits may be expensive but it is a fruitful investment and the expenditure on kit can be offset by the gains such as reduced crop losses and adoption of eco-friendly crop management measures. Both, public and private sectors in developing country like India should prioritize the development of diagnostic kits for the farmers.

KEYWORDS

- *Apple mosaic virus*
- *Bean yellow mosaic virus*
- *Carnation etched ring virus*
- *Chrysanthemum stem necrosis virus*
- fiber optic particle plasmon resonance
- serodiagnosis

REFERENCES

Aini, H. H., Omar, A. R., Hair-Bejo, M., & Aini, I., (2008). Comparison of SYBR green I, ELISA and conventional agarose gel-based PCR in the detection of infectious bursal disease virus. *Microbiol. Res., 163,* 556–563.

Babu, B., Brian, K., Washburn, T. S., Miller, S. H., Riddle, C. B., Knox, G. W., Ochoa-Corona, F. M., et al., (2017). A field-based detection method for Rose rosette virus using isothermal probe-based Reverse transcription-recombinase polymerase amplification assay. *J. Virol. Methods, 247,* 81–90.

Balodi, R., Bisht, S., Ghatak, A., & Rao, K. H., (2017). Plant disease diagnosis: Technological advancements and challenges. *Indian Phytopath., 70,* 275–281.

Boben, J., Mehle, N., Pirc, M., Mavric, I., & Ravnikar, M., (2007). New molecular diagnostic methods for detection of chrysanthemum stem necrosis virus (CSNV). *Acta Biologica Slovenica, 50,* 41–51.

Candresse, T., Macquaire, G., Monsion, M., & Dunez, J., (1988). Detection of chrysanthemum stunt viroid (CSV) using nick translated probes in a dot-blot hybridization assay. *J. Virol. Methods, 20,* 185–193.

Caruso, P., Gorris, M. T., Cambra, M., Palomo, J. L., Collar, J., & Lopez, M. M., (2002). Enrichment double-antibody sandwich indirect enzyme-linked immunosorbent assay that uses a specific monoclonal antibody for sensitive detection of *Ralstonia solanacearum* in asymptomatic potato tubers. *Appl. Environ. Microbiol., 68,* 3634–3638.

Chinestra, S. C., Facchinetti, C., Curvetto, N. R., & Marinangeli, P. A., (2010). Detection and frequency of lily viruses in Argentina. *Plant Dis., 94,* 1188–1194.

Choi, H., Jo, Y., Lian, S., Jo, K. M., Chu, H., Yoon, J. Y., Choi, S. K., Kim, K. H., & Cho, W. K., (2015). Comparative analysis of chrysanthemum transcriptome in response to three RNA viruses: Cucumber mosaic virus, Tomato spotted wilt virus and potato virus X. *Plant Molecular Biol., 88,* 233–248.

Clark, M. F., (1981). Immunosorbent assays in plant pathology. *Annu. Rev. Phytopathol., 19,* 83–106.

Cloyd, R. A., (2009). Western flower thrips (*Frankliniella occidentalis*) management on ornamental crops grown in greenhouses: Have we reached an impasse? *Pest Technol., 3,* 1–9.

Cloyd, R. A., (2016). In: McGovern, R., & Elmer, W., (eds.), *Insect Management for Disease Control in Florists Crops* (pp. 9–22). Springer, Cham.

Derks, A. F. L. M., & Hollinger, T. C., (1986). Similarities of and differences between potyviruses from bulbous and rhizomatous irises. *Acta Hortic., 177*, 555–562.

Dullemans, A. M., Verhoeven, J., Kormelink, R., & Vlugt, R. A. A., (2015). The complete nucleotide sequence of chrysanthemum stem necrosis virus. *Archives of Virol., 160*, 605–608.

Engelmann, J., & Hamacher, J., (2008). Plant virus diseases: Ornamental plants. *Enc. Virol.*, 207–229.

Fang, Y., & Ramasamy, R., (2015). Current and prospective methods for plant disease detection. *Biosensors, 5*, 537–561.

Foster, T. M., Lough, J. T., Emerson, S. J., Lee, R. H., Bowman, J. L., Forster, R. L. S., & Lucasa, W. J., (2002). A surveillance system regulates selective entry of RNA into the shoot apex. *The Plant Cell, 14*, 1497–1508.

Hammond, J., Lawson, R. H., & Hsu, H. T., (1985). Use of a monoclonal antibody reactive with several potyviruses for detection and identification in combination with virus-specific antisera. *Phytopathol., 75*, 1353.

Henson, J. M., & French, R., (1993). The polymerase chain reaction and plant disease diagnosis. *Ann. Rev. Phytopathol., 31*, 81–109.

Hsing-Ying, L., Chen-Han, H., Sin-Hong, L., I-Ting, K., & Lai-Kwan, C., (2014). Direct detection of orchid viruses using nanorod-based fiber optic particle plasmon resonance immunosensor. *Biosensors and Bioelectronics., 51*, 371–378.

Hull, R., (2002). *Matthews' Plant Virology.* Academic Press, San Diego, California.

Jordan, R., Guaragna, M. A., & Hammond, J., (2018). Development of polyclonal and monoclonal antibodies to rose rosette virus nucleoprotein. *Acta Horti., 193*, 77–82.

Katoch, M., Addin, M. Z., Ram, R., & Zaidi, A. A., (2003). An overview of diagnostics for viruses infecting gladiolus. *Crop Protection, 22*, 153–156.

Khurana, P. S. M., (2006). Detection of plant pathogens: Development and applications. *Indian Phytopath., 59*, 1–15.

Lau, H. Y., & Botella, J. R., (2017). Advanced DNA-based point-of-care diagnostic methods for plant diseases detection. *Front. Plant Sci., 8.*

Lockhart, B., Zlesak, D., & Fetzer, J., (2011). Identification and partial characterization of six new viruses of cultivated roses in the USA. *Acta Horti., 901*, 139–147.

Matsuura, S., Kubota, K., & Okuda, M., (2007). First report of chrysanthemum stem necrosis virus on chrysanthemums in Japan. *Plant Dis., 91*, 468.

Milosevic, S., Cingel, A., Jevremovic, S., Stankovic, I., Bulajic, A., Krstic, B., & Subotic, A., (2012). Virus elimination from ornamental plants using in vitro culture techniques. *Pestic. Phytomed. (Belgrade), 27*, 203–211.

Mitrofanova, I. V., Zakubanskiy, A. V., & Mitrofanova, O. V., (2018). Viruses infecting main ornamental plants: An overview Ornam. *Hortic., 24*, 2.

Pham, K. T. K., De Kock, M. J. D., Lemmers, M. E. C., & Derks, A. F. L. M., (2011). Molecular identification of potyviruses infecting bulbous ornamentals by the analysis of coat protein (CP) sequences. In: Derks, et al., (eds.), *Virus Diseases of Ornamental Plants* (p. 901). Proceed. XII. Acta Hort.

Raikhy, G., Hallan, V., Kulshrestha, S., & Zaidi, A., (2007). Polyclonal antibodies to the coat protein of carnation etched ring virus expressed in bacterial system: Production and use in immunodiagnosis. *J. Phytopathol., 155*, 616–622.

Raikhy, G., Hallan, V., Kulshrestha, S., Ram, R., & Zaidi, A., (2006a). Multiplex PCR and genome analysis of *Carnation mottle virus* Indian isolate. *Current Science, 90*, 74–82.

Raikhy, G., Hallan, V., Kulshrestha, S., Sharma, M. L., Verma, N., Ram, R., & Zaidi, A., (2006b). Detection of *carnation ringspot* and *carnation vein mottle viruses* in carnation cultivars in India. *Acta Horti., 722*, 247–258.

Raj, S., Srivastava, A., Chandra, G., & Singh, B., (2002). Characterization of cucumber mosaic virus isolate infecting gladiolus cultivars and comparative evaluation of serological and molecular methods for sensitive diagnosis. *Current Science, 83*, 1132–1137.

Ram, R., Verma, N., Singh, A. K., Singh, L., Hallan, V., & Zaidi, A. A., (2005). Indexing and production of virus-free chrysanthemums. *Biologia Plantarum, 49*, 149–152.

Sanchez-Navarro, J. A., Canizares, M. C., Cano, E. A., & Pallas, V., (1999). Simultaneous detection of five carnation viruses by non-isotopic molecular hybridization. *J. Virol. Methods, 82*, 167–175.

Sharma, A., Mahinghara, B. K., Singh, A. K., & Zaidi, A., (2005b). Identification, detection and frequency of lily viruses in Northern India. *Scientia Horti., 106*, 213–227.

Sharma, A., Zaidi, A. A., Sood, A., Sharma, A., Vij, S. P., Pathak, P., & Ahuja, P. S., (2005a). Virus analysis of indigenous orchid germplasm collection. *J. Orchid Soc. India, 19*, 41–46.

Sherpa, A. R., Hallan, V., & Zaidi, A. A., (2004). Cloning and sequencing of coat protein gene of an Indian *odontoglossum ringspot virus* isolate. *Acta Virol., 48*, 267–269.

Sherpa, A. R., Hallan, V., Pathak, P., & Zaidi, A. A., (2006). Coat protein gene of *cymbidium mosaic virus*: Characterization of geographical isolates from India. *J. Phytopathol., 154*, 275–280.

Sherpa, A. R., Hallan, V., Ram, R., Vij, S. P., Pathak, P., Garg, I. D., & Zaidi, A. A., (2003). First report of *cymbidium mosaic virus* on cymbidiums in India. *Plant Pathol., 52*, 788.

Singh, H. P., Hallan, V., Raikhy, G., Kulshrestha, S., Sharma, M. L., Ram, R., Garg, I. D., & Zaidi, A. A., (2005). Characterization of an Indian isolate of Carnation mottle virus infecting carnations. *Current Science, 88*, 594–601.

Song, A., You, Y., Chen, F., Li, P., Jiand, J., & Chen, S., (2012). A multiplex RT-PCR for rapid and simultaneous detection of viruses and viroids in chrysanthemum. *Letters in Applied Microbiol., 56*, 8–13.

Stella, W., Sek-Man, W. R., & Manjunatha, K., (2000). Rapid simultaneous detection of two orchid viruses using LC- and/or MALDI-mass spectrometry. *J. Virol. Methods, 85*, 93–99.

Thomas, B. J., (1980). The detection by serological methods of viruses infecting the rose. *Ann. Appl. Biol., 94*, 91–101.

Van, S. A. R., Derks, A. F. L. M., Lemmers, M. E. C., & Hollinger, T. C., (1988). Detection of iris mild mosaic virus in bulbs and leaves of bulbous iris by ELISA. *Acta Horti., 234*, 199–206.

Vazquez-Iglesias, I., Adams, I. P., Hodgetts, J., Fowkes, A., Forde, S., Ward, R., Buxton-Kirk, A., et al., (2019). High throughput sequencing and RT-qPCR assay reveal the presence of rose cryptic virus-1 in the United Kingdom. *J. Plant Pathol., 101*, 1171–1175.

Verhoeven, J. T. J., Roenhorst, J. W., Cortes, I., & Peters, D., (1996). Detection of a novel tospovirus in chrysanthemum. *Acta Horti., 432*, 44–51.

Verma, N., Sharma, A., Ram, R., Hallal, V., Zaidi, A., & Garg, D., (2003). Detection, identification and incidence of chrysanthemum B carlavirus in chrysanthemum in India. *Crop Protection, 22*, 425–429.

Webster, C. G., Wylie, S. J., & Jones, G. K., (2004). Diagnosis of plant viral pathogens. *Current Science, 86*, 12.

Wu, P. R., Chien, W. C., Okuda, M., Takeshita, M., Yeh, S. D., Wang, Y. C., & Chen, T. C., (2015). Genetic and serological characterization of chrysanthemum stem necrosis virus, a member of the genus tospovirus. *Archives of Virol., 160*, 529–536.

Xing-liang, L., Xi-ting, Z., Imtiaz, M., Bei-Bei, G., & Bo, H., (2014). Multiplex reverse transcription loop-mediated isothermal amplification for the simultaneous detection of CVB and CSVd in chrysanthemum. *J. Virol. Methods, 210*, 26–31.

Yong-Gil, S., & Jae-Young, R., (2014). Development of a PCR diagnostic system for iris yellow spot tospovirus in quarantine. *Plant Pathol. J., 30*, 440–444.

Yusuke, S., Moritsugu, O., Kenji, K., Takashi, H., Nana, S., Shuichi, U., Kenji, F., & Yuji, F., (2018). Use of reverse transcription loop-mediated isothermal amplification assay for detection of carnation ringspot virus in dianthus. *Res. Bull. Pl. Prot. Japan, 54*, 49–54.

Zaayen, A., Eijk, C., & Versluijs, J., (1992). Production of high quality, healthy ornamental crops through meristem culture. *Acta Botanica Neerlandica, 41*, 425–433.

Zaidi, A., Hallan, V., Raikhy, G., Singh, A., & Raja, R., (2011). Viruses of ornamental plants in India-current research status and future prospects. *Acta Horti., 901*, 7.

Zhao, X., Liu, X., Ge, B., Li, M., & Hong, B., (2015). A multiplex RT-PCR for simultaneous detection and identification of five viruses and two viroids infecting chrysanthemum. *Archives Virol., 160*, 1145–1152.

Zhiyong, G., Dan, W., Aiping, S., Fadi, C., Sumei, C., & Weimin, F., (2017). A highly sensitive method for the detection of chrysanthemum virus B. *Electronic J. Biotechnol., 26*, 64–68.

CHAPTER 9

Global Status on Diagnosis, Geographical Distribution, and Integrated Disease Management Strategies for Major Viruses Infecting Cucurbitaceous Crops

SOMNATH K. HOLKAR,[1] ATUL KUMAR,[1,2] and R. K. JAIN[3]

[1]*ICAR-Indian Institute of Sugarcane Research, Biological Control Center, Pravaranagar, Ahmednagar, Maharashtra–413712, India, E-mail: somnathbhu@gmail.com (S. K. Holkar)*

[2]*Amity Institute of Biotechnology, Amity University, Lucknow Campus, Lucknow–226028, Uttar Pradesh, India*

[3]*Division of Plant Pathology, ICAR-Indian Agricultural Research Institute, Pusa Campus, New Delhi–110012, India*

ABSTRACT

Biotic stresses caused due to fungi, bacteria, phytoplasma, viruses, viroids, nematodes are the major constraints in the cucurbits production globally. Among these pathogens, viruses are one of the major constraints in reducing quality and yield contributing parameters of cucurbits. Cucurbits are known to infect by >60 viral diseases which are caused by the members belonging to various genera *viz.*, begomo-, carmo-, como-, crini-, cucumo-, polero-, poty-, tobamo-, and orthotospoviruses. Viruses cause up to 100% crop loss in cucurbits. Due to severity of viral diseases, marketable quality deteriorates and growers facing remarkable financial loss worldwide. Therefore, accurate diagnosis and integrated disease management (IDM) strategies including cultural, chemical, and biological control of insect-vectors is essential to cope up the significant economic losses. In India, during the last couple of decades advancement in the virus diagnostics played an important role in exploring the possibility of associated viruses in newer cucurbits which were recorded from elsewhere. Due to the changes in the climatic conditions, outbreak of major insect-vectors, non-availability of the main hosts to these insect-vectors leads their widespread occurrence in new cucurbitaceous hosts and distribution across regions and countries of the world. Depsite the various efforts in identification of the resistant sources against different plant viruses infecting cucurbits, virus-resistance breeding needs further strengthening in exploitation of these resistant sources in the conventional and molecular breeding approaches. CRISPR/Cas system mediated resistance has to be developed in different cucurbits against the reported viruses.

9.1   INTRODUCTION

Cucurbits belonging to the family Cucurbitaceae which include >118 genera and 825 species. Worldwide, the major cucurbits cultivated including cucumber (*Cucumis sativus*), bottle gourd (*Lagenaria siceraria*), muskmelon (*Cucumis melo* L.), squash (*Cucurbita pepo*), bitter gourd (*Momordica charantia*), pumpkin (*Cucurbita moschata*), ridge gourd [*Luffa acutangula* (Roxb.) L.], snapmelon (*Cucumis melo* var. *momordica*), longmelon (*Cucumis melo* var. *utilissimus*), roundmelon (*Praecitrullus fistulosus* pang. Syn. *Citrullus vulgaris* var. *fistulolu*), snake gourd (*Trichosanthes cucumerina*), ivy gourd (*Coccinia indica*), pointed gourd (*Trichosanthes dioica*), ash gourd (*Beninasca hispida*), sponge gourd (*Luffa cylindrica*) and watermelon [*Citrulus lanatus* (Thunb.) Mansf]. Cucurbits are consumed in various forms, i.e., salad (cucumber, gherkins, long melon), sweet (ash gourd, pointed gourd), pickles (gherkins), and deserts (melons).

In India, vegetable production during 2017–2018 was recorded of about 184.39 million tons from the total area of about 10.259 million ha (Anonymous, 2019). In India, muskmelon, watermelon, bitter gourd, bottle gourd, cucumber, and pumpkins are the major cucurbits grown and consumed. During 2015–2016, bottle gourd alone was cultivated in an area of about 1.49 million ha with total production of 24.58 million tons. This was followed by watermelon 0.95 million ha with a production of 23.25 million tons. Bitter gourd, cucumber, pumpkin, and muskmelon cultivated on 0.93, 0.71, 0.68, and 0.45 million ha, respectively (NHB, 2018).

Globally, biotic, and abiotic stresses are the major constraints in the cucurbit production. Therefore, major emphasis on cucurbit research has been given on the improvement in the productivity by developing the resistant varieties against biotic stresses and quality parameters. Cucurbits are known to be affected by >200 diseases caused by fungi, bacteria, viruses, phytoplasma, viroids, and nematodes worldwide and cause economic losses (Zitter et al., 1996). Besides these pathogens, cucurbits are known to be susceptible for >60 different viruses (Provvidenti, 1996; Zitter et al., 1996; Lecoq et al., 2001; Lecoq and Desbiez, 2012) and more than 39 viruses are the major limiting factors affecting commercial cucurbit production worldwide (Provvidenti, 1996; Lecoq et al., 1998; Locoq, 2003). Of which 32 plant viruses caused major economic impact on production and quality parameters of cucurbits including, cucumber (*Cucumis sativus*), melon (*Cucumis melo*), pumpkins (*Cucurbita moschata*), watermelon (*Citrulus lanatus*) and zucchini squash (*Cucurbita pepo*) (Zitter et al., 1996). Plant viruses induce different symptoms in cucurbitaceous crops, of which mosaic is the first and foremost symptoms induced by frequently occurring and economically important viruses which have been reported from different regions of the world (Zitter et al., 1996; Lecoq et al., 2001). Viruses inducing yellowing symptoms in cucurbitaceous crops are transmitted either by aphids (Lecoq et al., 1992) and or by whiteflies (Wisler et al., 1998). Production of cucurbitaceous crops is severely affected by single or multiple infections of begomo-, como-, crini-, cucumo-, polero-, poty-, tobamo-, and orthotospoviruses (Table 9.1). As like other cucurbits, watermelon production is also severely affected by single or multiple infections of the various plant viruses as mentioned in Table 9.2.

In this chapter, the most important five groups of viruses belonging to begomo-, crini-, tobamo-, poty-, and orthotospoviruses have been discussed in detail on impact, species, transmission, diagnosis, distribution, host range, and integrated disease management (IDM) strategies.

9.1.1 BEGOMOVIRUSES (FAMILY: GEMINIVIRIDAE)

Members of this family (type member: *Bean golden yellow mosaic virus*) are the major threat to the global food security especially in the countries of the tropical and sub-tropical regions (Rybicki and Pietersen, 1999; Varma and Malathi, 2003; Rey et al., 2012). Up to 100% yield loss due to geminiviruses has been estimated (Dasgupta et al., 2003). *Begomovirus* is the largest genus among the nine genera of the family *Geminiviridae*. Currently, there are >350 species in this family of which approximately 320 virus species reported so far are transmitted by whiteflies (*Bemisia tabaci*) cryptic species complex (Brown et al., 2015). Begomviruses infecting cucurbits have been reported to have a bipartite genome (Faquet et al., 2005) and are phloem limited. Globally, cucurbits are known to infect by >10 begomovirus species including, *Chayote yellow mosaic virus* (ChaYMV) (Leke et al., 2019),

TABLE 9.1 List of Major Viruses Recorded on Cucurbitaceous Crops and Their Distribution

Virus Group	Virus Species	Distribution Profile	Vectored by	References
Begomovirus	*Beet curly top virus (BCTV)* Squash leaf curl virus (SLCV)	USA Egypt, Israel, Jordan, Lebanon, Saudi Arabia, Taiwan, USA	Leaf hopper Whitefly	Chen and Gilbertson (2009) Jawhari et al. (2012) (unpublished); Lapidot et al. (2014); Isakeit et al. (1994); Al-Musa et al. (2008)
	Watermelon chlorotic stunt virus (WmCSV)	Israel, Iran, Jordan, Lebanon, Saudi Arabia, Sudan, Oman, Palestine, Yemen	Grafting, Whitefly	Ali et al. (2012); Ali-Shtayeh et al. (2014); Al-Musa et al. (2011); Khan et al. (2012); Kheyr-Pour et al. (2000); Alhudaib et al. (2018); Jones et al. (1988); Samsatly et al. (2012)
	Squash leaf curl China virus (SLCCHV)	Australia, Malaysia, India, Pakistan, China, Philippines,	Whitefly	Maina et al. (2017); Saritha et al. (2011);
	Squash leaf curl Philippines virus (SLCPHV)	Vietnam	Whitefly	Mohammad Riyaz et al. (2013); Tahir et al. (2010); Sawangjit (2009); Wu et al. (2020); Kon et al. (2003)
	Squash leaf curl Yunnan virus (SLCuYV)	Taiwan China, Thailand	Mechanical, Whitefly Whitefly	(2008); Singh et al. (2008); Ito et al. (2008) Liao et al. (2007); Tsai et al. (2011)
	Squash yellow leaf curl virus (SYLCV)	Oman	Whitefly	Xie et al. (2003, 2006); Butnut et al. (2019)
		USA	Whitefly	Zouba et al. (1998)
	Melon chlorotic leaf curl virus (MCLCV)	France Benin, Nigeria, Togo	Whitefly Whitefly	Idris et al. (2007); Brown et al. (2011)
	Melon chlorotic mosaic virus (MCMV)	USA	Whitefly	Romay et al. (2019)
		USA	Whitefly	Leke et al. (2019)
	Chayote yellow mosaic virus (ChYMV)	USA India, Spain, Thailand, Pakistan, Italy, Tunisia, Indonesia, Central Java, Iran, Morocco	Whitefly	Zouba et al. (1998) Brown et al. (2001); Cohen et al. (1983); Guzman et al. (2000)
	Cucurbit leaf curl virus (CuLCV) or *Cucurbit leaf crumple virus* (CuLcrV)	India, Iran, Pakistan		Ito et al. (2008); Sohrab et al. (2003); Juarez et al. (2014); Tahir et al. (2005); Panno et al. (2016); Tiwari et al. (2010); Mnari-Hattab et al. (2015);
	Tomato leaf curl New			Mizutani et al. (2011); Phaneendra et al. (2012);

TABLE 9.1 *(Continued)*

Virus Group	Virus Species	Distribution Profile	Vectored by	References
	Delhi virus (ToLCNDV) *Tomato leaf curl Palampur Virus* (ToLCPMV)			Yazdani-Khameneh et al. (2016); Sifres et al. (2018) Namrata et al. (2010); Heydarnejad et al. (2009); Ali et al. (2010)
Carmovirus	*Cucumber leaf spot virus* (CLSV)	Bulgaria, Canada, Germany, Greece, Great Britain, Iran, Israel, Japan, Jordan, Poland, Saudi Arabia, Spain, UK, USA	*Olpidium bornovanus* zoospores	Bananej et al. (2014); Brunt et al. (1996); Campbell et al. (1991); Ghoshal_et al. (2014); Kostova et al. (2001); Miller et al. (1997); Pospieszny and Cajza (2004); Reade et al. (2003); Segundo et al. (2001)
	Cucumber soil-borne virus (CuSBV)	Germany	Fungi, mechanical	Koening et al. (1983)
	Melon necrotic spot virus (MNSV)	France, Israel, Japan, USA, Spain	Fungi (*Olpidium* spp.), seeds, contact	Barkan et al. (2006) (unpublished); Diaz et al. (2004); Mochizuki et al. (2009); Ohki et al. (2010); Riviere et al. (1989); Yakoubi et al. (2008)
Comovirus	*Squash mosaic virus* (SqMV)	China, Czech Republic, Japan, Korea, USA	beetles, seeds, contact	Han et al. (2002); Hu et al. (2009); Svoboda and Leisova-Svobodova (2011)
Crinivirus	*Beet pseudo-yellows virus* (BPYV)	Crete, Greece, Italy, Japan, New Zealand, South Africa, UK, USA	Whitefly	Rubio et al. (2001); Boubourakas et al. (2006); Ibaba et al. (2015); Berdiales et al. (1999); Livieratos et al. (1998); Ramirez et al. (2008); Tzanetakis and Martin (2004); Yamashita et al. (1979)
	Cucurbit yellow stunting disorder virus (CYSDV) *Cucurbit chlorotic yellows virus* (CCYV)	Egypt, France, Italy, Israel, Japan, Jordan, Lebanon, Morocco, Portugal, Saudi Arabia, Spain, Syria, UAE, Mexico, Turkey, USA, UK Greece, Cyprus	Whitefly	Abou-Jawdah et al. (2000); Decoin (2003); Manglli et al. (2016); Desbiez et al. (2000); Rubio et al. (2001); Gyoutoku et al. (2009); Hassan and Duffus (1991); Hourani and Abou-Jawdah (2003); Kao et al. (2000); Louro et al. (2000); Rubio et al. (1999); Wisler et al. (1998); Coffin and Coutts (1990) Orfanidou et al. (2014, 2019)

TABLE 9.1 *(Continued)*

Virus Group	Virus Species	Distribution Profile	Vectored by	References
	Lettuce infectious yellows virus (LIYV)	Japan, Mexico, USA	Whitefly	Duffus et al. (1986); Hartono et al. (2003); Brown and Nelson (1986)
Cucumovirus	Cucumber mosaic virus (CMV)	Austria, China, India, Iran, South Korea, Malaysia, Japan, Spain	Aphid	Balazs et al. (2011) (unpublished); Kim et al. (2010) (unpublished); Kumari et al. (2012) (unpublished); Lee et al. (2011) (unpublished); Lopez et al. (2007); Milojevic et al. (2012) (2013); Ren et al. (2007) (unpublished); Sohn et al. (2007) (unpublished); Sokhandan et al. (2007) (unpublished); Sun (2008) (unpublished); Suzuki et al. (2011) (unpublished)
Ipomovirus	Cucumber vein yellowing virus (CVYV)	France, Iran, Italy, Jordan, Lebanon, Spain, USA	Whitefly	Abrahamian et al. (2013); Galipienso et al. (2012); Janssen et al. (2005, 2007); Kianfar et al. (2013) (unpublished); Louro et al. (2004); Yakoubi et al. (2008)
Necrovirus	Tobacco necrosis virus (TNV)	Switzerland	Fungi	Molders et al. (1996)
Nepovirus	Artichoke yellow ringspot virus (AYRSV)	Turkey	Seed, mechanical, pollen	Paylan et al. (2013)
	Tobacco ringspot virus (TRSV)	USA	Xiphinema spp.	Abdalla et al. (2011)
	Tomato black ring virus (TBRV)	Poland	Xiphinema spp.	Pospieszny and Borodynko (2005)
	Tomato ringspot virus (ToRSV)	USA	Xiphinema spp.	Converse and Ramsdell (1982)
Ourmiavirus	Melon Ourmia virus (OuMV)	Iran	Whitefly	Gholamalizadeh et al. (2008); Lisa et al. (1988)

TABLE 9.1 *(Continued)*

Virus Group	Virus Species	Distribution Profile	Vectored by	References
Polerovirus	*Cucurbit aphid borne yellows virus* (CABYV)	China, Iran, India, Taiwan, Tunisia, Germany, Korea	Aphids (P)	Abkhoo (2012); Knierim et al. (2010) (unpublished); Rao et al. (2013); Xiang et al. (2008); : Suvedita et al. (2017); Sangeetha et al. (2019); Kumar et al. (2020); Menzel et al. (2020); Choi et al. (2015)
Potyvirus	*Algerian watermelon mosaic virus* (AWMV)	Algeria	Aphids	Yakoubi et al. (2008)
	Cucurbit vein-banding virus (CVBV)	USA	Aphids	Providenti and Gonsalves (1984)
	Melon vein-banding mosaic virus (MVBMV)	Taiwan	Aphids	Chen et al. (2012) (unpublished)
	Papaya ringspot virus-W (PRSV-W)	Brazil, Cuba, China, India, Iran, Korea, Taiwan, USA	Sap, aphids	Abdalla and Ali (2012); Mederos et al. (2017); Ali et al. (2004); Basavaraj and Jain (2013) (unpublished); Della-Vecchia et al. (2003); Gu et al. (2006) (unpublished); Liu et al. (2002); Mangrauthia et al. (2008); Naiimifar et al. (2014); Sohn et al. (2007) (unpublished)
	Telfairia mosaic virus (TeMV)	Nigeria	Aphids	Anno-Nyako (1988); Atiri and Varma (1983)
	Moroccan watermelon mosaic virus (MWMV)	Abkhazia, Albania, Algeria, Armenia, Andorra, France, Greece, Iran, Kenya, Nigeria, South Africa, Sudan, USA, Italy, Tunisia, Congo	Aphids	Miras et al. (2019); Lecoq et al. (2001); Malandraki et al. (2014); Bananej et al. (2018); Read et al. (2020); Ibaba et al. (2016b); Owolabi et al. (2012); Roggero et al. (1998); Yakoubi et al. (2008); Arocha et al. (2008)
	Watermelon mosaic virus 2 (WMV-2)	China, Egypt, France, India, Iran, Italy, Japan, Pakistan, Serbia, South Africa, Slovakia, South Korea, Turkey, USA	Mechanical, aphids (NP)	Ali et al. (2004); Desbiez et al. (2009); Finetti-Sialer et al. (2012); Gara et al. (1997); Ghasemzade et al. (2012); Glasa et al. (2012); Laney et al. (2012); Sharifi et al. (2008)

TABLE 9.1   (Continued)

Virus Group	Virus Species	Distribution Profile	Vectored by	References
	Zucchini shoestring virus (ZSSV) Zucchini yellow fleck virus (ZYFV)	South Africa, Zimbabwe Italy, France	Aphids Aphids	Ibaba et al. (2016)a; Karavina et al. (2020) Desbiez et al. (2007); Tomassoli et al. (2010)
	Zucchini yellow mosaic virus (ZYMV)	Brazil, China, Iran, France, Poland, Portugal, Serbia, Slovakia, USA, Venezuela	Aphids	Boiteux et al. (2013); Cardoso et al. (2004); Dunham et al. (2014); Lecoq et al. (2014); Massumi et al. (2009) (unpublished); Novakova et al. (2014); Pospieszny et al. (2009); Romay et al. (2014); Vucurovic et al. (2014) (unpublished)
Rhabdovirus	*Cucumber toad-skin virus* (CTSV)	France	NA	Katis et al. (1995); Lecoq (1983)
Tobamovirus	*Cucumber green mottle mosaic virus* (CGMMV) *Kyuri green mottle mosaic virus* (KGMMV) *Zucchini green mottle mosaic virus* (ZGMMV)	China, Greece, India, Israel, Japan, Spain, Ukraine, USA Japan, Korea, Indonesia Korea, China	Sap, seed, aphids, fungus Sap, Seed Sap	Boubourakas et al. (2004); Jabeen et al. (2004) (unpublished); Janssen et al. (2005); Liu et al. (2009); Rudnieva et al. (2009); Tan et al. (2000); Tian et al. (2014) Tan et al. (2000); Ryu et al. (2000); Daryono et al. (2005) Ryu et al. (2000); Li et al. (2018)
Tombusvirus	*Cucumber necrosis virus* (CNV)	New Zealand, USA	*Olpidium bornovanu*	Harris et al. (2006) (unpublished); Obermeier et al. (2005) (unpublished)
Tospovirus	*Groundnut bud necrosis virus* (GBNV)	India	Sap, Thrips	Kunkalikar et al. (2011); Nagendran et al. (2018); Krishnan et al. (2019)
	Melon severe mosaic virus (MeSMV)	Italy, Mexico	Sap, Thrips	Ciuffo et al. (2009, 2017)
	Watermelon silver mott e virus (WSMoV)	Japan, Thailand, Taiwan, China, Ecuador	Sap, Thrips	Kato et al. (2000a); Chiemsombat et al. (2008); Chen et al. (2008); Gu et al. (2012); Quito-Avila et al. (2014)

TABLE 9.1 *(Continued)*

Virus Group	Virus Species	Distribution Profile	Vectored by	References
	Watermelon bud necrosis virus (WBNV)	China, Japan, Russia, Taiwan, Thailand India	Sap, thrips (NC; PC) Sap, Thrips	Chiemsombat and Choi (2007) (unpublished); Mnari-Hattab et al. (2009); Okuda et al. (2001); Prihodko et al. (2011) (unpublished); Seepiban et al. (2009) (unpublished); Yeh et al. (1995) Jain et al. (1998)
	Tomato spotted wilt virus (TSWV) *Zucchini lethal chlorosis virus* (ZLCV)	USA Argentina, Brazil	Sap, Thrips (CP) Sap, Thrips	Bhat et al. (1999) Pozzi et al. (2019); Yuki et al. (2000); Camelo-Garcia et al. (2014)
Tymovirus	*Chayote mosaic virus* (ChMV)	Germany, Costa Rica	Mechanical	Bernal et al. (2000)
	Melon rugose mosaic virus (MRMV)	Sudan, Yemen	Seed	Jones et al. (1986); Mahgoub et al. (1997)
Unassigned	*Squash necrosis virus* (SqNV) *Cucumber vein yellowing virus* (CVYV)	Switzerland Spain, Jordan, Israel, Turkey, Tunisia, Portugal, Iran, France, Lebanon, Sudan	Seed Whitefly	Koenraadt and Remeeus (2014) Cuadrado et al. (2001); Mansour and Al-Musa (1993); Al-Musa et al. (1985); Cohen and Nitzany (1960); Yilmaz et al. (1989); Yakoubi et al. (2007); Louro et al. (2004); Bananej et al. (2006); Lecoq et al. (2007); Abrahamian et al. (2013); Desbiez et al. (2019)

Note: Unpublished references were obtained from the NCBI Genbank accession numbers and are not presented in the reference list.

TABLE 9.2 Viruses Recorded on Watermelon (*Citrullus* spp.) and Their Distribution

Virus group	Virus Species	Distribution Profile	Vectored by	References
Begomovirus	*Melon chlorotic leaf curl virus* (MCLCuV)	Mexico, USA	Sap, Whitefly (NP)	Brown et al. (2001); Hernandez-Zepeda et al. (2007)
	Watermelon chlorotic stunt virus (WmCSV)	Jordan, Saudi Arab, Oman, Palestine,	Grafting, Whitefly (CNP)	Ali et al. (2012); Al-Muska et al. (2011); Ali-Shtayeh et al. (2012) (2014); Khan et al. (2012); Saleh et al. (2014)
Comovirus	*Squash mosaic virus* (SqMV)	USA	Sap, beetle, seed	Ali et al. (2012)
Cucumovirus	*Cucumber mosaic virus* (CMV)	China, Serbia, South Korea	Sap, aphids (NP), seed	Milojevic et al. (2012) (2013); Ren et al. (2007) (unpublished); Sohn et al. (2007) (unpublished)
Polerovirus	*Cucurbit aphid borne yellows virus* (CABYV)	China, Iran, Taiwan, Tunisia	Aphids (P)	Abkhoo (2012); Knierim et al. (2010) (unpublished); Mnari-Hattab et al. (2009); Xiang et al. (2008)
Potyvirus	*Papaya ring spot virus-W* (PRSV-W)	Brazil, China, India, Korea, USA	Sap, aphids	Abdalla and Ali (2012); Ali et al. (2012); Basavaraj and Jain (2013) (unpublished); Della-Vecchia et al. (2002); Gu et al. (2006) (unpublished); Sohn et al. (2007) (unpublished)
	Watermelon mosaic virus-2 (WMV-2)	China, Egypt, France, India, Iran, Italy, Japan, Serbia, South Africa, South Korea, Turkey, USA	Mechanical, aphids (NP)	Choi et al. (2007) (unpublished); Desbiez et al. (2009); Finetti-Sialer et al. (2012); Gara et al. (1997); Ghasemzadeh et al. (2008); Laney et al. (2012); Salem et al. (2007); Trkulja et al. (2013); Verma et al. (2014) (unpublished); Wu et al. (2006) (unpublished)
	Zucchini yellow mosaic virus (ZYMV)	China, Italy, Serbia, Syria, Taiwan, USA	Sap, aphids (NP)	Ali et al. (2013); Trkulja et al. (2013); Turina et al. (2012); Tsai et al. (2010); Usher et al. (2012); Zhao et al. (2003)
Tobamovirus	*Cucumber green mottle mosaic virus* (CGMMV)	China, Greece, Israel Japan, USA	Sap, seed, aphids, fungus	Boubourakas et al. (2004); Liu et al. (2009); Reingold et al. (2013); Tan et al. (2000); Tian et al. (2014)

TABLE 9.2 *(Continued)*

Virus group	Virus Species	Distribution Profile	Vectored by	References
Tospovirus	Melon yellow spot virus (MYSV)	China, Ecuador, Japan, Taiwan, Thailand	Sap, thrips (CP)	Chen et al. (2012); Gu et al. (2006) (unpublished); Gu et al. (2012); Kato et al. (2000); Mound (2011); Quito-Avila et al. (2014) (unpublished); Seepiban et al. (2009) (unpublished)
	Groundnut bud necrosis virus (GBNV)	India	Sap, Thrips	Kunkalikar et al. (2011)
	Tomato spotted wilt virus (TSWV)	USA	Sap, thrips (CP)	Bhat et al. (1999)
	Watermelon bud necrosis virus (WBNV)	India	Sap, thrips (CP)	Jain et al. (1998); Mandal et al. (2012); Pappu et al. (2009)
	Watermelon silver mottle virus (WSMoV)	China, Japan, Russia, Taiwan, Thailand	Sap, thrips (NC and PC)	Chiemsombat and Maneechoat (2010); Mnari-Hattab et al. (2009); Okuda et al. (2001); Prihodko et al. (2011) (unpublished); Rao et al. (2013); Seepiban et al. (2009) (unpublished); Yeh et al. (1995)

Note: CP: circulative propagative; CNP: circulative non-propagative; PC: propagative circulative; NC: non-circulative; P: persistent; NP: non-persistent.

Note: Unpublished references were obtained from the NCBI Genbank accession numbers and are not presented in the reference list.

Cucurbit leaf crumple virus (CuLCrV; also known as *Cucurbit leaf curl virus* (Brown et al., 2000, 2002; Guzman et al., 2000); *Luffa yellow mosaic virus* (LYMV; Revill et al., 2003), *Melon chlorotic leaf curl virus* (MCLuCV; Brown et al., 2001), *Squash leaf curl China virus* (SLCCV), *Squash leaf curl Yunnan virus* (SLCYV; Xei et al., 2003), *Squash leaf curl virus* (SLCV; Brown et al., 1986, 1989), *Squash leaf curl mild virus* (SLCMV; Brown et al., 2002), *Squash leaf curl Philippines virus* (SLCPPV; Kon et al., 2003), and *Watermelon chlorotic stunt virus* (WmCSV; Hanley-Bowdoin et al., 1999; Kheyr-Pour et al., 2000) (Table 9.1).

In India, >30 begomviruses are known to infect several dicotyledonous plant species and this is reviewed by Bohrah and Dasgupta (2012). In India, seven begomoviruses have been reported to infect cucurbitaceous crops *viz.*, bottle gourd, chayote, cucumber, muskmelon, pumpkin, ridge gourd, ivy gourd and sponge gourd which includes, ChYMV (Mandal et al., 2004), *Pumpkin yellow vein mosaic virus* (PYVMV; Muniyappa et al., 2003), SLCCV-*India* (Singh et al., 2008), *Tomato leaf curl virus* (ToLCV; Varma and Giri, 1998; Raj et al., 2005), *Tomato leaf curl New Delhi virus* (ToLCNDV; Muniyappa et al., 2003; Sohrab et al., 2003, 2010; Jyothsna et al., 2013), *Tomato leaf curl Palampur virus* (ToLCPMV; Namrata et al., 2010), *Indian cassava mosaic virus* (ICMV) in bitter gourd (Rajinimala and Rabindran, 2007) and SLCYNV was reported by Maruthi et al. (2007).

9.1.2   CRINIVIRUSES (FAMILY: CLOSTEROVIRIDAE)

Criniviruses belongs to the family *Closteriviridae*. Members of this genus having linear ssRNA with the genome of up to 20 kb in size. These are one of the largest RNA viruses infecting plants. Genome is encapsidated and composed of two RNA molecules *viz.*, RNA-1 and RNA-2 (Livieratos et al., 2004). Family *Closteroviridae* composed of clostero- and ampeloviruses which are transmitted by aphids (Martelli et al., 2002). *Crinivirus* is the only genus in this family transmitted by whiteflies (*B. tabaci, Trialeurodes vaporariarum,* and *T. abutiloneus*) in semi-persistent manner (Wisler et al., 1998; Martelli et al., 2012).

Members of the genus *Crinivirus* are one of the emerging constraints to cucrbitaceous crop production (Navas-Castillo et al., 2011). Of the 14 criniviruses reported so far only four species are known to infect cucurbits (Table 9.1) which includes, *Beet pseudo-yellows virus* (BPYV), *Cucurbit chlorotic yellows virus* (CCYV), *Cucurbit yellow stunting disorder virus* (CYSDV), and *Lettuce infectious yellows virus* (LIYV). CCYV and CYSDV belongs to the genus *Crinivirus* and are associated with the cucurbit yellows disease (CYD) and causes significant yield losses in different cucurbitaceous crops worldwide (Wintermant et al., 2009; Okuda et al., 2010). Both of these viruses are naturally transmitted by *Bemisia tabaci* MEAM 1 (Fomerely known as biotype B) and MED (earlier known as biotype Q) (Celix et al., 1996; Berdiales et al., 1999; Okuda et al., 2010).

All the criniviruses cause economic losses and cause significant reductions in the yield and plant growth. It was reported that, significant reduction in the fruit weight due to the incidence of BPYV in pumpkin in California (Wintermantel, 2004). Subsequently, CYSDV causes 30–50% and 10–20% yield loss in cucumber and melon, respectively (Abou-Jawdah et al., 2000). Similarly, CCYV alone causes 10–20% yield loss in melon (Gu et al., 2011).

9.1.3 POTYVIRUSES (FAMILY: POTYVIRIDAE)

Potyviruses belong to family *Potyviridae* (type species *Potato virus* Y) having global distribution. Members belonging to the *Potyvirus* genus in the family *Potyviridae* are the second largest group of plant viruses infecting wide range of host plants in agricultural and horticultural crops worldwide. The family *Potyviridae* encompasses host range of >30% of the total plants (Berger et al., 2005). *Potyvirus* is the largest genus in the family comprising of >100 species. The members of potyviruses infect large numbers of monocot and dicot plants. Potyviruses are transmitted by aphids in a non-persistent manner. Virions are of flexuous filaments rods of 680–900 nm x 11–13 nm in size. Virions are positive sense single stranded (ss) RNA of approximately 10 kb in size. Potyviruses have been transmitted by aphids, and mechanical inoculation. The genus consisting of >140 distinct species and 30 tentative species (King et al., 2011).

In India, aprox.: 40 potyvirus species have been known to infect plant species belonging to solanaceae, leguminaceae, and cucurbitaceae (Sharma et al., 2014). Among the known potyviruses 13 species are known to infect cucurbits including, *Algerian watermelon mosaic virus* (AWMV), *Clover yellow vein virus* (ClYVV), *Cucurbit vein-banding virus* (CVBV), *Melon vein-banding mosaic virus* (MVBMV), *Papaya ringspot virus*-W (PRSV-W), *Telfairia mosaic virus* (TeMV), *Turnip mosaic virus* (TuMV), *Watermelon leaf mottle virus* (WLMV), *Watermelon mosaic virus Morocco strain* (WMV-MO), *Watermelon mosaic virus 2* (WMV-2), *Zucchini yellow fleck virus* (ZYFV), *Zucchini shoestring virus* (ZSSV), and *Zucchini yellow mosaic virus* (ZYMV).

9.1.4   TOBAMOVIRUSES (FAMILY: VIRGAVIRIDAE)

Tobamoviruses belongs to the family *Virgaviridae,* currently there are >37 species (Type species: *Tobacco mosaic virus* (TMV)) are reported in the genus. Cucurbits can severely be affected when infected with different tobamoviruses such as *Cucumber green mottle mosaic virus* (CGMMV), *Cucumber fruit mottle mosaic virus* (CFMMV), *Kyuri green mottle mosaic virus* (KGMMV), and *Zucchini green mottle mosaic virus* (ZGMMV), (Antignus et al., 2001; Choi et al., 2001; Yoon et al., 2001, 2002). CGMMV is rod-shaped, microscopic (300 X 18 nm) and it is the most stable virus and survives for a long period in the infected crop debris. The CGMMV is known to be seed transmitted and found associated with seed coat, due to the seed contamination very serious infection occur at an early crop stage (Coutts and Jones, 2005). CGMMV is considered to have a narrow host range that is primarily limited to cucurbit species including, watermelon, melon, cucumber, pumpkin, squash, gourds, etc. It causes severe damage to the host plant and fruit resulting in substantial yield losses. Therefore, it is an important threat to fresh market, export, and cucurbit seed industries in the areas where CGMMV is known to occur (Tian et al., 2014).

Of the 37 species, in India, only 10 species have been reported to infect cucurbits including CGMMV, *Frangipani mosaic virus* (FrMV), *Odontoglossum ringspot virus* (ORSV), *Pepper mild mottle virus* (PMMoV), *Plumeria mosaic virus* (PluMV), *Sunn-hemp mosaic virus* (SHMV), TMV and *Tomato mosaic virus* (ToMV), *Zucchhini green mottle mosaic virus* (ZGMMV). In India, CGMMV is most commonly infecting several cucurbits like bottle gourd, cucumber, watermelon, muskmelon, gherkin, ash gourd, snake gourd, and sponge gourd (Table 9.1).

9.1.5   ORTHOTOSPOVIRUSES (FAMILY: TOSPOVIRIDAE)

Orthotospoviruses (type member: *Tomato spotted wilt virus*; TSWV), the major constraints of approx.: 15 monocots and 69 dicot plant species (Parella et al., 2003). Orthotospoviruses belongs to family *Tospoviridae*. Orthotospoviruses are one of the most economically important plant viruses affecting wide range of horticultural plant species globally (Pappu et al., 2009). Orthotospoviruses are naturally spread by several species of thrips in a circulative and propagative manner (Ullman et al., 1997; Jones et al., 2005). Worldwide, approx.: 20 orthtospoviruses have been identified and are known to be transmitted by >14 different thrips species (Jones, 2005; Pappu et al., 2009; Ciuffio et al., 2009; Hassani-Mehraban et al., 2010). Among the 20 species only seven species *viz.*, *Groundnut bud necrosis viru*s (GBNV), *Melon yellow spot virus* (MYSV), *Melon severe mosaic virus* (MeSMV), TSWV, *Watermelon bud necrosis virus* (WBNV), *Watermelon silver mottle virus* (WSMoV) and *Zucchhini lethal chlorosis virus* (ZLCV) are known to infect cucurbitaceous crops.

In India, six tospoviruses are known to occur including, GBNV during 1968 on groundnut (Reddy et al., 1992) and WBNV on watermelon during 1991–1992 (Krishnareddy and Singh, 1993; Singh and Krishnareddy, 1996). Both the WBNV and GBNV are transmitted by *Thrips palmi* (Vijayalakshmi, 1994; Rabijit et al., 2012) and are most prevalent and infecting watermelon, muskmelon, ridge gourd, bitter gourd, etc., (Jain et al., 2007; Mandal et al., 2012; Holkar et al., 2017, 2018a, 2019). WBNV is most devastating in watermelon in India causing up to 90–100% crop loss (Reddy et al., 1995; Singh and Krishnareddy, 1996; Jain et al., 1998, 2007). TSWV has been reported to occur in India on chrysanthemum (Renukadevi et al., 2015) but its occurrence on cucurbits has not yet studied. However, infections of MSMV, WSMoV, and ZLCV on cucurbits in Indian subcontinents cannot be ignored.

9.2 DIAGNOSIS

9.2.1 *BEGOMOVIRUSES*

Since the inception of polymerase chain reaction (PCR), it has been utilized for the detection of begomoviruses (Varma and Malathi, 2003). Coat protein gene specific primers successfully detected the association of begomoviruses in different cucucrbits by PCR (Sohrab et al., 2010; Tiwari et al., 2010). By PCR with CP genes *viz.*, AC1 and AV2 specific primers of ToLCNDV, detected the association of begomoviruses in different cucurbits including bitter gourd, cucumber, muskmelon, winter squash, bottle gourd, sponge gourd, ridge gourd, ivy gourd, pumpkin, and watermelon (Varma and Giri, 1998; Mandal et al., 2004; Raj et al., 2005; Sohrab et al., 2006).

The association of begomovirus with yellow mosaic disease (YMD) was detected by PCR with CP gene specific primers and southern hybridization. In addition, the associated virus was named as *Bitter gourd yellow mosaic virus* (BGYMV; Raj et al., 2005). Subsequently, the whitefly transmissibility of BGYMV was studied by Rajinimala et al. (2005). Similarly, by serological and nucleic acid based diagnostics, the association ICMV was detected in bittergourd (Rajinimala and Rabindran, 2007). Later, an improved and economical protocol for DNA isolation and PCR reaction was optimized for efficient detection of begomoviruses infecting legumes which are rich in polyohenols, tannins, and polysaccharides (Rouhibakhsh et al., 2008). Detection begomoviruses are primarily effected by PCR, but other methods include enzyme-linked immunosorbent assay (ELISA), lateral flow immuno chromatographic assays (LFA), dot blot hybridization, rolling circle amplification (RCA) and loop-mediated isothermal amplification (LAMP) (Fukuta et al., 2003; Polston and Lapidot, 2008; Kushwaha et al., 2010; Almasi et al., 2013a, b). Recently, Nagendran et al. (2017) detected the presence of ToLCNDV and SLCCNV with the 98.6% incidence in different cucurbits cultivated in Tamil Nadu state in India.

Recently, another isothermal amplification method widely used is known as recombinase polymerase amplification (RPA). RPA was first developed in 2006 and based on the principle of extension of primers induced by recombination proteins (Piepenburg et al., 2006). This rapid and sensitive RPA assay was used for the detection of *BGYMV*, *Tomato mottle virus* and TYLCV (Londono et al., 2016).

9.2.2 CRINIVIRUSES

Criniviruses are phloem limited and low titer virus in the infected plants (Wisler et al., 1998). Therefore, detection of criniviruses is difficult and needs sensitive nucleic acid methods. Earlier, diagnosis based on serology was more cumbersome due to purification of the virions for antibody production, but due to recombinant technology has changed the way of diagnostics. Several workers developed antibodies for CYSDV and CCYV based on the CP gene expressed in *E. coli* cells (Hourani and Abou-Jawdah, 2003). These antibodis have been widely used in different formats of ELISA bot blot assays including DAS-ELISA, plate trapped antibody ELISA (PTA-ELISA), tissue immuno blot assay (TIBA), Dot immuno binding assay (DIBA) and immuno-capture reverse transcription PCR (RT-PCR) and immuno-electron microscopy (IEM; Hourani and Abou-Jawdah, 2003). Similarly, multiplex PCR (M-PCR) and RT-qPCR have been extensively used for the sensitive and reliable detection of these two viruses (Gil-Salas et al., 2007; Abou-Jawdah et al., 2008; Abrahamian et al., 2013).

9.2.3   POTYVIRUSES

As like other plant viruses, several methods have been practiced for diagnosis of potyviruses. Dignsotic protocols based on serological (DAS-ELISA) and nucleic acid-based techniques (RT-PCR) have been routinely used for potyviruses including WMV, ZYMV, PRSV in different cucurbits (Hosseini et al., 2007; Zheng et al., 2010; Kamberoglu et al., 2016; Amer, 2015; Ghanem et al., 2016; Perotto et al., 2016; Verma et al., 2016; Ali, 2017; Mederos et al., 2017; Singh et al., 2017; Santosa et al., 2018; Topkaya et al., 2019). Simultaneous detection of WMV, ZYMV, and PRSV through multiplex-RT-PCR assay has been studied (Rajbanshi et al., 2019). Recently, Perotto et al. (2018) have characterized the complete genome of CVBV by Illumina next-generation sequencing.

9.2.4   TOBAMOVIRUSES

The CGMMV have been detected earlier by electron microscopy (EM) and DAC-ELISA (Antignus et al., 1990) and by Real-time TaqMan qRT-PCR (Hongyun et al., 2008). Recently, serological-based detection of KGMMV through DAS-ELISA has also been done by Daryono et al. (2016) and ZGMMV have diagnosed by RT-PCR (Li et al., 2018).

9.2.5   ORTHOTOSPOVIRUSES

Globally, diagnosis of orthotospoviruses has accomplished notable achievements during the last two-three decades; which led to the identification of orthotospoviruses in different new plant species. Diagnostic techniques *viz.*, EM, biological, immno-based, and nucleo-based has been developed worldwide. Initially, during early 80s biological and immuno-based techniques were widely used for the diagnosis of virus species, later 90s due to improvement in the nucleo-based techniques *viz.*, PCR, RT-PCR, duplex PCR, M-PCR, qRT-PCR, LAMP, RPA were extensively used for the diagnosis of orthotospoviruses.

In India, cowpea (*Vigna unguiculata*) cvs. Pusa Komal and C-152 have been studied as one of the best indicator hosts for the local and systemic symptoms by sap inoculations of GBNV, WBNV, and CaCV (Ghanekar et al., 1979; Singh and Krishnareddy, 1996; Jain et al., 2007; Kunkalikar et al., 2011; Holkar et al., 2017; Basavaraj et al., 2018). Orthotospoviruses are known to produce varying symptoms and are difficult to distinguish by symptomatology alone. Physical properties of GBNV in crude sap are known (Ghanekar et al., 1979). ELISA have been successfully used for the detection of these three orthotospoviruses *viz*., GBNV, WBNV, and CaCV from different plant species using polyclonal antibodies (PAb) specific to N protein of WSMoV and GBNV (Jain et al., 2002, 2005, 2007; Anjaneya Reddy et al., 2008). PAb was unable to distinguish these three viruses therefore, in order to get differential detection of GBNV and WBNV monoclonal antibodies (MAb) were developed to N protein of GBNV (Jain et al., 2007) Likewise, MAb were developed against N protein of GBNV which could distinguish GBNV and WBNV isolates (Hemalatha et al., 2008). More recently, PAb against NSs protein of WBNV was able to differentiate these two species (Basavaraj et al., 2020). Reverse transcription-polymerase chain reaction (RT-PCR) by species-specific N gene primers were optimize and utilized (Jain et al., 1998). Subsequently, Kunkalikar et al. (2011) standardized single tube one-step RT-PCR method by conserved common forward and virus-specific reverse primers based on N gene for specific detection of GBNV, WBNV, and CaCV. Moreover, duplex RT-PCR have been developed for the specific detection of GBNV and WBNV using virus-specific forward and degenerate, common single reverse primers (Holkar et al., 2017). Duplex RT-PCR has been widely used in India which led to the identification of three new hosts of GBNV *viz*., muskmelon, dahlia, and chrysanthemum and WBNV chrysanthemum (Holkar et al., 2017). However, other rapid techniques like LAMP, RPA, and lateral flow assay (LFA) have not yet standardized for orthotospoviruses in the Indian subcontinent.

9.3   DISTRIBUTION AND HOST RANGE

9.3.1   *BEGOMOVIRUSES*

In the recent past, the cucurbit infecting begomoviruses has become one of the major constraints in the production and is widely distributed across the globe. Begomoviruses infecting cucurbits including SLCV was first time recorded on winter and summer squash in the Southern and Northern U.S. during 1977 and 1978, respectively (Flock and Mayhew, 1981; Lazarowitz and Lazdins, 1991). Since then, association SLCV was recorded on watermelon, melon, cucumber, pumpkin, and squashes from various countries like Lebanon, Israel, Jordan, Saudi Arabia, Egypt, and Taiwan (Table 9.1). Likewise, WmCSV is known to infect cucurbits mainly in the Eastern

Mediterranean and Western hemispheres. WmCSV was first time recorded from Yemen in 1982 (Jones et al., 1988). Since then, it was reported from Jordan, Israel, Iran, Lebanon, Sudan, Palestine, Saudi Arabia, and Oman on watermelon, squash, cucumber, bottle gourd, zucchini, melons, and pumpkins (Kheyr-Pour et al., 2011; Ali et al., 2012; Al-Musa et al., 2011; Khan et al., 2012; Ali-Shtayeh et al., 2014; Alhudaib et al., 2018). The third important species of begomovirus includes SLCCNV which has limited cucurbit host range and known to infect mainly squash and pumpkin. SLCCNV has been reported from Australia, Malaysia, India, Pakistan, China, Philippines, and Vietnam (Table 9.1; Kon et al., 2003; Ito et al., 2008; Singh et al., 2008; Sawangjit, 2009; Tahir et al., 2010; Saritha et al., 2011; Mohammad-Riyaz et al., 2013; Maina et al., 2017; WU et al., 2020). Whereas, SLCPHV was known to occur in Taiwan on chayote, pumpkin, and wax gourd (Liao et al., 2007; Tsai et al., 2011). Likewise, SLCYNV were reported from China and Thailand on pumpkin, wax gourd and ridge gourd (Xie and Zou et al., 2003; Xie et al., 2006; Butnut et al., 2019). MCLCV was recorded on different cucurbits hosts like cantaloupe, cucumber, pumpkin, watermelon, and muskmelon in Zacapa Valley, Guatemala, USA (Guzman et al., 2000; Brown et al., 2001, 2011; Idris et al., 2008) and MeCMV was reported from Venezuela on melon (Ramirez et al., 2004) and on watermelon, squash, and cucumber (Romay et al., 2010, 2014, 2015).

ChaYMV has been known to infect bitter gourd/bitter melon from some African countries like Benin, Nigeria, and Togo (Leke et al., 2016). Similarly, a new and unexplored member belonging to the SLCV lineage, i.e., CuLCV was occurred in the Western USA and Northern Mexico (Cohen et al., 1983). Similarly, CuLCrV was reported in the USA in different cucurbits (Kuo et al., 2007). LYMV is known to infect cucurbits in the Philippines and Vietnam (Revill et al., 2003). ToLCNDV is widely reported from different countries of Asia, the Middle East, North Africa, and Europe. ToLCNDV is known to occur on different cucurbits *viz*., chayote, sponge gourd, zucchini, cucumber, bottle gourd, muskmelon, bitter gourd, pumpkin, ash gourd and snake gourd from various countries like India, Spain, Thailand, Pakistan, Italy, Indonesia, Tunisia, Iran, and Morocco (Table 9.1; Sohrab et al., 2003; Tahir et al., 2005; Ito et al., 2008; Namrata et al., 2010; Tiwari et al., 2010; Mizutani et al., 2011; Phaneendra et al., 2012; Juarez et al., 2014; Mnari-Hattab et al., 2015; Panno et al., 2016; Yazdani-Khameneh et al., 2016; Nagendran et al., 2017; Siferes et al., 2018) and Whereas, ToLCPMV on cucurbits is reported from India, Iran, and Pakistan (Heydarnejad et al., 2009; Ali et al., 2010; Namrata et al., 2010).

9.3.2   CRINIVIRUSES

CYSDV was known to be confined only with cucurbitaceous crops (Wisler, 1998), now due to the advancement in the different detection techniques it has been recorded from non-cucurbitaceous, and certain weed hosts (Wintermantel et al., 2009). CYSDV is reported from Egypt, France, Italy, Israel, Japan, Jordan, Lebanon, Morocco, Portugal, Saudi Arabia, Spain, Syria, UAE, Mexico, Turkey, the USA, the UK on cucurbits *viz*., cucumber, melon, and zucchini (Coffin and Coutts, 1990; Wisler et al., 1998; Desbiez et al., 2000; Louro et al., 2000; Rubio et al., 2001; Al-Saleh et al., 2015; Manglli et al., 2016; Orfanidou et al., 2019). Whereas, CCYV was found to infect only cucurbits like cucumber, melon, watermelon, and squash and reported from Greece and Cyprus (Okuda et al., 2010; Gu et al., 2011; Hamed et al., 2011; Bananej et al., 2013; Al-Saleh et al., 2015; Orphanidou et al., 2014). BPYV is naturally affecting cucurbits with yellowing symptoms (Duffus, 1965). BPYV was believed to have a monopartite genome (Woudt et al., 1993) in the genus *Closterovirus* (Fauquet and Mayo, 1999; Wisler et al., 1998). Later, it was found that it has bipartite genome which is similar in size and organization to that of LIYV (Hartono et al., 2003). BPYV reported from Crete, Italy, Spain, Greece, New Zealand, and the USA on different cucurbits (Berdiales et al., 1999; Rubio et al., 1999; Tomassoli et al., 2003; Wintermantel, 2004; Hammond et al., 2005; Boubourakas et al., 2006).

9.3.3   POTYVIRUSES

AWMV having a very narrow cucurbit host range and therefore reported on squash from Algeria in 1986 (Yakoubi et al., 2008), so far no any information available on infecting cucurbits from different countries. Likewise, CVBV is reported from the USA (Providenti et al., 1984) and MVBMV a distinct potyvirus reported in pumpkin from Taiwan (Huang et al., 1993). MWMV is reported from many countries Abkhazia, Algeria, Albania, Armenia, Andorra, France, Greece, Iran, Kenya, Nigeria, South Africa, Sudan, the USA, Italy, Tunisia, and Congo on different cucurbitaceous crops (Roggero et al., 1998; Lecoq et al., 2001; Arocha et al., 2008; Yakoubi et al., 2008; Owolabi et al., 2012; Malandraki et al., 2014; Ibaba et al., 2016; Bananej et al., 2018; Miras et al., 2019; Read et al., 2020). PRSV-W isolate is infecting crop plants belonging to the cucurbitaceae and chenopodiace families (Tripathi et al., 2008). PRSV-W isolate has been well studied from different countries including Brazil, Cuba, China, India, Iran, Korea, Taiwan, and the USA on different cucurbits (Liu et al., 2002; Della-Vecchia et al., 2003; Ali et al., 2004; Mangruthia et al., 2008; Abdalla and Ali, 2012). Watermelon is affected by WMV-2, belongs to the same group and known to occur in different countries *viz.*, China, Egypt, France, India, Iran, Italy, Japan, Pakistan, Serbia, South Africa, Slovakia, South Korea, Turkey, and the USA (Table 9.1). Similarly, ZSSV, ZYFV, and ZYMV are known to infect primarily on Zucchini and reported from various countries *viz.*, South Africa, Zimbabwe, and Italy, France, and Brazil, China, Iran, France, Poland, Portugal, Serbia, Slovakia, USA, Venezuela, respectively (Table 9.1).

9.3.4　TOBAMOVIRUSES

All the tobamoviruses have a similar natural host range and confined mainly to cucurbitaceous crops (Yoon et al., 2002). Among tobamoviruses, CGMMV has been reported to infect watermelon from China, Greece, Japan, and the USA which was originally described during 1930s. CGMMV has been found to be responsible for heavy losses of cucurbitaceous crops in European and Asian countries. This virus has been known since 1935, it was first reported from the United Kingdom, since then it has been identified from other European countries, Thailand (Noda et al., 1993), India (Varma and Giri, 1998), Japan (Lee et al., 1996), China (Zhang et al., 2009), Greece (Varveri et al., 2002), Korea (Yoon et al., 2008), Myanmar (Kim et al., 2010), and in Ukraine (Budzanivska et al., 2007). In the United States, it was first reported from California in 2013 (Baker, 2013). Recently, an elaborated review on global distribution, etiology, and management of CGMMV from its first record to 2016 has been compiled by Dombrovsky et al. (2017).

In India, CGMMV is known to infect different cucurbits including, cucumber (Mandal et al., 2008), bottle gourd, watermelon, gherkin, and muskmelon (Varma and Giri, 1998), ridge gourd (Sharma et al., 2014), snake gourd (Nagendran et al., 2015).

9.3.5　ORTHOTOSPOVIRUSES

Among the known orthotospovirus species MYSV, MeSMV, TSWV, WSMoV, and ZLCV are known to infect cucurbits and has widely distributed globally. GBNV and WBNV are the distinct species predominantly occurring in the Indian subcontinent, GBNV has wide host range in general and narrow host range as far as cucurbits is concerned, and vice versa in WBNV. The distribution of 20 orthotospoviruses is known in different continents including Africa, Asia, Australasia, Europe, North America, and South America.

MYSV was the first time was reported from Japan on melon (Kato et al., 1999, 2000). Since then, it was reported on different cucurbits from different countries including on watermelon from Taiwan (Chen et al., 2008), on melon and cucumber from Thailand (Chiemsombat et al., 2008), on cucumber from Taiwan (Chao et al., 2010), on melon from China (Gu et al., 2012), from Ecuador (Quito-Avila et al., 2014), and on pumpkin from Thailand (Supakitthana-Kom et al., 2018). MeSMV is a new orthotospovirus was recorded from Italy and Mexico on different cucurbits (Ciuffo et al., 2009, 2017). WSMoV was first reported from Taiwan on watermelon and was identified as a distinct serogroup IV infecting cucurbits (Yeh and Chang, 1995). Subsequently, reported in India (Kunkalikar et al., 2007) and Taiwan (Zheng et al., 2008), and in recent past in Hawaii (Melzer et al., 2014).

WBNV was first recorded in India during 1991–1992 on watermelon at IIHR experimental fields (Krishnareddy and Singh, 1993). Subsequently, Jain et al. (1998) characterized it as a new distinct orthotospovirus from Delhi, India, and also found its natural occurrence in cucumber and ridge guard (Jain et al., 1998, 2007; Li et al., 2011; Kumar et al., 2011; Mandal et al., 2003). Symptoms of WBNV on watermelon and muskmelon were characterized by bud, stem, and leaf tip necrosis (Figure 9.1(a–c)). Likewise, GBNV is the most studied orthotospovirus in the Indian subcontinent (Mandal et al., 2012). Earlier till 2010, it was considered that WBNV was confined only with the cucurbitaceous crops whereas, GBNV with the crops belonging to Leguminoceae, Solanaceae, Euphorbiaceae, and Asteraceae families. Recently, Holkar et al. (2017) detected natural infection of WBNV on chrysanthemum and Kunkalikar et al. (2011) detected infection of GBNV on watermelon and muskmelon (Holkar et al., 2017).

In India, GBNV, and WBNV are broadly spread and prevalent in various states comprising Andhra Pradesh, Gujarat, Haryana, Himachal Pradesh, Karnataka, Kerala, Madhya Pradesh, Maharashtra, Orissa, Punjab, Rajasthan, Tamil Nadu, Uttar Pradesh, and West Bengal (Mandal et al., 2012). In Bangladesh, bud necrosis disease of watermelon is known but the causal viruses were not yet identified (*Personal communication with Dr. Shamim Akhter, BARI, Bangladesh*).

FIGURE 9.1 Symptoms of Watermelon bud necrosis virus (WBNV) on watermelon and muskmelon recorded in India. (A) Watermelon plant showing leaf and stem necrosis; (B) watermelon plant showing typical bud and leaf tip necrosis; and (C) muskmelon plant showing chlorotic spots all over the leaf surface.

9.4 INTEGRATED DISEASE MANAGEMENT (IDM)

Management of plant viruses infecting cucurbits is very crucial and important due to their huge economic loss to the growers. Management of plant viruses is becoming more challenging day by day due to changes in climatic conditions, emergence of new insect-vector species transmitting plant viruses, changes in host preference by these insect-vectors leading to the occurrence of viruses in newer hosts. In order to manage plant viruses, integrated pest management strategies need to be followed which including cultural, chemical, biological methods for effective management of insect-vectors. Moreover, plant virus-resistance breeding program needs to be emphasized and non-conventional methods including pathogen-derived resistance (PDR) needs to be adopted. Recently, CRISPR/cas system approach of genome editing must be taken into consideration against the cucurbit viruses because this technology has certain advantages over PDR.

Recently, CRISPR/Cas9 system has been widely used as a robust and efficient genome editing technology to develop host resistance against ssDNA viruses (Ali et al., 2015, 2016; Baltes et al., 2015; Ji et al., 2015). Zhang et al. (2018), for the first time, applied FnCas9 and RNA-targeting sgRNAs specific for *Cucumber mosaic virus* (CMV) and TMV in *N. benthamiana* and Arabidopsis plants. Similarly, Aman et al. (2018) utilized the RNA-guided ribonuclease to edit the TuMV by targeting four different viral genomic portions. Chandrasekaran et al. (2016) developed cucumber plants resistant to CVYV, ZYMV, and PRSV-W. Subsequently, Pyott et al. (2016) utilized a similar approach for conferring resistance against TuMV. Recently, CRISPR/Cas13a system has been used to engineer interference with TuMV in plants (Aman et al., 2018). Therefore, application of CRISPR/Cas9 and CRISPR/Cas13a machinery for the development of resistance against DNA and RNA viruses, respectively in cucurbits by targeting the viral genomic regions could be emphasized.

9.4.1 BEGOMOVIRUSES AND CRINIVIRUSES

Management of the begomoviruses is more challenging, once the plant are diseased it is very difficult to manage therefore, chemical control only reduce the population of whitefly insect-vector. Thus, keeping these things in mind preventive measures and use of resistant varieties play an important role in managing the widespread occurrence of begomoviruses in different cucurbits. The well-known preventive measures including the proper cultural practices viz., clean cultivation by burning or disposal of crop residue, rouging, use of non-preference crops as an intercrop, use of barrier crops, etc.

The management of begomoviruses of cucurbits is dependent on control of whiteflies and use of resistant varieties (Babitha, 1996). Further studies need to emphasize on the development of resistant cucurbit genotypes against begomoviruses and *B. tabaci* biotypes. The prophylactic practices including whitefly control using chemicals, biological agents, development of virus-resistant and virus-free plants. The use of non-host "trap crops" needs to be practiced to attract whitefly to reduce possible the spread of the virus (Bragard et al., 2013).

9.4.2 POTYVIRUSES

In order to reduce the population of insect vectors and possible spread of virus, the various methods had been used including use of resistant cultivars, improvement in plant resistant through use of Plant growth-promoting rhizobacteria (PGPR) (Murphy et al., 2003; Jones, 2006; Barakat et al., 2012; Ruwanthi et al., 2014). The PGPR strains secrete Chitosan which is a nontoxic and biodegradable biopolymer having the ability to induce resistance in plants through its eliciting behavior (Xing et al., 2015). Recently, it has been used to control PRSV-W in Florida, USA (Abdalla et al., 2017). Moreover, the use of different chemicals *viz.*, salicyclic acid and acetyl salicylic acid (SA) play an important role in inducing defense response against virus (Zhang et al., 2007; Lewsey et al., 2009; Madhusudhan et al., 2011). In recent past, Elsharkawy, and Mousa (2015) reported the importance of use of silicon against PRSV-W in cucurbits. Similarly, several root extracts of *Boerhaavia diffusa* and leaf extract of *Clerodendrum aculeatum* has been demonstrated for effective management of PRSV (Awasthi and Singh, 2009).

9.4.3   TOBAMOVIRUSES

In order to avoid the initial spread of the CGMMV in field conditions, the traditional methods play an important role and which include the use of resistant varieties, use of healthy see material (van Koot and van Dorst, 1959; Komuro et al., 1971), expose the seed materials to a high temperature in compost (Avgelis and Manios, 1992), avoiding the use of contaminated irrigation water (Vani and Varma, 1993), heat treatment at 75°C for 3 days (Kim and Lee, 2000), Soaking the seed in 10% solution of sodium phosphate for 20 min and by soil fumigation (80 gm methyl bromide/m^2) we can effectively control the virus transmission to some extent (Komuro, 1971).

9.4.4   ORTHOTOSPOVIRUSES

Due to the severe economic impact of the orthotospoviruses in different crops proper attention must be given to the IDM approaches. IDM approaches like, cultural, chemical, biological, host-plant resistance and phytosanitory measures need to involve in managing the tospoviruses (Pappu et al., 2009). These approaches have been found efficient in cucurbits in minimizing the economic losses due to the infection of GBNV (Sreekanth et al., 2003). On IDM approaches for management of orthotospoviruses there are excellent reviews published on orthotospoviruses by Pappu et al. (2009); Mandal et al. (2012); Holkar et al. (2019). Genome editing using CRISPR/Cas a robust, reliable, and novel technology so far has not yet utilized against orthotospovirus resistance in experimental hosts and crop plants. This is challenging and gray area of research in the *Orthotospovirus* group.

Management of insect-vectors by chemicals would certainly help to reduce the virus infection. Therefore, chemical treatment of seed with the systemic insecticide, i.e., imidacloprid 70 WS @ 10 g/Kg in combination of two-three foliar sprays of imidacloprid 17.8 SL @ 0.25 ml/L was found to have reduced infection of WBNV in watermelon (Kamanna et al., 2010). Similarly, application of imidacloprid for seed treatment @ 5 g/Kg of seed, along with the silver color UV reflective mulch and foliar sprays of spinosad (0.3 ml/L) or imidacloprid (0.3 ml/L) or thimithoxam 0.2 g/L has found to have reduced WBNV infection in watermelon (Rajasekharam, 2010).

Extensive studies have been found to enhance host resistance using truncated or full nucleocapsid protein (N) gene sequences for orthotospoviruses which has led to heritable resistance against homologous virus species. RNA silencing or RNA interference (RNAi) approach has been demonstrated and reviewed for development of transgenics against orthotospoviruses (Raja, 2005; Raja and Jain, 2006; Pappu et al., 2009; Runo, 2011; Duan et al., 2012; Catoni et al., 2013; Holkar et al., 2018b, 2019; Kumar et al., 2019). Agro-infiltartion technique using partial N gene of WBNV has effectively conferred resistance in the watermelon plants when sap inoculated with the virus (Figure 9.2(a–f)).

In order to enhance level of resistance, transgenic resistance could be energized. *Agrobacterium* mediated transformation is one of the best method which is routinely utilized (Table 9.3). In Watermelon, the first successful *Agrobacterium* mediated transformation against CGMMV using CP gene was reported by Park et al. (2005). Subsequently, genetic transformation of wild watermelon plants has been achieved by Akashi et al. (2005). In the recent past, the *Agrobacterium* mediated genetic transformation using CP gene, transgenic *Citrullus* spp. have been developed against different plant viruses viz., CGMMV, CMV, PRSV-W, WMV, WSMoV, and ZYMV (Choi et al., 2008; Ibrahim et al., 2009; Yu et al., 2010; Huang et al., 2011; Lin et al., 2012). In India, efforts have been made to develop N protein mediated resistance in watermelon against WBNV (Kumar and Mandal, 2012; Holkar et al., 2018b; Kumar et al., 2019). Identification of the resistant genes in different cucurbits against various viruses has been worked out (Table 9.4). In India, WBNV is the most studied orthotospovirus in cucurbits over the past three decades which is presented schematically starting from its first record on watermelon till its management including non-conventional methods (Figure 9.3).

FIGURE 9.2 Validation of the partial N gene construct conferring resistance through transient expression for the proof of concept against Watermelon bud necrosis virus (WBNV): (A) healthy watermelon seedlings of 30 days old; (B) WBNV inoculated watermelon of 30 DAS showing necrotic spots at 6 days post inoculation; (C) symtomatic plants were subjected to agroinfiltation using partial N gene construct; (D) upper young leaves of agroinfiltrated plant was remained healthy; (E) agroinfiltrated leaf was showing severe necrosis at 6 days of post agroinfiltration; (F) agroinfiltrated part of the leaf was showing necrosis by hypersensitive reaction (HR) at 3 dpai and thus conferring resistance against WBNV.

TABLE 9.3 *Agrobacterium* Mediated Genetic Transformation of Cucurbitaceous Crops

Explant	Transgene	Virus Targeted*	Agrobacterium Strain	References
Cucumber (*Cucumis sativus*)				
Cotyledon	CP	CMV (1.1)	EHA105	Gonsalves et al. (1992)
Cotyledon	NPT II and Bar	–(1.1)	EHA105, LBA4404	Vengadesan et al. (2005)
Leaf	HC-Pro	PRSV-W, WMV	Not mentioned	Leibman et al. (2011)
Squash (*Cucurbita pepo* L.)				
Cotyledon	CP	CMV	Not mentioned	Fusch et al. (1995)
Cotyledon	CP	WMV, ZYMV	Not mentioned	Claugh and Hamm (1995)
Cantaloupe/Melons (*Cucumis melo* L.)				
Cotyledon	CP	WMV, ZYMV	Not mentioned	Claugh and Hamm (1995)
Cotyledon	CP	ZYMV	LBA4404	Wu et al. (2009)
Leaf and	HC-Pro	ZYMV	Not mentioned	Leibman et al. (2011)
Bottle Gourd (*Lagenaria ciceria*)				
Cotyledon	Glufosinate ammonium	–(1.9)	LBA4404/EHA105	Han et al. (2005)
Watermelon (*Citrullus lanatus*)				
Cotyledon	NPT II	–(Up to 16%)	LBA4404,	Choi et al. (1994)
Cotyledon	CP	CCGMV (0.1)	LBA4404, EHA105	Park et al. (2005)
Cotyledon	NPT II and bar	–(1.16)	LBA4404, GV3101, EHA101	Cho et al. (2008)
Cotyledon	NPT II and bar	–(1.16)	EHA 105LBA4404, GV3101	Ibrahim et al. (2009)
Cotyledon	CP	CMV, CGMMV, WMV	LBA4404	Lin et al. (2012)

*Values in parenthesis represent transformation efficiency.

TABLE 9.4 Identification of Resistant Genes Against Different Viruses in Various Cucurbit Genotypes and Accessions

SL. No.	Genotype/Accession	Crop	Resistant against virus species	Resistant gene	Country	Reference
1.	PI 371795	Melon	*Zucchini yellow mosaic virus* / *Watermelon mosaic virus*	–	India	Kolez (1948)
2.	PI 161375	Melon	*Cucumber mosaic virus*	*cmv-1*	Korea	Essafi et al. (2009)
3.	IC 267353, IC 267384, IC 274010	Snapmelon	*Cucumber mosaic virus*	–	India	Dhilon et al. (2007)
4.	IC 274007, IC 274014	Snapmelon	*Zucchini yellow mosaic virus*	–	India	Dhilon et al. (2007)
5.	IC 274014, SM 67, SM 72, SM 3, SM 82	Snapmelon	*Cucumber mosaic virus*	*cmv-1*	India	Dillon et al. (2009)
6.	PI 414723-4	Snapmelon	*Zucchini yellow mosaic virus*	*Zym-1, Zym-2 Zym-3*	–	Provvidenti (1996)
7.	Faizabadi Phoont, 90625, PI 124112, PI 124440, PI 414723	Melon	*Cucurbit aphid borne yellows virus*	*cab-1, cab-2*	India	Dogimont et al. (1997)
8.	PI 180280, PI 180283	Snapmelon	*Papaya ringspot virus*	*Prv1* and *Prv2*	India	Dhilon et al. (2007)
9.	PI 313970	Melon	*Lettuce infectious yellows virus and Cucurbit leaf crumple virus*	*Liy*	US, Mexico	McCreight and Wintermantel (2008)
10.	PI 313970, Ames 20203, PI 614185, PI 614213	Melon	*Cucurbit yellow stunting disorder virus*	–	US, Mexico	McCreight and Wintermantel (2008)
11.	90625, PI 124112, PI 4144723	Melon	*Watermelon chlorotic stunt virus*	–	India	Yousif et al. (2007)

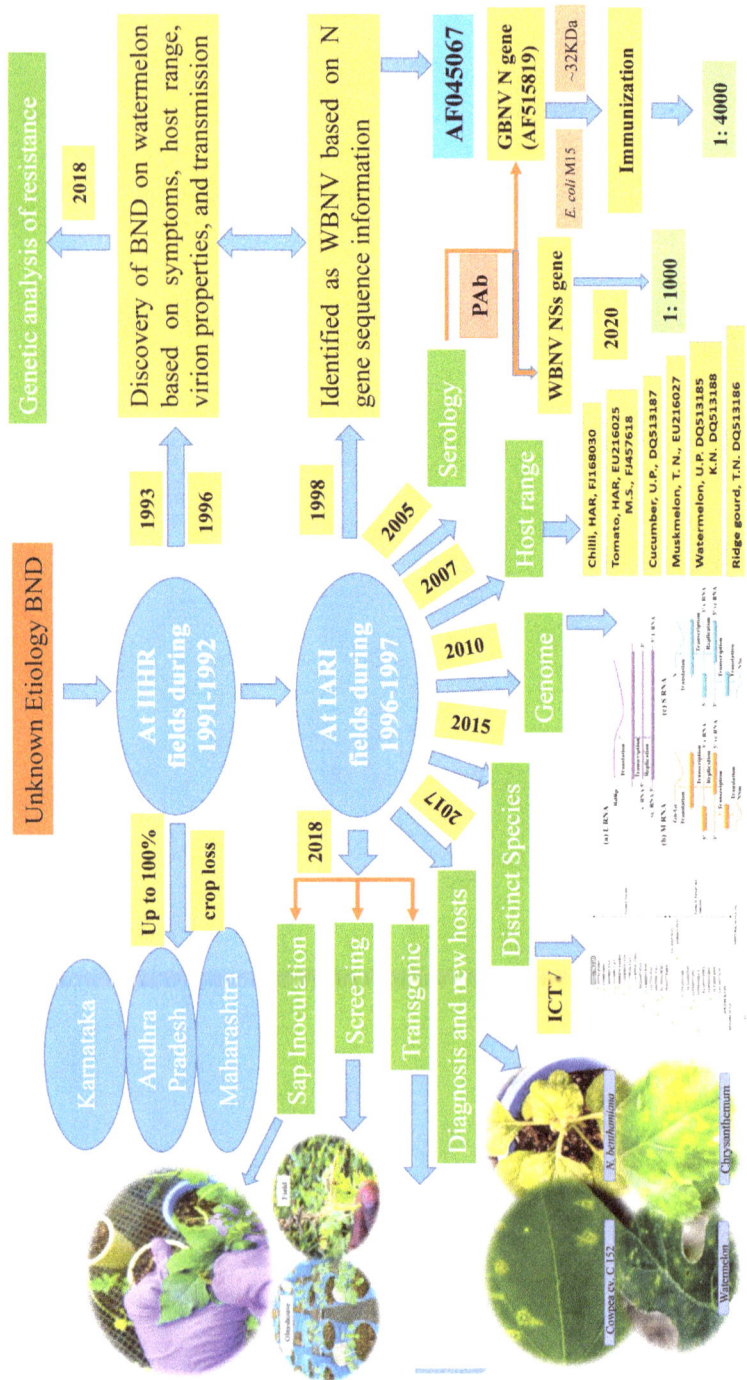

FIGURE 9.3   Schematic re-presentation of the progress achieved during last three decades on *Watermelon bud necrosis virus* (WBNV), a distinct orthotospovirus species since its first occurrence on watermelon (*Citrullus lanatus*) in India.

9.5 CONCLUSION

Rapid diagnosis of the disease is the key to devise management options. Therefore, rapid, robust, and reliable diagnostic techniques of plant viruses infecting cucurbits is well known and easily available in public domain. These diagnostic techniques are required to explore in different crop plants and this possibly help in identification of new hosts and new insect-vector species. Management of cucurbit viruses poses a tough challenge and this has received considerable attention due to their serious economic impact on production and productivity. Effective and appropriate strategy for managing cucurbit viruses need to involve the epidemiological principles and IDM approaches, *viz.*, cultural, chemical, biological, host-plant resistance and phytosanitary measures. Management of virus-infected plants is more challenging than the preventive measures, therefore much emphasis to be given on development the virus and insect resistant cucurbit cultivars and cultural and mechanical practices. Development of the host-plant resistance against plant viruses using the conventional and molecular breeding approaches definitely help in reducing the huge economical losses in the country. CRISPR/Cas is one of the novel and robust technology which can be applied in different cucurbits for developing resistance against the members belonging to begomo-, crini-, poty-, tobamo-, and orthotospoviruses.

KEYWORDS

- **Cucurbitaceae**
- **diagnosis**
- *Geminiviridae*
- **integrated disease management**
- *Potyviridae*
- *Tospovirirdae*
- **viruses**

REFERENCES

Abdalla, O. A., & Ali, A., (2012). Genetic diversity in the 3'-terminal region of *papaya ringspot virus* (PRSV-W) isolates from watermelon in Oklahoma. *Arch. Virol., 157*(3), 405–412.

Abdalla, O. A., Bruton, B. D., Fish, W. W., & Ali, A., (2012). First confirmed report of *tobacco ringspot virus* in cucurbits crops in Oklahoma. *Plant Dis., 96*(11), 1705.

Abdalla, O. A., Shagufta, B., & Shouan, Z., (2017). Integration of chitosan and plant growth-promoting rhizobacteria to control *papaya ringspot virus* and *tomato chlorotic spot virus*. *Arch. Phytopathol. Plant Prot., 50*(19/20), 997–1007.

Abkhoo, J., (2012). Serological and molecular detection and prevalence of *cucurbit aphid-borne yellow virus* in the Sistan region *Afr. J. Biotechnol., 11*, 13119–13122.

Abou-Jawdah, Y., Eid, S. G., Atamian, H. S., & Michael, H., (2008). Assessing the movement of *cucurbit yellow stunting disorder virus* in susceptible and tolerant cucumber germplasms using serological and nucleic acid-based methods. *J. Phytopathol., 156*(7/8), 438–445.

Abou-Jawdah, Y., Sobh, H., Fayad, A., Lecoq, H., Delecolle, B., & Trad-Ferre, J., (2000). *Cucurbit yellow stunting disorder virus*: A new threat to cucurbits in Lebanon. *J. Plant Pathol., 1*, 55–60.

Abrahamian, P. E., Sobh, H., Seblani, R., Samsatly, J., Jawhari, M., & Abou-Jawdah, Y., (2013). First report of *cucumber vein yellowing virus* on cucumber in Lebanon. *Plant Dis., 97*(11), 1516–1516.

Akashi, K., Morikawa, K., & Yokota, A., (2005). *Agrobacterium*-mediated transformation system for the drought and excess light stress-tolerant wild watermelon (*Citrullus lanatus*). *Plant Biotechnol., 22*(1), 13–18.

Alhudaib, K. A., Rezk, A. A., & Soliman, A. M., (2018). Current status of *watermelon chlorotic stunt virus* (WmCSV) on some cucurbit plants (Cucurbitaceae) in Al-Ahsa region of Saudi Arabia. *Sci. J. King Faisal Univ., 19*(2), 1440H.

Ali, A., (2017). First complete genome sequence of *papaya ringspot virus*-W isolated from a gourd in the United States. *Genome Announc., 5*(2), 14–16.

Ali, A., Mohammad, O., & Khattab, A., (2012). Distribution of viruses infecting cucurbit crops and isolation of potential new virus-like sequences from weeds in Oklahoma. *Plant Dis., 96*(2), 243–248.

Ali, A., Natsuaki, T., & Okuda, S., (2004). Identification and molecular characterization of viruses infecting cucurbits in Pakistan. *J. Phytopathol., 152*(11, 12), 677–682.

Ali, I., Malik, A. H., & Mansoor, S., (2010). First report of *tomato leaf curl Palampur virus* on bitter gourd in Pakistan. *Plant Dis., 94*(2), 276.

Ali, Z., Abulfaraj, A., Idris, A., Ali, S., Tashkandi, M., & Mahfouz, M. M., (2015). CRISPR/Cas9-mediated viral interference in plants. *Genome Biol., 16*(1), 238.

Ali, Z., Ali, S., Tashkandi, M., Zaidi, S. S., & Mahfouz, M. M., (2016). CRISPR/Cas9-mediated immunity to Gemini viruses: Differential interference and evasion. *Sci. Rep., 6*, 26912.

Ali Ali, E. M., Al Hachach, H., Al-Aqeel, H., & Hejji, A. B., (2013). Multiple important plant viruses are present on vegetables crops in Kuwait. *J. Clin. Trials, 3*(136), 2167–0870.

Ali-Shtayeh, M. S., Jamous, R. M., Omar, B., Mallah, O. B., & Abu-Zeitoun, S. Y., (2014). Molecular characterization of *watermelon chlorotic stunt virus* (WmCSV) from Palestine. *Viruses, 6*, 2444–2462.

Almasi, M. A., Aghapour-Ojaghkandi, M., & Saeedeh-Aghaei, S., (2013b). Visual detection of curly top virus by the colorimetric loop-mediated isothermal amplification. *J. Plant Pathol. Microbiol., 4*, 198.

Almasi, M. A., Aghapour-Ojaghkandi, M., Hemmatabadi, A., Hamidi, F., & Aghae, S., (2013a). Development of colorimetric loop-mediated isothermal amplification assay for rapid detection of the *tomato yellow leaf curl virus*. *J. Plant Pathol. Microbiol., 4*, 153.

Al-Musa, A., Anfoka, G., Al-Abdulat, A., Misbeh, S., Ahmed, F. H., & Otri, I., (2011). *Watermelon chlorotic stunt virus* (WmCSV): A serious disease threatening watermelon production in Jordan. *Virus Genes, 43*(1), 79–89.

Al-Musa, A., Anfoka, G., Misbeh, S., Abhary, M., & Ahmad, F. H., (2008). Detection and molecular characterization of *squash leaf curl virus* (SLCV) in Jordan. *J. Phytopathol., 156* (5), 311–316.

Al-Musa, A., Qusus, M. S. J., & Mansour, A. N., (1985). *Cucumber vein yellowing virus* on cucumber in Jordan. *Plant Dis., 69,* 361.

Al-Saleh, M. A., Al-Shahwan, I. M., Amer, M. A., Shakeel, M. T., Abdalla, O. A., Orfanidou, C. G., & Katis, N. I., (2015). First report of cucurbit chlorotic yellows virus in cucumber in Saudi Arabia. *Plant Dis., 99*(5), 734–734.

Al-Saleh, M. A., Al-Shahwan, I. M., Brown, J. K., & Idris, A. M., (2014). Molecular characterization of a naturally occurring intraspecific recombinant begomovirus with close relatives widespread in Southern Arabia. *Virol. J., 11*(1), 103.

Aman, R., Ali, Z., Butt, H., Mahas, A., Aljedaani, F., Khan, M. Z., Ding, S., & Mahfouz, M., (2018). RNA virus interference via CRISPR/Cas13a system in plants. *Genome Biol., 19*(1), 1.

Amer, M. A., (2015). Serological and molecular characterization of *cucurbit chlorotic yellows virus* affecting cucumber plants in Egypt. *Intl. J. Virol., 11*(1), 1–11.

Anjaneya, R. B., Krishna, R. M., Salil, J., Patil, M. S., & Usharani, T. R., (2008). Detection of a tospovirus infecting tomato (*Solanum lycopersicon*). *Indian J. Virol., 19*(1), 32–35.

Anno-Nyako, F. O., (1988). Seed transmission of *telfairia mosaic virus* in fluted pumpkin (*Telfairia occidentalis* Hook f.) in Nigeria. *J. Phytopathol., 121*(1), 85–87.

Anonymous, (2017). *Annual Report 2016–2017* (p. 164). National Horticulture Board, Ministry of Agriculture and Framers Welfare Govt. of India.

Anonymous, (2019). *Annual Report 2018–2019* (p. 188). ICAR-Indian Institute of Vegetable Research, Varanasi, Uttar Pradesh, India.

Antignus, Y., Pearlsman, M., Yoseph, R. B., & Cohen, S., (1990). Occurrence of a variant of *cucumber green mottle mosaic virus* in Israel. *Phytoparasitica, 18*(1), 50–56.

Antignus, Y., Wang, Y., Pearlsman, M., Lachman, O., Lavi, N., & Gal-On, A., (2001). Biological and molecular characterization of a new cucurbit-infecting tobamovirus. *Phytopathology, 91,* 565–571.

Arocha, Y., Vigheri, N., Nkoy-Florent, B., Bakwanamaha, K., Bolomphety, B., Kasongo, M., Betts, P., et al., (2008). First report of the identification of *Moroccan watermelon mosaic virus* in Papaya in the Democratic Republic of Congo. *Plant Pathol., 57*(2), 387–387.

Atiri, G. I., & Varma, A., (1983). Development of improved lines of *Telfairia occidentalis* Hook. f. resistant to mosaic disease. *Tropical Agric.*

Avgelis, A. D., & Manios, V. I., (1992). Elimination of cucumber green mottle mosaic tobamo virus by composting infected cucumber residues. *Acta Hort., 302,* 311–314.

Awasthi, L. P., & Singh, S., (2009). Management of ringspot disease of papaya through plant products. *Indian Phytopath., 62*(3), 369–375.

Babitha, C. R., (1996). *Transmission, Host Range and Serodiagnosis of Pumpkin Yellow Mosaic Gemini Virus* (p. 123). MSc (Agri.) Thesis, University of Agricultural Sciences, Bangalore, India.

Baker, C., (2013). *Cucumber Green Mottle Mosaic Virus (CGMMV) Found in the United States (California) in Melon.* Pest Alert Florida Department of Agriculture and Consumer Services, Division of Plant Industry. DACS, P-01863.

Bananej, K., Desbiez, C., Girard, M., Wipf-Scheibel, C., Vahdat, I., Kheyr-Pour, A., Ahoonmanesh, A., & Lecoq, H., (2006). First report of *cucumber vein yellowing virus* on cucumber, melon, and watermelon in Iran. *Plant Dis., 90*(8), 1113–1113.

Bananej, K., Kianfar, N., Winter, S., & Menzel, W., (2014). The status of *cucumber vein yellowing virus* in Iran. *Phytopathol. Mediterr., 53*(2), 269–276.

Bananej, K., Menzel, W., Kianfar, N., Vahdat, A., & Winter, S., (2013). First report of *cucurbit chlorotic yellows virus* infecting cucumber, melon, and squash in Iran. *Plant Dis., 97*(7), 1005.

Bananej, K., Orfanidou, C. G., Maliogka, V. I., & Katis, N. I., (2018). First report of *Moroccan watermelon mosaic* virus in zucchini in Iran. *Plant Dis., 102*(10), 2047.

Barakat, O. S., Goda, H. A., Mahmoud, S. M., & Emara, K. S., (2012). Induction of systemic acquired resistance in watermelon against *watermelon mosaic virus*-2. *Arab. J. Biotech., 15*(2), 23–44.

Basavaraj, Mandal, B., Gawande, S. J., Renukadevi, P., Holkar, S. K., Krishnareddy, M., Ravi, K. S., & Jain, R. K., (2017). The occurrence, biology, serology and molecular biology of tospoviruses in Indian agriculture. In: Mandal, B., (ed.), *A Century of Plant Virology in India* (pp. 445–474). Springer; Singapore.

Basavaraj, Y. B., Siwach, J., Rawat, S., Kumar, A., Yadav, M., Patel, S., Jain, R. K., & Mandal, B., (2020). Non-structural protein based recombinant polyclonal antibodies for specific detection of *watermelon bud necrosis virus* in plants and thrips vectors. In: *Proceedings of 7ᵗʰ International Conference on Phytopathology in Achieving UN Sustainable Development Goals* (p. 66). Indian Agricultural Research Institute: New Delhi, India.

Berdiales, B., Bernal, J. J., Saez, E., Woudt, B., Beitia, F., & Rodriguez-Cerezo, E., (1999). Occurrence of *cucurbit yellow stunting disorder virus* (CYSDV) and *beet pseudo-yellows virus* in cucurbit crops in Spain and transmission of CYSDV by two biotypes of *Bemisia tabaci*. *Eur. J. Plant Pathol., 105*, 211–215.

Berger, P. H., (2005). In: Fauquet, C. M., Mayo, M. A., Maniloff, J., Desselberger, U., & Ball, L. A., (eds.), *Virus Taxonomy: Eighth Report of the International Committee on the Taxonomy of Viruses* (pp. 819–841). Elsevier Academic Press: San Diego.

Bernal, J. J., Jimenez, I., Moreno, M., Hord, M., Rivera, C., Koenig, R., & Rodriguez-Cerezo, E., (2000). *Chayote mosaic virus*, a new tymovirus infecting cucurbitaceae. *Phytopathology, 90*(10), 1098–1104.

Bhargava, K. S., & Joshi, R. D., (1960). Detection of *watermelon mosaic virus* in Uttar Pradesh. *Curr. Sci., 29*, 11.

Bhat, A. I., Pappu, S. S., Pappu, H. R., Deom, C. M., & Culbreath, A. K., (1999). Analysis of the intergenic region of tomato spotted wilt tospovirus medium RNA segment. *Virus Res., 61*(2), 161–170.

Boiteux, L. S., Spadotti, D. M. D. A., Rezende, J. A. M., & Kitajima, E. W., (2013). *Fevillea trilobata* as a natural host of *zucchini yellow mosaic virus* in Brazil. *Plant Dis., 97*(9), 1261.

Borah, K. B., & Dasgupta, I., (2012). PCR-RFLP analysis indicates that recombination might be a common occurrence among the cassava infecting begomo viruses in India. *Virus Genes, 45*(2), 327–332.

Boubourakas, I. N., Avgelis, A. D., Kyriakopoulou, P. E., & Katis, N. I., (2006). Occurrence of yellowing viruses (*beet pseudo-yellows virus, cucurbit yellow stunting disorder virus* and *cucurbit aphid-borne yellows virus*) affecting cucurbits in Greece. *Plant Pathol., 55*(2), 276–283.

Boubourakas, I. N., Hatziloukas, E., Antignus, Y., & Katis, N. I., (2004). Etiology of leaf chlorosis and deterioration of the fruit interior of watermelon plants. *J. Phytopathol., 152* (10), 580–588.

Bragard, C., Caciagli, P., Lemaire, O., Lopez-Moya, J. J., MacFarlane, S., Peters, D., Susi, P., & Torrance, L., (2013). Status and prospects of plant virus control through interference with vector transmission. *Ann. Rev. Phytopathol., 51,* 177–201.

Brown, J. K., & Merritt, R. N., (1989). Characterization of *watermelon curly mottle virus,* a Gemini virus distinct from squash leaf curl virus. *Ann. Appl. Biol., 115*(2), 243–252.

Brown, J. K., & Nelson, M. R., (1986). Whitefly-borne viruses of melons and lettuce in Arizona. *Phytopathology, 76*(2), 236–239.

Brown, J. K., (2000). Molecular markers for the identification and global tracking of whitefly vector-begomovirus complexes. *Virus Res., 71*(1, 2), 233–260.

Brown, J. K., Idris, A. M., Alteri, C., & Stenger, D. C., (2002). Emergence of a new cucurbit-infecting begomovirus species capable of forming viable reassortants with related viruses in the squash leaf curl virus cluster. *Phytopathology, 92*(7), 734–742.

Brown, J. K., Idris, A. M., Rogan, D., Hussein, M. H., & Palmieri, M., (2001). *Melon chlorotic leaf curl virus,* a new begomovirus associated with *Bemisia tabaci* infestations in Guatemala. *Plant Dis., 85*(9), 1027–1027.

Brown, J. K., Mills-Lujan, K., & Idris, A. M., (2011). Phylogenetic analysis of *melon chlorotic leaf curl virus* from Guatemala: Another emergent species in the *squash leaf curl virus* clade. *Virus Res., 158*(1/2), 257–262.

Brown, J. K., Zerbini, F. M., Navas-Castillo, J., Moriones, E., Ramos-Sobrinho, R., Silva, J. C., Fiallo-Olive, E., et al., (2015). Revision of begomovirus taxonomy based on pairwise sequence comparisons. *Arch. Virol., 160*(6), 1593–1619.

Brunt, A. A., Crabtree, K., Dallwitz, M. J., Gibbs, A. J., & Watson, L., (1996). *Viruses of Plants, Descriptions and Lists from the VIDE Database.* CAB International, Wallingford, UK.

Budzanivska, I. G., Rudneva, T. O., Shevchenko, T. P., Boubriak, I., & Polischuk, V. P., (2007). Investigation of Ukrainian isolates of *cucumber green mottle mosaic virus. Arch. Phytopathol. Plant Prot., 40*(5), 376–380.

Butnut, N., (2019). *Squash leaf curl Yunnan virus. Thai Agril. Res. J., 37*(1), 2–13.

Camelo-Garcia, V. M., Lima, E. F. B., Mansilla-Cordova, P. J., Rezende, J. A. M., Kitajima, E. W., & Barreto, M., (2014). Occurrence of *groundnut ringspot virus* on Brazilian peanut crops. *J. Gen. Plant Pathol., 80*(3), 282–286.

Campbell, R. N., Lecoq, H., Wipf-Scheibel, C., & Sim, S. T., (1991). Transmission of cucumber leaf spot virus by *Olpidium radicale. J. Gen. Virol., 72,* 3115–3119.

Cardoso, J. M. S., Felix, M. R., Oliveira, S., & Clara, M. I. E., (2004). A *tobacco necrosis virus* D isolate from olea Europaea L.: Viral characterization and coat protein sequence analysis. *Arch. Virol., 149*(6), 1129–1138.

Catoni, M., Lucioli, A., Doblas-Ibanez, P., Accotto, G. P., & Vaira, A. M., (2013). From immunity to susceptibility: Virus resistance induced in tomato by a silenced transgene is lost as TGS overcomes PTGS. *The Plant J., 75*(6), 941–953.

Celix, A., Lopez-Sese, A., Almarza, N., Gomez-Guillamon, M. L., & Rodriguez-Cerezo, E., (1996). Characterization of *cucurbit yellow stunting disorder virus,* a *Bemisia tabaci*-transmitted closterovirus. *Phytopathology, 86,* 1370–1376.

Chandrasekaran, J., Brumin, M., Wolf, D., Leibman, D., Klap, C., Pearlsman, M., Sherman, A., Arazi, T., & Gal-On, A., (2016). Development of broad virus resistance in non-transgenic cucumber using CRISPR/Cas9 technology. *Mol. Plant Pathol., 17*(7), 1140–1153.

Chao, C. H., Chen, T. C., Kang, Y. C., Li, J. T., Huang, L. H., & Yeh, S. D., (2010). Characterization of melon yellow spot virus infecting cucumber (*Cucumis sativus* L.) in Taiwan. *Plant Pathol. Bull., 19*(1), 41–52.

Chen, L. F., & Gilbertson, R. L., (2009). Curtovirus-cucurbit interaction: Acquisition host plays a role in leafhopper transmission in a host-dependent manner. *Phytopathology, 99*(1), 101–108.

Chen, T. C., Li, J. T., Lin, Y. P., Yeh, Y. C., Kang, Y. C., Huang, L. H., & Yeh, S. D., (2012). Genomic characterization of *calla lily chlorotic spot virus* and design of broad-spectrum primers for detection of tospoviruses. *Plant Pathol., 61*(1), 183–194.

Chen, T. C., Lu, Y. Y., Cheng, Y. H., Chang, C. A., & Yeh, S. D., (2008). Melon yellow spot virus in Watermelon: A first record from Taiwan. *Plant Pathol., 57*(4), 765.

Chiemsombat, P., Gajanandana, O., Warin, N., Hongprayoon, R., Bhunchoth, A., & Pongsapich, P., (2008). Biological and molecular characterization of tospoviruses in Thailand. *Arch. Virol., 153*(3), 571–577.

Cho, M. A., Moon, C. Y., Liu, J. R., & Choi, P. S., (2008). *Agrobacterium*-mediated transformation in *Citrullus lanatus. Biol. Plant., 52*, 365–369.

Choi, G. S., Kim, J. H., Chung, B. N., Kim, H. R., & Choi, Y. M., (2001). Simultaneous detection of three tobamoviruses in cucurbits by rapid immunofilter paper assay. *The Plant Pathol. J., 17*, 106–109.

Choi, P. S., Soh, W. Y., Kim, Y. S., Yoo, O. J., & Liu, J. R., (1994). Genetic transformation and plant regeneration of watermelon using *Agrobacterium tumefaciens. Plant Cell Rep., 13*, 344–348.

Choi, S. K., Yoon, J. Y., & Choi, G. S., (2015). Biological and molecular characterization of a Korean isolate of *cucurbit aphid-borne yellows virus* infecting *Cucumis* species in Korea. *The Plant Pathol. J., 31*(4), 371.

Ciuffo, M., Kurowski, C., Vivoda, E., Copes, B., Masenga, V., Falk, B. W., & Turina, M., (2009). A new tospovirus sp. in cucurbit crops in Mexico. *Plant Dis., 93*(5), 467–474.

Ciuffo, M., Nerva, L., & Turina, M., (2017). Full-length genome sequence of the tospovirus *melon severe mosaic virus. Arch. Virol., 162*(5), 1419–1422.

Claough, G. H., & Hamm, P. B., (1995). Coat protein transgenic resistance to watermelon mosaic and *zucchini yellows mosaic virus* in squash and cantaloupe. *Plant Dis., 79*(11), 1107–1109.

Coffin, R. S., & Coutts, R. H. A., (1990). The occurrence of *beet pseudo-yellows virus* in England. *Plant Pathol., 39*(4), 632–635.

Cohen, S., & Nitzany, F. E., (1960). A whitefly transmitted virus of cucurbits in Israel. *Phytopathol. Mediterr.,* 44–46.

Cohen, S., Duffus, J. E., Larsen, R. C., Liu, H. Y., & Flock, R. A., (1983). Purification, serology, and vector relationships of *squash leaf curl virus*, a whitefly-transmitted Gemini virus. *Phytopathology, 73*(12), 1669–1673.

Converse, R. H., & Ramsdell, D. C., (1982). Occurrence of tomato and tobacco ringspot viruses and of dagger and other nematodes associated with cultivated highbush blueberries in Oregon. *Plant Dis., 66*(8), 710–712.

Cuadrado, I. M., Janssen, D., Velasco, L., Ruiz, L., & Segundo, E., (2001). First report of *cucumber vein yellowing virus* in Spain. *Plant Dis., 85*(3), 336–336.

Daryono, B. S., Alaydrus, Y., Natsuaki, K. T., & Somowiyarjo, S., (2016). Inheritance of resistance to *kyuri green mottle mosaic virus* in Melon. *Sabrao J. Breed. Genet., 48*(1), 33–40.

Daryono, B. S., Somowiyarjo, S., & Natsuaki, K. T., (2005). Biological and molecular characterization of melon-infecting *kyuri green mottle mosaic virus* in Indonesia. *J. Phytopathol., 153*(10), 588–595.

Dasgupta, I., Malathi, V. G., & Mukherjee, S. K., (2003). Genetic engineering for virus resistance. *Curr. Sci., 84*(3), 341–354.

Decoin, M., (2003). Tomatoes and cucumbers watch out there are five new viruses about. *Phytoma,* 558, 27–29.

Della-Vecchia, M. G., Camargo, L. E., & Rezende, J. A., (2003). Nucleotide sequence comparison of the capsid protein gene of severe and protective mild strains of *papaya ringspot virus. Fitopatol. Brasil., 28,* 678–681.

Desbiez, C., Caciagli, P., Wipf-Scheibel, C., Millot, P., Ruiz, L., Marian, D., Dafalla, G., & Lecoq, H., (2019). Evidence for long-term prevalence of *cucumber vein yellowing virus* in Sudan and genetic variation of the virus in Sudan and the Mediterranean basin. *Plant Pathol., 68*(7), 1268–1275.

Desbiez, C., Joannon, B., Wipf-Scheibel, C., Chandeysson, C., & Lecoq, H., (2009). Emergence of new strains of *watermelon mosaic virus* in south-eastern France: Evidence for limited spread but rapid local population shift. *Virus Res., 141,* 201–208.

Desbiez, C., Justafre, I., & Lecoq, H., (2007). Molecular evidence that *zucchini yellow fleck virus* is a distinct and variable potyvirus related to *papaya ringspot virus* and *Moroccan watermelon mosaic virus. Arch. Virol., 152*(2), 449–455.

Desbiez, C., Lecoq, H., Aboulama, S., & Peterschmitt, M., (2000). First report of *cucurbit yellow stunting disorder virus* in Morocco. *Plant Dis., 84*(5), 596–596.

Dhillon, N. P. S., Ranjana, R., Singh, K., Eduardo, I., Monforte, A. J., Pitrat, M., Dhillon, N. K., & Singh, P. P., (2007). Diversity among landraces of Indian snapmelon (*Cucumis melo* var. *momordica*). *Genet. Res. Crop Evol., 54*(6), 1267–1283.

Dhillon, N. P. S., Singh, J., Fergany, M., Monforte, A. J., & Sureja, A. K., (2009). Phenotypic and molecular diversity among landraces of snapmelon (*Cucumis melo* var. *momordica*) adapted to the hot and humid tropics of Eastern India. *Plant Genet. Res., 7*(3), 291–300.

Diaz, J. A., Nieto, C., Moriones, E., Truniger, V., & Aranda, M. A., (2004). Molecular characterization of a *melon necrotic spot virus* strain that overcomes the resistance in melon and non-host plants. *Mol. Plant Microbe Int., 17*(6), 668–675.

Dogimont, C., Bussemakers, A., Martin, J., Slama, S., Lecoq, H., & Pitrat, M., (1997). Two complementary recessive genes conferring resistance to *cucurbit aphid borne yellows luteovirus* in an Indian melon line (*Cucumis melo* L.). *Euphytica, 96*(3), 391–395.

Dombrovsky, A., Tran-Nguyen, L. T., & Jones, R. A., (2017). *Cucumber green mottle mosaic virus*: Rapidly increasing global distribution, etiology, epidemiology, and management. *Ann. Rev. Phytopathol., 55,* 231–256.

Duan, C. G., Wang, C. H., & Guo, H. S., (2012). Application of RNA silencing to plant disease resistance. *Silence, 3*(1), 5.

Duffus, J. E., & Gold, A. H., (1965). Transmission of *beet western yellows virus* by aphids feeding through a membrane. *Virology, 27*(3), 388–390.

Duffus, J. E., Larsen, R. C., & Liu, H. Y., (1986). Lettuce infectious yellows virus: A new type of whitefly-transmitted virus. *Phytopathology, 76*(1), 97–100.

Dunham, J. P., Simmons, H. E., Holmes, E. C., & Stephenson, A. G., (2014). Analysis of viral (*zucchini yellow mosaic virus*) genetic diversity during systemic movement through a *Cucurbita pepo* vine. *Virus Res., 191,* 172–179.

Elsharkawy, M. M., & Mousa, K. M., (2015). Induction of systemic resistance against *papaya ring spot virus* (PRSV) and its vector *Myzus persicae* by *Penicillium simplicissimum* GP17-2 and silica (SiO2) nanopowder. *Int. J. Pest Manag., 61*(4), 353–358.

Essafi, A., Diaz-Pendon, J. A., Moriones, E., Monforte, A. J., Garcia-Mas, J., & Martin-Hernandez, A. M., (2009). Dissection of the oligogenic resistance to *cucumber mosaic virus* in the melon accession PI 161375. *Theo. Appl. Genet., 118*(2), 275–284.

Fauquet, C. M., Mayo, M. A., Maniloff, J., Desselberger, U., & Ball, L. A., (2005). *Virus Taxonomy: VIIIth Report of the International Committee on Taxonomy of Viruses*. Academic Press.

Fauquet, M. C., & Mayo, M. A., (1999). Abbreviations for plant virus names-1999. *Arch. Virol., 144*, 1249–1273.

Finetti-Sialer, M. M., Mascia, T., Cillo, F., Crisostomo, V. C., & Gallitelli, D., (2012). Biological and molecular characterization of a recombinant isolate of *watermelon mosaic virus* associated with a watermelon necrotic disease in Italy. *Eur. J. Plant Pathol., 132*, 317–322.

Flock, R. A., & Mayhew, D. E., (1981). Squash leaf curl, a new disease of cucurbits in California. *Plant Dis., 65*(1), 75–76.

Fuchs, M., Ferreira, S., & Gonsalves, D., (1997). Management of virus diseases by classical and engineered protection. *Mol. Plant Pathol.* Available at: http://www.bspp.org.uk/mppol/]1997/0116fuchs (accessed on 9 February 2021).

Fukuta, S., Kato, S., Yoshida, K., Mizukami, Y., Ishida, A., Ueda, J., Kanbe, M., & Ishimoto, Y., (2003). Detection of *tomato yellow leaf curl virus* by loop-mediated isothermal amplification reaction. *J. Virol. Meth., 112*, 35–40.

Galipienso, L., Rubio, L., Aramburu, J., Velasco, L., & Janssen, D., (2012). Complete nucleotide sequence of a severe isolate of *cucumber vein yellowing virus* from Jordan. *Arch. Virol., 157*(6), 1189–1192.

Gara, I. W., Kondo, H., Maeda, T., Inouye, N., & Tamada, T., (1997). Stunt disease of *Habenaria radiata* caused by a strain of *watermelon mosaic virus*. *Ann. Phytopath. Soc. Japan., 63*, 113–117.

Ghanekar, A. M., Reddy, D. V. R., Iizuka, N., Amin, P. W., & Gibbons, R. W., (1979). Bud necrosis of groundnut (*Arachis hypogaea*) in India caused by *tomato spotted wilt virus*. *Ann. Appl. Biol., 93*, 173–179.

Ghanem, G. A. M., Noura-Hassan, M., Kheder, A. A., Mazyad, H. M., & Abdel-Alim, A., (2016). Antiserum production, biological and serological detection of *cucurbit yellow stunting disorder crinivirus* (CYSDV) in Egypt. *Intl. J. Adv. Res., 4*(4), 1116–1128.

Ghasemzade, A., Bashir, N. S., & Masoudi, N., (2012). Sequencing part of *watermelon mosaic virus* genome and phylogenetical comparison of 5 isolates with other isolates from world. *J. Agric. Food Chem., 2*, 93–101.

Ghasemzadeh, A., Sokhandan, B. N., & Khakvar, R., (2012). Identification of Important viruses for legominosae family with using of universal primers from east-Azarbaijan province. *Proc. 12th Iranian Genetics Congress* (pp. 92–97). Tehran, Iran.

Gholamalizadeh, R., Vahdat, V., Keshavarz, T., Elahinia, A., & Bananej, K., (2008). Occurrence and distribution of ten viruses infecting cucurbits in guilan province, Iran. *Acta Virol., 52*, 113–118.

Ghoshal, K., Theilmann, J., Reade, R., Sanfacon, H., & Rochon, D. A., (2014). The *cucumber leaf spot virus* p25 auxiliary replicase protein binds and modifies the endoplasmic reticulum via N-terminal transmembrane domains. *Virology, 468*, 36–46.

Gil-Salas, F. M., Morris, J., Colyer, A., Budge, G., Boonham, N., Cuadrado, I. M., & Janssen, D., (2007). Development of real-time RT-PCR assays for the detection of *cucumber vein yellowing virus* (CVYV) and *cucurbit yellow stunting disorder virus* (CYSDV) in the whitefly vector *Bemisia tabaci*. *J. Virol. Methods, 146*(1/2), 45–51.

Glasa, M., Prichodko, Y., Zhivaeva, T., Shneider, Y., Predajňa, L., Subr, Z., & Candresse, T., (2012). Complete and partial genome sequences of the unusual *plum pox virus* (PPV) isolates from sour cherry in Russia suggest their classification to a new PPV strain. In: *22ⁿᵈ International Conference on Virus and Other Transmissible Diseases of Fruit Crops* (ICVF) (pp. 3–8). Rome, Italy.

Gonsalves, D., Chee, P., Provvidenti, R., Seem, R., & Slightom, J. L., (1992). Comparison of coat protein-mediated and genetically-derived resistance in cucumbers to infection by *cucumber mosaic virus* under field conditions with natural challenge inoculations by vectors. *Nat. Biotechnol., 10*(12), 1562–1570.

Gu, Q. S., Liu, Y. H., Wang, Y. H., Huangfu, W. G., Gu, H. F., Xu, L., Song, F. M., & Brown, J. K., (2011). First report of cucurbit chlorotic yellows virus in cucumber, melon, and watermelon in China. *Plant Dis., 95*(1), 73.

Gu, Q. S., Wu, H. J., Chen, H. Y., Zhang, X. J., Wu, M. Z., Wang, D. M., Peng, B., Kong, X. Y., & Liu, T. J., (2012). Melon yellow spot virus identified in China for the first time. *New Dis. Rep., 25*(7), 2044–2588.

Guzman, P., Sudarshana, M. R., Seo, Y. S., Rojas, M. R., Natwick, E., Turini, T., Mayberry, K., & Gilbertson, R., (2000). A new bipartite Gemini virus (*Begomovirus*) causing leaf curl and crumpling in cucurbits in the imperial valley of California. *Plant Dis., 84*(4), 488.

Gyoutoku, Y., Okazaki, S., Furuta, A., Etoh, T., Mizobe, M., Kuno, K., Hayashida, S., & Okuda, M., (2009). Chlorotic yellows disease of melon caused by *cucurbit chlorotic yellows virus*, a new crinivirus. *Jpn. J. Phytopathol., 75*, 109–111.

Hamed, K., Menzel, W., Dafalla, G., Gadelseed, A. M. A., & Winter, S., (2011). First report of cucurbit chlorotic yellows virus infecting muskmelon and cucumber in Sudan. *Plant Dis., 95*(10), 1321.

Hammond, R. W., Hernandez, E., Mora, F., & Ramirez, P., (2005). First report of beet pseudo-yellows virus on *Cucurbita moschata* and *C. pepo* in costa Rica. *Plant Dis., 89* (10), 1130.

Han, J. S., Kim, C. K., Park, S. H., Hirschi, K. D., & Mok, I. G., (2005). *Agrobacterium*-mediated transformation of bottle gourd (*Lagenaria siceraria* Standl.). *Plant Cell Rep., 23*(10/11), 692–698.

Han, S. S., Yoshida, K., Karasev, A. V., & Iwanami, T., (2002). Nucleotide sequence of a Japanese isolate of *squash mosaic virus. Arch. Virol., 147*(2), 437–443.

Hanley-Bowdoin, L., Settlage, S. B., Orozco, B. M., Nagar, S., & Robertson, D., (1999). Gemini viruses: Models for plant DNA replication, transcription, and cell cycle regulation. *Critical Rev. Plant Sci., 18*(1), 71–106.

Hartono, S., Natsuaki, T., Genda, Y., & Okuda, S., (2003). Nucleotide sequence and genome organization of *cucumber yellows virus*, a member of the genus crinivirus. *J. Gen. Virol., 84*(4), 1007–1012.

Hassan, A. A., & Duffus, J. E., (1991). A review of a yellowing and stunting disorder of cucurbits in the United Arab Emirates. *Emir. J. Agric. Sci., 2*, 1–16.

Hassani-Mehraban, A., Botermans, M., Verhoeven, J. T. J., Meekes, E., Saaijer, J., Peters, D., Goldbach, R., & Kormelink, R., (2010). A distinct tospovirus causing necrotic streak on *Alstroemeria* sp. in Colombia. *Arch. Virol., 155*(3), 423–428.

Hemalatha, V., Pradnya, G., Anjali, A. K., Krishnareddy, M., & Savithri, H. S., (2008). Monoclonal antibodies to the recombinant nucleocapsid protein of a *groundnut bud necrosis virus* infecting tomato in Karnataka and their use in profiling the epitopes of Indian tospovirus isolates. *Curr. Sci., 95*, 952–957.

Hernandez-Zepeda, C., Idris, A. M., Carnevali, G., Brown, J. K., & Moreno-Valenz-uela, O. A., (2007). Molecular characterization and experimental host range of euphorbia mosaic virus-Yucatan peninsula, a begomovirus species in the *squash leaf curl virus* clade. *Plant Pathol., 56*(5), 763–770.

Heydarnejad, J., Mozaffari, A., Massumi, H., Fazeli, R., Gray, A. J., Meredith, S., Lakay, F., Shepherd, D. N., Martin, D. P., & Varsani, A., (2009). Complete sequences of *tomato leaf curl Palampur virus* isolates infecting cucurbits in Iran. *Arch. Virol., 154*(6), 1015–1018.

Holkar, S. K., Kumar, R., Yogita, M., Katiyar, A., Jain, R. K., & Mandal, B., (2017). Diagnostic assays for two closely related tospovirus species, *watermelon bud necrosis virus* and *groundnut bud necrosis virus* and identification of new natural hosts. *J. Plant Biochem. Biotechnol., 26*(1), 43–51.

Holkar, S. K., Mandal, B., & Jain, R. K., (2018). Development and validation of marker-free constructs based on nucleocapsid protein gene of *watermelon bud necrosis orthotospovirus* in watermelon. *Curr. Sci., 114*(8), 1742.

Holkar, S. K., Mandal, B., Reddy, M. K., & Jain, R. K., (2019). *Watermelon bud necrosis orthotospovirus*: An emerging constraint in the Indian subcontinent: An overview. *Crop Prot., 117*, 52–62.

Hongyun, C., Wenjun, Z., Qinsheng, G., Qing, C., Shiming, L., & Shuifang, Z., (2008). Real time TaqMan RT-PCR assay for the detection of *cucumber green mottle mosaic virus. J. Virol. Methods, 149*(2), 326–329.

Hosseini, S., Mosahebi, G. H., Koohi, Habibi, M., & Okhovvat, S. M., (2007). Characterization of the *zucchini yellow mosaic virus* from squash in Tehran province. *J. Agric. Sci. Technol., 9*, 137–143.

Hourani, H., & Abou-Jawdah, Y., (2003). Immunodiagnosis of *cucurbit yellow stunting disorder virus* using polyclonal antibodies developed against recombinant coat protein. *J. Plant Pathol.*, 197–204.

Hu, J., Zhou, T., Liu, L., Peng, B., Li, H., Fan, Z., & Gu, Q., (2009). The genomic sequence of a Chinese isolate of *squash mosaic virus* with novel 5′ conserved ends. *Virus Genes, 38*(3), 475–477.

Huang, C. H., Liang, S. C., Deng, T. C., & Hseu, S. H., (1993). Comparison of diagnostic host and serological tests for four cucurbit potyviruses. *Plant Pathol. Bull., 2*(3), 169–176.

Huang, Y. C., Chiang, C. H., Li, C. M., & Yu, T. A., (2011). Transgenic watermelon lines expressing the nucleocapsid gene of *watermelon silver mottle virus* and the role of thiamine in reducing hyperhydricity in regenerated shoots. *Plant Cell Tissue Org. Cult., 106*(1), 21–29.

Ibaba, J. D., Laing, M. D., & Gubba, A., (2015). Incidence and phylogeny of viruses infecting cucurbit crops in Kwa Zulu-natal, republic of South Africa. *Crop Prot., 75*, 46–54.

Ibaba, J. D., Laing, M. D., & Gubba, A., (2016a). *Zucchini shoestring virus*: A distinct potyvirus in the *papaya ringspot virus* cluster. *Arch. Virol., 161*(8), 2321–2323.

Ibaba, J. D., Laing, M. D., & Gubba, A., (2016b). Genome sequence analysis of two South African isolates of *Moroccan watermelon mosaic virus* infecting cucurbits. *Virus Genes, 52*(6), 896–899.

Ibrahim, I. A., Nower, A. A., Badr-Elden, A. M., & Elaziem, T. M. A., (2009). High efficiency plant regeneration and transformation of watermelon (*Citrulus lanatus* cv. Giza1). *Res. J. Agric. Biol. Sci., 5*(5), 689–697.

Idris, A. M., Mills-Lujan, K., Martin, K., & Brown, J. K., (2008). *Melon chlorotic leaf curl virus*: Characterization and differential reassortment with closest relatives reveal adaptive virulence in the squash leaf curl virus clade and host shifting by the host-restricted *bean calico mosaic virus*. *J. Virol., 82*(4), 1959–1967.

Isakeit, T., Robertson, N. L., Brown, J. K., & Gilbertson, R. L., (1994). First report of *squash leaf curl virus* on watermelon in Texas. *Plant Dis., 78*(10), 1010.

Ito, T., Ogawa, T., Samretwanich, K., Sharma, P., & Ikegami, M., (2008). Yellow leaf curl disease of pumpkin in Thailand is associated with *squash leaf curl China virus*. *Plant Pathol., 57*(4), 766.

Jain, R. K., Bag, S., Umamaheswaran, K., & Mandal, B., (2007). Natural infection by tospovirus of cucurbitaceous and fabaceous vegetable crops. *J. Phytopathol., 155*, 22–25.

Jain, R. K., Pandey, N. A., Reddy, M. K., & Mandal, B., (2005). Immunodiagnosis of groundnut and watermelon bud necrosis viruses using polyclonal antiserum to recombinant nucleocapsid protein of *groundnut bud necrosis virus*. *J. Virol. Methods, 130*, 162–164.

Jain, R. K., Pappu, H. R., Pappu, S. S., Krishanareddy, M., & Vani, A., (1998). *Watermelon bud necrosis tospovirus* is a distinct virus species belonging to serogroup IV. *Arch. Virol., 143*, 1637–1644.

Janssen, D., Saez, E., Segundo, E., Martin, G., Gil, F., & Cuadrado, I. M., (2005). *Capsicum annuum*: A new host of *Parietaria mottle virus* in Spain. *Plant Pathol., 54*(4), 567–567.

Janssen, D., Velasco, L., Martin, G., Segundo, E., & Cuadrado, I. M., (2007). Low genetic diversity among *cucumber vein yellowing virus* isolates from Spain. *Virus Genes, 34*(3), 367–371.

Ji, X., Zhang, H., Zhang, Y., Wang, Y., & Gao, C., (2015). Establishing a CRISPR-Cas-like immune system conferring DNA virus resistance in plants. *Nat. Plants, 1*(10), 1–4.

Jones, D. R., (2005). Plant viruses transmitted by thrips. *Eur. J. Plant Pathol., 113*(2), 119–157.

Jones, P., Angood, S., & Carpenter, J. M., (1986). *Melon rugose mosaic virus*, the cause of a disease of watermelon and sweet melon. *Ann. Appl. Biol., 108*(2), 303–307.

Jones, P., Sattar, M. H. A., & Al-Kaff, N., (1988). The incidence of virus disease in watermelon and sweet melon crops in the people democratic republic of Yemen and its impact on cropping policy. *Ann. Appl. Biol., 17*, 203–207.

Jones, R. A., (2006). Control of plant virus diseases. *Adv. Virus Res., 67*, 205–244.

Juarez, M., Tovar, R., Fiallo-Olive, E., Aranda, M. A., Gosalvez, B., Castillo, P., Moriones, E., & Navas-Castillo, J., (2014). First detection of *tomato leaf curl New Delhi virus* infecting zucchini in Spain. *Plant Dis., 98*(6), 857–857.

Jyothsna, P., Haq, Q. M. I., Singh, P., Sumiya, K. V., Praveen, S., Rawat, R., Briddon, R. W., & Malathi, V. G., (2013). Infection of *tomato leaf curl New Delhi virus* (ToLCNDV), a bipartite begomovirus with beta satellites, results in enhanced level of helper virus components and antagonistic interaction between DNA B and beta satellites. *Appl. Microbiol. Biotechnol., 97*(12), 5457–5471.

Kamanna, B. C., Jadhav, S. N., & Shankarappa, T. H., (2010). Evaluation of insecticides against thrips vector for the management of *watermelon bud necrosis virus* (WBNV) disease. *Karnataka J. Agric. Sci., 23*, 172–173.

Kamberoglu, M. A., Caliskan, A. F., & Desbiez, C., (2016). Current status of some cucurbit viruses in Cukurova region (Adana and Mersin provinces) of turkey and molecular characterization of *zucchini yellow mosaic virus* Isolates. *Romanian Biotechnol. Letters, 21*(4), 11709.

Kao, J., Jia, L., Tian, T., Rubio, L., & Falk, B. W., (2000). First report of *cucurbit yellow stunting disorder virus* (genus *Crinivirus*) in North America. *Plant Dis., 84*(1), 101.

Karavina, C., Ibaba, J. D., & Gubba, A., (2020). High-throughput sequencing of virus-infected *Cucurbita pepo* samples revealed the presence of *zucchini shoestring virus* in Zimbabwe. *BMC Res. Notes, 13*(1), 53.

Kato, K., Hanada, K., & Kameya-Iwaki, M., (1999). Transmission mode, host range and electron microscopy of a pathogen causing a new disease of melon (*Cucumis melo*) in Japan. *Ann. Phytopathol. Soci. Jpn., 65*, 624–627.

Kato, K., Hanada, K., & Kameya-Iwaki, M., (2000). *Melon yellow spot virus*: A distinct species of the genus tospovirus isolated from Melon. *Phytopathology, 90*, 422–426.

Kawchuk, L. M., Martin, R. R., & McPherson, J., (1990). Resistance in transgenic potato expressing the potato leafroll virus coat protein gene. *Mol. Plant Microbe Interact., 3*, 301–307.

Khan, A. J., Akhtar, S., Briddon, R. W., Ammara, U., Al-Matrooshi, A. M., & Mansoor, S., (2012). Complete nucleotide sequence of *watermelon chlorotic stunt virus* originating from Oman. *Viruses, 4*(7), 1169–1181.

Kheyr-Pour, A., Bananej, K., Dafalla, G. A., Caciagli, P., Noris, E., Ahoonmanesh, A., Lecoq, H., & Gronenborn, B., (2000). *Watermelon chlorotic stunt virus* from the Sudan and Iran: Sequence comparisons and identification of a whitefly-transmission determinant. *Phytopathology, 90*(6), 629–635.

Kim, D. H., & Lee, J. M., (2000). Seed treatment for *cucumber green mottle mosaic virus* (CGMMV) in Gourd (*Lagenaria siceraria*) seeds and its detection. *J. Korean Soc. Hort. Sci., 41*, 1–6.

Kim, O. K., Mizutani, T., Natsuaki, K. T., Lee, K. W., & Soe, K., (2010). First report and the genetic variability of *cucumber green mottle mosaic virus* occurring on bottle gourd in Myanmar. *J. Phytopathol., 158*(7, 8), 572–575.

King, A. M., Lefkowitz, E., Adams, M. J., & Carstens, E. B., (2011). *Virus Taxonomy: Ninth Report of the International Committee on Taxonomy of Viruses* (p. 9). Elsevier.

Koenig, R., Lesemann, D. E., Huth, W., & Makkouk, M. M., (1983). Comparison of a new soilborne virus from cucumber with tombus-, diantho-, and other similar viruses. *Phytopathology, 73*, 515–520.

Koenraadt, H. M. S., & Remeeus, P. M., (2010). Detection of *squash mosaic virus, cucumber green mottle mosaic virus* and *melon necrotic spot virus* in Cucurbits. International Seed Testing Association (ISTA), International Rules for Seeding Testing, Bassersdorf, Switzerland. In: *Seed Health Methods* (pp. 7-026-3).

Komuro, Y., (1971). *Cucumber green mottle mosaic virus* on cucumber and watermelon and *melon necrotic spot virus* on muskmelon. *Jpn. Agric. Res. Quart., 6*, 11 15.

Kon, T., Dolores, L. M., Bajet, N. B., Hase, S., Takahashi, H., & Ikegami, M., (2003). Molecular characterization of a strain of *squash leaf curl China virus* from the Philippines. *J. Phytopathol., 151*(10), 535–539.

Koot, Y. V., & Dorst, H. J. M., (1959). Virusziekten Van de Komko mer in Nederland (virus disease of cucumber in Netherlands). *Tijdschrift Plziekt., 65*, 257–271.

Kostova, D., Masenga, V., Milne, R. G., & Lisa, V., (2001). First report of *eggplant mottled dwarf virus* in cucumber and pepper in Bulgaria. *Plant Pathol., 50*(6), 804–804.

Krishnan, N., Kumari, S., Dubey, V., Rai, A. B., Meena, B. R., Singh, A. K., Chinnappa, M., & Singh, B., (2019). First report of natural occurrence of *watermelon bud necrosis virus* in round melon (*Praecitrullus fistulosus*) in India. *Plant Dis., 103*(4), 781.

Krishnareddy, M., & Singh, S. J., (1993). Immunological and molecular based diagnosis of tospovirus infecting watermelon. In: *Golden Jubilee Symposium on Horticultural Research:*

Changing Scenario (pp. 247, 248). Indian Institute of Horticultural Research: Bangalore, India.

Kumar A., Basavaraj, Y. B., Bhattarai, A., Rathore, A. S, Behera, T., Renukadevi, P., Baranwal, V. K., and Jain, R. K., (2020). Natural occurrence of cucurbit aphid-borne yellows virus on cucurbitacious hosts from India. In: *VIROCON 2020: International Conference on Evolution of Viruses and Viral Diseases.* (pp. 153). Indian National Science Academy: New Delhi, India.

Kumar, R., & Mandal, B., (2013). Transgenic constructs and transformation of watermelon cv. sugar baby for developing transgenic resistance against *watermelon bud necrosis virus. Indian J. Virol., 24*(1), 99–149.

Kumar, R., Geetanjali, A. S., Krishnareddy, M., Jaiwal, P. K., & Mandal, B., (2019). Standardization of regeneration, agrobacterium-mediated transformation, and introduction of nucleocapsid gene of *watermelon bud necrosis virus* in watermelon. *Proc. Natl. Acad. Sci. India Sec. B: Biol. Sci., 1*–8.

Kumar, R., Mandal, B., Geetanjali, A. S., Jain, R. K., & Jaiwal, P. K., (2010). Genome organization and sequence comparison suggest intraspecies incongruence in mRNA of *watermelon bud necrosis virus. Arch. Virol., 155*(8), 1361–1365.

Kunkalikar, S. R., Sudarsana, P., Arun, B. M., Rajagopalan, P., Chen, T. C., Yeh, S. D., Naidu, R. A., Zehr, U. B., & Ravi, K. S., (2011). Importance and genetic diversity of vegetable-infecting tospoviruses in India. *Phytopathology, 101*, 367–376.

Kunkalikar, S., Poojari, S., Rajagopalan, P., Zehr, U. B., Naidu, R. A., & Kankanallu, R. S., (2007). First report of *capsicum chlorosis virus* in tomato in India. *Plant Health Progr., 8*(1), 37.

Kuo, Y. W., Rojas, M. R., Gilbertson, R. L., & Wintermantel, W. M., (2007). First report of cucurbit yellow stunting disorder virus in California and Arizona, in association with *cucurbit leaf crumple virus* and *squash leaf curl virus. Plant Dis., 91*(3), 330–330.

Kushwaha, N., Singh, A. K., Chattopadhyay, B., & Chakraborty, S., (2010). Recent advances in Gemini virus detection and future perspectives. *J. Plant Prot. Sci., 2*, 1–18.

Laney, A. G., Avanzato, M. V., & Tzanetakis, I. E., (2012). High incidence of seed transmission of *papaya ringspot virus* and *watermelon mosaic virus*, two viruses newly identified in *Robinia pseudoacacia. Eur. J. Plant Pathol., 134*(2), 227–230.

Lapidot, M., Gelbart, D., Galon, A., Sela, N., Anfoka, G., Ahmed, F. H., & Duffy, S., (2014). Frequent migration of introduced cucurbit-infecting begomoviruses among middle eastern Countries. *Virol. J., 11*(1), 181.

Lazarowitz, S. G., & Inara, B. L., (1991). Infectivity and complete nucleotide sequence of the cloned genomic components of a bipartite *squash leaf curl Gemini virus* with a broad host range phenotype. *Virology, 180*(1), 58–69.

Lecoq, H., & Desbiez, C., (2012). Viruses of cucurbit crops in the Mediterranean region: An ever-changing picture. *Adv. Virus Res., 84*, 67–126.

Lecoq, H., Bourdin, D., Wipf-Scheibel, C., Bon, M., Lot, H., Lemaire, O., & Herrbach, E., (1992). A new yellowing disease of cucurbits caused by a luteovirus, cucurbit aphid-borne yellows virus. *Plant Pathol., 41*(6), 749–761.

Lecoq, H., Dafalla, G., Desbiez, C., Wipf-Scheibel, C., Delecolle, B., Lanina, T., Ullah, Z., & Grumet, R., (2001). Biological and molecular characterization of *Moroccan watermelon mosaic virus* and a potyvirus isolate from eastern Sudan. *Plant Dis., 85*(5), 547–552.

Lecoq, H., Desbiez, C., Wipf-Scheibel, C., & Girard, M., (2003). Potential involvement of melon fruit in the long-distance dissemination of cucurbit potyviruses. *Plant Dis., 87*(8), 955–959.

Lecoq, H., Dufour, O., Wipf-Scheibel, C., Girard, M., Cotillon, A. C., & Desbiez, C., (2007). First report of *cucumber vein yellowing virus* in melon in France. *Plant Dis., 91*(7), 909–909.

Lecoq, H., Lisa, V., & Dellavalle, G., (1983). Serological identity of *muskmelon yellow stunt* and *zucchini yellow mosaic viruses. Plant Dis., 67,* 824–825.

Lecoq, H., Wipf-Scheibel, C., Nozeran, K., Millot, P., & Desbiez, C., (2014). Comparative molecular epidemiology provides new insights into *zucchini yellow mosaic virus* occurrence in France. *Virus Res., 186,* 135–143.

Lecoq, H., Wisler, G., & Pitrat, M., (1998). Cucurbit viruses: The classics and the emerging. In: *Cucurbitaceae, 98,* 126–142.

Leibman, D., Wolf, D., Saharan, V., Zelcer, A., Arazi, T., Yoel, S., Gaba, V., & Galon, A., (2011). A High level of transgenic viral small RNA is associated with broad potyvirus resistance in cucurbits. *Mol. Plant Microbe Interact., 24*(10), 1220–1238.

Leke, W. N., Brown, J. K., & Fondong, V., (2020). First report of *Chayote yellow mosaic virus* (ChaYMV) and its associated beta satellite infecting papaya (*Carica papaya*) in Cameroon. *Plant Dis.* Published online: https://doi.org/10.1094/PDIS-11-19-2293-PDN.

Leke, W. N., Mignouna, D. B., Brown, J. K., & Fondong, V. N., (2016). First report of chayote yellow mosaic virus infecting bitter melon (*Momordica charantia*) exhibiting yellow mosaic symptoms in Benin, Nigeria, and Togo. *Plant Disease, 100*(5), 1031.

Lewsey, M. G., & Carr, J. P., (2009). Effects of DICER-like proteins 2, 3 and 4 on *cucumber mosaic virus* and *tobacco mosaic virus* infections in salicylic acid-treated plants. *J. Gen. Virol., 90*(12), 3010–3014.

Li, J. T., Yeh, Y. C., Yeh, S. D., Raja, J. A., Rajagopalan, P. A., Liu, L. Y., & Chen, T. C., (2011). Complete genomic sequence of *watermelon bud necrosis virus. Arch. Virol., 156*(2), 359–362.

Li, Z., Yang, S., Qin, B., Xie, H., Cui, L., Su, Q., Cai, J., & Gu, Q., (2018). First report of natural infection of *zucchini green mottle mosaic virus* on bottle gourd in Guangxi, China. *Plant Dis., 102*(11), 2384.

Liao, J. Y., Hu, C. C., Lin, T. K., Chang, C. A., & Deng, T. C., (2007). Identification of *squash leaf curl Philippines virus* on *Benincasa hispida* in Taiwan. *Plant Pathol. Bull., 16,* 11–18.

Lin, C. Y., Ku, H. M., Chiang, Y. H., Ho, H. Y., Yu, T. A., & Jan, F. J., (2012). Development of transgenic watermelon resistant to *cucumber mosaic virus* and *watermelon mosaic virus* by using a single chimeric transgene construct. *Transgenic Res., 21*(5), 983–993.

Lisa, V., Milne, R. G., Accotto, G. P., Boccardo, G., Caciagli, P., & Parvizy, R., (1988). *Ourmia melon virus,* a virus from Iran with novel properties. *Ann. Appl. Biol., 112,* 291–302.

Liu, Y., Schiff, M., & Dinesh-Kumar, S. P., (2002). Virus-induced gene silencing in tomato. *Plant J., 31,* 777–786.

Liu, Y., Wang, Y., Wang, X., & Zhou, G., (2009). Molecular characterization and distribution of *cucumber green mottle mosaic* virus in China. *J. Phytopathol., 157*(7/8), 393–399.

Livieratos, I. C., Eliasco, E., Muller, G., Olsthoorn, R. C. L., Salazar, L. F., Pleij, C. W. A., & Coutts, R. H., (2004). Analysis of the RNA of potato yellow vein virus: Evidence for a tripartite genome and conserved 3′-terminal structures among members of the genus crinivirus. *J. Gen. Virol., 85*(7), 2065–2075.

Livieratos, I. C., Katis, N., & Coutts, R. H. A., (1998). Differentiation between *cucurbit yellow stunting disorder virus* and *beet pseudo-yellows virus* by a reverse transcription polymerase chain reaction assay. *Plant Pathol., 47*(3), 362–369.

Londono, A., Capobianco, H., Zhang, S., & Polston, J. E., (2012). First record of *tomato chlorotic spot virus* in the USA. *Trop. Plant Pathol., 37,* 333–338.

Londono, M. A., Harmon, C. L., & Polston, J. E., (2016). Evaluation of recombinase polymerase amplification for detection of begomoviruses by plant diagnostic clinics. *Virol. J., 13*(1), 48.

Lopez, C., Aramburu, J., Galipienso, L., & Nuez, F., (2007). Characterization of several heterogeneous species of defective RNAs derived from RNA 3 of *cucumber mosaic virus. Arch. Virol., 152*(3), 621–627.

Louro, D., Quinot, A., Neto, E., Fernandes, J. E., Marian, D., Vecchiati, M., Caciagli, P., & Vaira, A. M., (2004). Occurrence of *cucumber vein yellowing virus* in cucurbitaceous species in Southern Portugal. *Plant Pathol., 53*(2), 241.

Louro, D., Vicente, M., Vaira, A. M., Accotto, G. P., & Nolasco, G., (2000). *Cucurbit yellow stunting disorder virus* (Genus *Crinivirus*) associated with the yellowing disease of cucurbit crops in Portugal. *Plant Dis., 84*(10), 1156.

Madhusudhan, K. N., Vinayarani, G., Deepak, S. A., Niranjana, S. R., Prakash, H. S., Singh, G. P., Sinha, A. K., & Prasad, B. C., (2011). Antiviral activity of plant extracts and other inducers against tobamoviruses infection in bell pepper and tomato plants. *Intl. J. Plant Pathol., 2*, 35–42.

Mahgoub, H. A., Desbiez, C., Wipf-Scheibel, C., Dafalla, G., & Lecoq, H., (1997). Characterization and occurrence of *zucchini yellow mosaic virus* in Sudan. *Plant Pathol., 46*(5), 800–805.

Maina, S., Edwards, O. R., De Almeida, L., Ximenes, A., & Jones, R. A., (2017). First complete *squash leaf curl China virus* genomic segment DNA-A sequence from east Timor. *Genome Announc., 5*(24), e00483–00417.

Malandraki, I., Vassilakos, N., Xanthis, C., Kontosfiris, G., Katis, N. I., & Varveri, C., (2014). First report of *Moroccan watermelon mosaic virus* in zucchini crops in Greece. *Intl. J. Virol., 10*, 253–262.

Mandal, B., Jain, R. K., Chaudhary, V., & Varma, A., (2003). First report of natural infection of *Luffa acutangula* by *watermelon bud necrosis virus* in India. *Plant Dis., 87*, 598.

Mandal, B., Jain, R. K., Reddy, M. K., Kumar, N. K., Ravi, K. S., & Pappu, H. R., (2012). Emerging problems of tospoviruses (Bunyaviridae) and their management in the Indian subcontinent. *Plant Dis., 96*(4), 468–479.

Mandal, B., Mandal, S., Sohrab, S. S., Pun, K. B., & Varma, A., (2004). A new yellow mosaic disease of chayote in India. *Plant Pathol., 53*, 797.

Maneechoat, P., Takeshita, M., Uenoyama, M., Nakatsukasa, M., Kuroda, A., Furuya, N., & Tsuchiya, K., (2015). A single amino acid at n-terminal region of the 2b protein of *cucumber mosaic virus* strain m1 has a pivotal role in virus attenuation. *Virus Res., 197*, 67–74.

Manglli, A., Murenu, M., Sitzia, M., & Tomassoli, L., (2016). First report of *cucurbit yellow stunting disorder virus* infecting cucurbits in Italy. *New Dis. Rep., 34*, 23–23.

Mangrauthia, S. K., Parameswari, B., Jain, R. K., & Praveen, S., (2008). Role of genetic recombination in the molecular architecture of *papaya ringspot virus. Biochem. Genet., 46*(11/12), 835–846.

Mansour, A., & Al-Musa, A., (1993). *Cucumber vein yellowing virus*; Host range and virus vector relationships. *J. Phytopathol., 137*(1), 73–78.

Martelli, G. P., Agranovsky, A. A., Bar-Joseph, M., Boscia, D., Candresse, T., Coutts, R. H. A., Dolja, V. V., et al., (2002). The family closteroviridae revised. *Arch. Virol., 147*, 2039–2044.

Martelli, G. P., Ghanem-Sabanadzovic, N. A., Agranovsky, A. A., Rwahnih, M. A., Dolja, V. V., Dovas, C. I., Fuchs, M., et al., (2012). Taxonomic revision of the family closteroviridae with special reference to the grapevine leafroll-associated members of the genus *Ampelovirus* and the putative species unassigned to the family. *J. Plant Pathol., 7–19.*

Maruthi, M. N., Rekha, A. R., & Muniyappa, V., (2007). Pumpkin Yellow vein mosaic disease is caused by two distinct begomoviruses: Complete viral sequences and comparative transmission by an indigenous *Bemisia tabaci* and the introduced B-biotype. *EPPO Bull., 37*(2), 412–419.

McCreight, J. D., Wintermantel, W. M., & Pitrat, M., (2008). Potential new sources of genetic resistance in melon to *cucurbit yellow stunting disorder virus*. In: Pitrat, M., (ed.), *Proceedings of the IX*th *EUCARPIA Meeting on Genetics and Breeding of Cucurbitaceae.* INRA, Avignon, France.

Mederos, D. C., Giolitti, F., Nome, C., Bejerman, N., & Portal, O., (2017). *Papaya ringspot virus* W infecting *Luffa aegyptiaca* in Cuba. *Austral. Plant Dis. Notes, 12*(1), 5.

Mehdi, S., & Sudhakar, D. W., (2008). Tomato breeding for resistance to *Tomato spotted wilt virus* (TSWV): An overview of conventional and molecular approaches. *Czech. J. Genet. Plant Breed., 44*(3), 83–92.

Melzer, M. J., Shimabukuro, J., Long, M. H., Nelson, S. C., Alvarez, A. M., Borth, W. B., & Hu, J. S., (2014). First report of capsicum chlorosis virus infecting waxflower (*Hoya calycina* Schlecter) in the United States. *Plant Dis., 98*(4), 571.

Menzel, W., Maeritz, U., & Seigner, L., (2020). First report of *cucurbit aphid-borne yellows virus* infecting cucurbits in Germany. *New Dis. Rep., 41*, 1.

Miller, J. S., Damude, H., Robbins, M. A., Reade, R. D., & Rochon, D. M., (1997). Genome structure of *cucumber leaf spot virus*: Sequence analysis suggests it belongs to a distinct species within the tombusviridae. *Virus Res., 52*(1), 51–60.

Milojevic, K., Stankovic, I., Vucurovic, A., Ristic, D., Milosevic, D., Bulajic, A., & Krstic, B., (2013). First report of *cucumber mosaic virus* infecting peperomia tuisana in Serbia. *Plant Dis., 97*(7), 1004.

Milojevic, K., Stankovic, I., Vucurovic, A., Ristic, D., Nikolic, D., Bulajic, A., & Krstic, B., (2012). First report of *cucumber mosaic virus* infecting watermelon in Serbia. *Plant Dis., 96*(11), 1706–1706.

Miras, M., Juarez, M., & Aranda, M. A., (2019). Resistance to the emerging *Moroccan watermelon mosaic virus* in squash. *Phytopathology, 109*(5), 895–903.

Mizutani, T., Daryono, B. S., Ikegami, M., & Natsuaki, K. T., (2011). First report of *tomato leaf curl New Delhi virus* infecting cucumber in central java, Indonesia. *Plant Dis., 95*(11), 1485–1485.

Mnari-Hattab, M., Gauthier, N., & Zouba, A., (2009). Biological and molecular characterization of the *cucurbit aphid-borne yellows virus* affecting cucurbits in Tunisia. *Plant Dis., 93*(10), 1065–1072.

Mnari-Hattab, M., Zammouri, S., Belkadhi, M. S., Dona, D. B., Ben, N. E., & Hajlaoui, M. R., (2015). First report of *tomato leaf curl New Delhi virus* infecting cucurbits in Tunisia. *New Dis. Rep., 31*(21), 2044–0588.

Mochizuki, T., Hirai, K., Kanda, A., Ohnishi, J., Ohki, T., & Tsuda, S., (2009). Induction of necrosis via mitochondrial targeting of *melon necrotic spot virus* replication protein p29 by its second trans-membrane domain. *Virology, 390*(2), 239–249.

Mohammed, R. S. U., Deepan, S., Dharanivasan, G., Jesse, M. I., Muthuramalingam, R., & Kathiravan, K., (2013). First report on a variant of *squash leaf curl China virus* (SLCCNV) infecting *Benincasa hispida* in India. *New Dis. Rep., 28*, 20.

Molders, W., Buchala, A., & Metraux, J. P., (1996). Transport of salicylic acid in *tobacco necrosis virus*-infected cucumber plants. *Plant Physiol., 112*(2), 787–792.

Mound, L. A., (2011). Grass-dependent Thysanoptera of the family Thripidae from Australia. *Zootaxa, 3064*(1), 1–40.

Muniyappa, V., Maruthi, M. N., Babitha, C. R., Colvin, J., Briddon, R. W., & Rangaswamy, K. T., (2003). Characterization of *pumpkin yellow vein mosaic virus* from India. *Ann. App. Biol., 142*, 323–331.

Murphy, J. F., Reddy, M. S., Ryu, C. M., Kloepper, J. W., & Li, R. X., (2003). Rhizobacteria mediated growth promotion of tomato leads to protection against *cucumber mosaic virus*. *Phytopathology, 93*, 1301–1307.

Nagendran, K., Aravintharaj, R., Mohankumar, S., Manoranjitham, S. K., Naidu, R. A., & Karthikeyan, G., (2015). First report of *cucumber green mottle mosaic virus* in snake gourd (*Trichosanthes cucumerina*) in India. *Plant Dis., 99*(4), 559–559.

Nagendran, K., Kumari, S., Rai, A. B., Manimurugan, C., Singh, B., Karthikeyan, G., & Naidu, R. A., (2018). First report of *peanut bud necrosis virus* infecting bitter gourd (*Momordica charantia* L.) in India. *Plant Dis., 102*(3), 690–690.

Nagendran, K., Mohankumar, S., Aravintharaj, R., Balaji, C. G., Manoranjitham, S. K., Singh, A. K., Rai, A. B., Singh, B., & Karthikeyan, G., (2017). The occurrence and distribution of major viruses infecting cucurbits in Tamil Nadu State, India. *Crop Prot., 99*, 10–16.

Namrata, J., Saritha, R. K., Datta, D., Singh, M., Dubey, R. S., Rai, A. B., & Rai, M., (2010). Molecular characterization of *tomato leaf curl Palampur virus* and pepper leaf curl beta satellite naturally infecting pumpkin (*Cucurbita moschata*) in India. *Ind. J. Virol., 21*(2), 128–132.

Navas-Castillo, J., Fiallo-Olive, E., & Sanchez-Campos, S., (2011). Emerging virus diseases transmitted by whiteflies. *Annu. Rev. Phytopathol., 49*, 219–248.

Noda, C., Kittipakorn, K., Inchan, P., Wanapee, L., & Deema, N., (1993). Distribution of cucurbit viruses and reactions of some cucurbit species to certain viruses. *Proc. Kasetsart Univ. Annu. Conf., 31*[st] (pp. 341–347). Bangkok, Bangkok, Thail.: Kaestart Univ.

Novakova, S., Svoboda, J., & Glasa, M., (2014). Analysis of the Complete sequences of two biologically distinct *zucchini yellow mosaic virus* isolates further evidences the involvement of a single amino acid in the virus pathogenicity. *Acta. Virol., 58*, 364–367.

Ohki, T., Akita, F., Mochizuki, T., Kanda, A., Sasaya, T., & Tsuda, S., (2010). The protruding domain of the coat protein of *melon necrotic spot virus* is involved in compatibility with and transmission by the fungal vector *Olpidium bornovanus*. *Virol., 402*(1), 129–134.

Okuda, M., Okazaki, S., Yamasaki, S., Okuda, S., & Sugiyama, M., (2010). Host range and complete genome sequence of *cucurbit chlorotic yellows virus*, a new member of the genus *Crinivirus*. *Phytopathology, 100*, 560–566.

Okuda, M., Taba, S., Tsuda, S., Hidaka, S., Kameya-Iwaki, M., & Hanada, K., (2001). Comparison of the S RNA segments among Japanese isolates and Taiwanese isolates of *watermelon silver mottle virus*. *Arch. Virol., 146*(2), 389–394.

Orfanidou, C. G., Papayiannis, L. C., Pappi, P. G., Katis, N. I., & Maliogka, V. I., (2019). Criniviruses associated with cucurbit yellows disease in Greece and Cyprus: An ever-changing scene. *Plant Pathol., 68*(4), 764–774.

Orfanidou, C., Maliogka, V. I., & Katis, N. I., (2014). First report of *cucurbit chlorotic yellows virus* in cucumber, melon, and watermelon in Greece. *Plant Dis., 98*(10), 1446.

Owolabi, A. T., Rabenstein, F., Ehrig, F., Edgar, M. M., & Vetten, H. J., (2012). Strains of moroccan *watermelon mosaic virus* Isolated from *Lagenaria breviflorus* and *Coccinia barteri* in Calabar, South-Eastern Nigeria. *Intl. J. Virol., 8*, 258–270.

Panno, S., Iacono, G., Davino, M., Marchione, S., Zappardo, V., Bella, P., Tomassoli, L., Accotto, G. P., & Davino, S., (2016). First report of *tomato leaf curl New Delhi virus* affecting zucchini squash in an important horticultural area of Southern Italy. *New Dis. Rep., 33*(6), 2044–0588.

Pappu, H. R., Jones, R. A. C., & Jain, R. K., (2009). Global status of tospovirus epidemics in diverse cropping systems: Successes achieved and challenges ahead. *Virus Res., 141*(2), 219–236.

Park, S. M., Lee, J. S., Jegal, S., Jeon, B. Y., Jung, M., Park, Y. S., Han, S. L., et al., (2005). Transgenic watermelon rootstock resistant to CGMMV (*cucumber green mottle mosaic virus*) Infection. *Plant Cell Rep., 24*(6), 350–356.

Parrella, G., Gognalons, P., Gebre-Selassie, K., Vovlas, C., & Marchoux, G., (2003). An update of the host range of *tomato spotted wilt virus*. *J. Plant Pathol., 1*, 227–264.

Paylan, I. C., Ergun, M., & Erkan, S., (2013). First report of *artichoke yellow ringspot virus* in globe artichoke in Turkey. *Plant Dis., 97*(10), 1388–1388.

Perotto, M. C., Celli, M. G., Pozzi, E. A., Luciani, C. E., & Conci, V. C., (2016). Occurrence and characterization of a severe isolate of *watermelon mosaic virus* from Argentina. *Europ. J. Plant Pathol., 146*(1), 213–218.

Perotto, M. C., Pozzi, E. A., Celli, M. G., Luciani, C. E., Mitidieri, M. S., & Conci, V. C., (2018). Identification and characterization of a new potyvirus infecting cucurbits. *Arch. Virol., 163*(3), 719–724.

Phaneendra, C., Rao, K. R. S. S., Jain, R. K., & Mandal, B., (2012). *Tomato leaf curl New Delhi virus* is associated with pumpkin leaf curl: A new disease in Northern India. *Indian J. Virol., 23*(1), 42–45.

Piepenburg, O., Williams, C. H., Stemple, D. L., & Armes, N. A., (2006). DNA detection using recombination proteins. *PLoS Biol., 4*(7). e204, 1115–1121.

Polston, J. E., & Lapidot, M., (2008). *Tomato yellow leaf curl virus*: A whitefly-transmitted DNA virus. In: Rao, G. P., Kumar, L., & Holguin-Pena, R. J., (eds.), *Characterization, Diagnosis, and Management of Plant Viruses: Vegetable and Pulse Crops* (Vol. 3, pp. 141–161). Texas, USA: Stadium Press LLC.

Pospieszny, H., & Borodynko, N., (2005). First report of *tomato black ring virus* (TBRV) in the natural infection of zucchini (*Cucurbita Pepo* L. Convar Giromantiina) in Poland. *J. Plant Prot. Res., 45*(4), 321–325.

Pozzi, E. A., Luciani, C. E., Celli, M. G., Conci, V. C., & Perotto, M. C., (2019). First report of *zucchini lethal chlorosis virus* in Argentina infecting squash crops. *Plant Dis.*, PDIS–05.

Providenti, R., Gonsalves, D., & Humaydan, H. S., (1984). Occurrence of *zucchini yellow mosaic virus* in cucurbits from Connecticut, Florida, and California. *Plant Dis., 68*(1), 443–446.

Provvidenti, R., (1996). Diseases caused by viruses. *Comp. Cucurbit Dis.*, 37–45.

Pyott, D. E., Sheehan, E., & Molnar, A., (2016). Engineering of CRISPR/Cas9-mediated potyvirus resistance in transgene-free *Arabidopsis* plants. *Mol. Plant Pathol., 17*(8), 1276–1288.

Quito-Avila, D. F., Peralta, E. L., Martin, R. R., Ibarra, M. A., Alvarez, R. A., Mendoza, A., Insuasti, M., & Ochoa, J., (2014). Detection and occurrence of melon yellow spot virus in Ecuador: An emerging threat to cucurbit production in the region. *Eur. J. Plant Pathol., 140*(2), 193–197.

Raj, S. K., Singh, R., Pandey, S. K., & Singh, B. P., (2005). *Agrobacterium* mediated tomato transformation and regeneration of transgenic lines expressing *tomato leaf curl virus* coat protein gene for resistance against TLCV infection. *Curr. Sci., 88*, 1674–1679.

Raja, P., & Jain, R. K., (2008). Development of nucleocapsid gene mediated resistance in tomato against *groundnut bud necrosis virus*. *J. Plant Pathol., 90*(S2), 442.

Raja, P., (2005). *Development of Nucleocapsid Gene Mediated Resistance in Tomato Against Groundnut Bud Necrosis Virus*. Doctoral Dissertation, Indian Agricultural Research Institute, New Delhi.

Rajasekharam, T., (2010). *Biological and Molecular Characterization and Management of Watermelon Bud Necrosis Virus.* Doctoral dissertation, UAS Dharwad.

Rajinimala, N., & Rabindran, R., (2007). First report of Indian *cassava mosaic virus* on bitter gourd (*Momordica charantia*) in Tamil Nadu, India. *Australasian Plant Dis. Notes, 2*(1), 81, 82.

Rajinimala, N., Rabimdran, R. A., Kamalakannan, A., & Mareeswari, P., (2005). Virus-vector relationship of bitter gourd yellow mosaic virus and the whitefly *Bemisia tabaci* genn. *Acta Phytopathol. Entomol. Hung., 40*(1, 2), 23–30.

Ramirez, P., Chicas, M., Salas, J., Maxwell, D., & Karkashian, J., (2004). Identificacion de un Nuevo Begomovirus en melon (*Cucumis melo* L.) en Lara, Venezuela. *Rev. Man. Itegr. de Plag. Y Agroecol., 72*, 22–30.

Ramirez, P., Hernandez, E., Mora, F., Abraitis, R., & Hammond, R. W., (2008). Limited geographic distribution of *beet pseudo-yellows virus* in costa Rican cucurbits. *J. Plant Pathol.*, 331–335.

Rao, X., Wu, Z., & Li, Y., (2013). Complete genome sequence of a *watermelon silver mottle virus* isolate from China. *Virus Genes, 46*(3), 576–580.

Read, D. A., Muoma, J., & Thompson, G. D., (2020). Metaviromic analysis reveals coinfection of papaya in western Kenya with a unique strain of *Moroccan watermelon mosaic virus* and a novel member of the family alphaflexiviridae. *Arch. Virol.*, 1–4.

Reade, R., Miller, J., Robbins, M., Xiang, Y., & Rochon, D., (2003). Molecular analysis of the *cucumber leaf spot virus* genome. *Virus Res., 91*(2), 171–179.

Reddy, D. V. R., Buiel, A. A. M., Satyanarayana, T., Dwivedi, S. L., Reddy, A. S., Ratna, A. S., Vijaylakshmi, K., et al., (1995). *Peanut bud necrosis virus* disease: An overview. In: Buiel, A. A. M., Parlevliet, J. E., & Leenne, J. M., (eds.), *Recent Studies on Peanut bud Necrosis Disease* (pp. 3–7). ICRISAT Asia Centre, Hyderabad.

Reddy, D. V. R., Ratna, A. S., Sudarshana, M. R., Poul, F., & Kumar, I. K., (1992). Serological relationships and purification of bud necrosis virus, a tospovirus occurring in peanut (*Arachis hypogaea* L.) in India. *Ann. Appl. Biol., 120*(2), 279–286.

Reingold, V., Lachman, O., Koren, A., & Dombrovsky, A., (2013). First report of *cucumber green mottle mosaic virus* (CGMMV) symptoms in watermelon used for the discrimination of non-marketable fruits in Israeli commercial fields. *New Dis. Rep., 28*, 11.

Renukadevi, P., Nagendran, K., Nakkeeran, S., Karthikeyan, G., Jawaharlal, M., Alice, D., Malathi, V. G., & Pappu, H. R., (2015). First report of *tomato spotted wilt virus* infection of chrysanthemum in India. *Plant Dis., 99*(8), 1190.

Revill, P. A., Ha, C. V., Porchun, S. C., Vu, M. T., & Dale, J. L., (2003). The complete nucleotide sequence of two distinct Gemini viruses infecting cucurbits in Vietnam. *Arch. Virol., 148*(8), 1523–1541.

Rey, M. E., Ndunguru, J., Berrie, L. C., Paximadis, M., Berry, S., Cossa, N., Nuaila, V. N., et al., (2012). Diversity of dicotyledenous-infecting Gemini viruses and their associated DNA molecules in Southern Africa, including the South-west Indian Ocean Islands. *Viruses, 4*(9), 1753–1791.

Riviere, C. J., Pot, J., Tremaine, J. H., & Rochon, D. M., (1989). Coat Protein of melon necrotic spot carmovirus is more similar to those of tombusviruses than those of carmoviruses. *J. Gen. Virol., 70*(11), 3033–3042.

Roggero, P., Dellavalle, G., Lisa, V., & Stravato, V. M., (1998). First report of moroccan *watermelon mosaic Potyvirus* in zucchini in Italy. *Plant Dis., 82*(3), 351–351.

Romay, G., Chirinos, D., Geraud-Pouey, F., & Desbiez, C., (2010). Association of an atypical alpha satellite with a bipartite new world begomovirus. *Arch. Virol., 155*(11), 1843–1847.

Romay, G., Lecoq, H., & Desbiez, C., (2015). Melon chlorotic mosaic virus and associated alpha satellite from Venezuela: Genetic variation and sap transmission of a begomovirus-satellite complex. *Plant Pathol., 64*(5), 1224–1234.

Romay, G., Lecoq, H., Geraud-Pouey, F., Chirinos, D. T., & Desbiez, C., (2014). Current status of cucurbit viruses in Venezuela and characterization of Venezuelan isolates of *zucchini yellow mosaic virus. Plant Pathol., 63*(1), 78–87.

Romay, G., Pitrat, M., Lecoq, H., Wipf-Scheibel, C., Millot, P., Girardot, G., & Desbiez, C., (2019). Resistance against *melon chlorotic mosaic virus* and *tomato leaf curl New Delhi virus* in melon. *Plant Dis., 103*(11), 2913–2919.

Rouhibakhsh, A., Priya, J., Periasamy, M., Haq, Q. M. I., & Malathi, V. G., (2008). An improved DNA isolation method and PCR protocol for efficient detection of multicomponents of begomovirus in legumes. *J. Virol. Methods, 147*(1), 37–42.

Rubio, L., Abou-Jawdah, Y., Lin, H. X., & Falk, B. W., (2001). Geographically distant isolates of the crinivirus cucurbit yellow stunting disorder virus show very low genetic diversity in the coat protein gene. *J. Gen. Virol., 82*(4), 929–933.

Rubio, T., Borja, M., Scholthof, H. B., & Jackson, A. O., (1999). Recombination with host transgenes and effects on virus evolution: An overview and opinion. *Mol. Plant Microbe Interact., 12*(2), 87–92.

Rudnieva, T. A., (2009). *Characterization of Ukrainian Isolates of Cucumber Green Mottle Mosaic Virus Dissertation.* Kyiv, Taras Shevchenko National University of Kyiv, Ukrainian.

Runo, S., (2011). Engineering host-derived resistance against plant parasites through RNA interference: Challenges and opportunities. *Bioeng. Bugs, 2*(4), 208–213.

Ruwanthi, K. H. D., Ranasinghe, C., & De Costa, D. M., (2014). Identification of plant growth promoting rhizobacterial isolates as potential biocontrol agents of papaya ringspot virus disease. *Proc. Peradeniya Uni. Int. Res. Sess.*, 4–5.

Rybicki, E. P., & Gerhard, P., (1999). Plant virus disease problems in the developing world. In: *Advances in Virus Research* (Vol. 53, pp. 127–175). Academic Press.

Ryu, K. H., Min, B. E., Choi, G. S., Choi, S. H., Kwon, S. B., Noh, G. M., Yoon, J. Y., et al., (2000). *Zucchini green mottle mosaic virus* is a new tobamovirus; comparison of its coat protein gene with that of kyuri green mottle mosaic virus. *Arch. Virol., 145*(11), 2325–2333.

Sangeetha, B., Malathi, V. G. and Renukadevi, P., (2019). Emergence of cucurbit aphid-borne yellows virus in bitter gourd (Momordica charantia) in Tamil Nadu, India. *Plant Dis., 103*(6), 1441.

Salem, T. Z., El-Gamal, S. M., & Sadik, A. S., (2007). Use of helper component proteinase to identify a new Egyptian isolate of *watermelon mosaic potyvirus. Intl. J. Virol., 3,* 107–116.

Samsatly, J., Sobh, H., Jawhari, M., Najjar, C., Haidar, A., & Abou-Jawdah, Y., (2012). First report of *watermelon chlorotic stunt virus* in cucurbits in Lebanon. *Plant Dis., 96*(11), 1703–1703.

Santosa, A. I., Al-Shahwan, I. M., Abdalla, O. A., Al-Saleh, M. A., & Amer, M. A., (2018). Characterization of a *watermelon mosaic virus* isolate inducing a severe disease in watermelon in Saudi Arabia. *J. Agric. Sci. Tech., 8*(4), 221–230.

Saritha, R. K., Bag, T. K., Loganathan, M., Rai, A. B., & Rai, M., (2011). First report of *squash leaf curl China virus* causing mosaic symptoms on summer squash (*Cucurbita pepo*) grown in Varanasi district of India. *Arch. Phytopathol. Plant Prot., 44*(2), 179–185.

Sawangjit, S., (2009). The complete nucleotide sequence of squash leaf curl china virus-[Wax gourd] and its phylogenetic relationship to other Gemini viruses. *Sci. Asia, 35,* 131–136.

Segundo, E., Janssen, D., Velasco, L., Ruiz, L., & Cuadrado, I. M., (2001). First report of *cucumber leaf spot virus* in Spain. *Plant Dis., 85*(10), 1123.

Sharifi, M., Massumi, H., Heydarnejad, J., Pour, A. H., Shaabanian, M., & Rahi-mian, H., (2008). Analysis of the biological and molecular variability of *watermelon mosaic virus* isolates from Iran. *Virus Genes, 37*(3), 304–313.

Sharma, P., Verma, R. K., Mishra, R., Sahu, A. K., Choudhary, D. K., & Gaur, R. K., (2014). First report of cucumber green mottle mosaic virus association with the leaf green mosaic disease of a vegetable crop, *Luffa acutangula* L. *Acta Virol., 58*, 299–300.

Sifres, C. A. G., Saez-Sanchez, C., Ferriol, M. M., Selmani, E. A., Riado, J., Pico, S. M. B., & Lopez, D. R. C., (2018). First report of *tomato leaf curl New Delhi Virus* infecting zucchini in Morocco. *Plant Dis., 102*(5), 1045–1045.

Singh, R., Raj, S. K., & Prasad, V., (2008). Molecular characterization of a strain of *squash leaf curl china virus* from North India. *J. Phytopathol., 156*(4), 222–228.

Singh, S. J., & Krishnareddy, M., (1996). Watermelon bud necrosis: A new tospovirus disease. *Acta Hort., 431*, 68–77.

Singh, S., Jangre, A., & Kumar, P., (2017). Studies on the molecular variability in Indian isolates of papaya ringspot virus. *Virol. Res. J., 1*(1), 10–19.

Sobh, H., Samsatly, J., Jawhari, M., Najjar, C., Haidar, A., & Abou-Jawdah, Y., (2012). First report of *squash leaf curl virus* in cucurbits in Lebanon. *Plant Dis., 96*(8), 1231–1231.

Sohrab, S. S., Mandal, B., Ali, A., & Varma, A., (2006). Molecular diagnosis of emerging begomovirus diseases in cucurbits occurring in Northern India. *Indian J. Virol., 17*, 88–95.

Sohrab, S. S., Mandal, B., Ali, A., & Varma, A., (2010). Chlorotic curly stunt: A severe begomovirus disease of bottle gourd in Northern India. *Indian J. Virol., 21*(1), 56–63.

Sohrab, S. S., Mandal, B., Pant, R. P., & Varma, A., (2003). First report of association of *tomato leaf curl virus*-New Delhi with yellow mosaic disease of *Luffa cylindrica* in India. *Plant Dis., 87*(9), 1148.

Sreekanth, M., Sreeramulu, M., Rao, R. D. V. J., Babu, B. S., & Babu, T. R., (2003). Relative efficacy and economics of different imidacloprid schedules against *Thrips palmi*, the vector of *Peanut bud necrosis virus* on mung bean. *Indian J. Plant Prot., 31*, 43–47.

Supakitthanakorn, S., Akarapisan, A., & Ruangwong, O. U., (2018). First record of melon yellow spot virus in pumpkin and its occurrence in cucurbitaceous crops in Thailand. *Aus. Plant Dis. Notes, 13*(1), 32.

Suveditha, S., Bharathi, L. K., and Krishna Reddy, M., (2017). First report of Cucurbit aphid-borne yellows virus infecting bitter gourd (Momordica charantia) and teasel gourd (Momordica subangulata subsp. renigera) in India. *New Dis. Rep., 36*, 7.

Svoboda, J., & Leisova-Svobodova, L., (2011). First report of *squash mosaic virus* in ornamental pumpkin in the Czech republic. *Plant Dis., 95*(10), 1321.

Tahir, M., & Haider, M. S., (2005). First report of *tomato leaf curl New Delhi virus* infecting bitter gourd in Pakistan. *Plant Pathol., 54*(6), 807.

Tahir, M., Haider, M. S., & Briddon, R. W., (2010). First report of *squash leaf curl China virus* in Pakistan. *Aus. Plant Dis. Notes, 5*(1), 21–24.

Tamura, K., Peterson, D., Peterson, N., Stecher, G., Nei, M., & Kumar, (2011). MEGA5: Molecular evolutionary genetics analysis using maximum likelihood, evolutionary distance, and maximum parsimony methods. *Mol. Biol. Evol., 28*(10), 2731–2739.

Tan, S. H., Nishiguchi, M., Murata, M., & Motoyoshi, F., (2000). The genome structure of *kyuri green mottle mosaic tobamovirus* and its comparison with that of *cucumber green mottle mosaic tobamovirus*. *Arch. Virol., 145*(6), 1067–1079.

Tesoriero, L. A., Chambers, G., Srivastava, M., Smith, S., Conde, B., & Tran-Nguyen, L. T. T., (2016). First report of *cucumber green mottle mosaic virus* in Australia. *Aus. Plant Dis. Notes, 11*(1), 1–3.

Tian, T., Posis, K., Maroon-Lango, C. J., Mavrodieva, V., Haymes, S., Pitman, T. L., & Falk, B. W., (2014). First report of *cucumber green mottle mosaic virus* on melon in the United States. *Plant Dis., 98*(8), 1163.

Tiwari, A. K., Sharma, P. K., Khan, M. S., Snehi, S. K., Raj, S. K., & Rao, G. P., (2010). Molecular detection and identification of *tomato leaf curl New Delhi virus* isolate causing yellow mosaic disease in bitter gourd (*Momordica charantia*), a medicinally important plant in India. *Medicinal Plants, 2*, 117–123.

Tomassoli, L., Lumia, V., Siddu, G. F., & Barba, M., (2003). Yellowing disease of melon in Sardinia (Italy) caused by *beet pseudo yellows virus. J. Plant Pathol.,* 59–61.

Tomassoli, L., Tiberini, A., & Meneghini, M., (2010). *Zucchini yellow fleck virus* is an emergent virus on melon in Sicily (Italy). *J. Phytopathol., 158*(4), 314–316.

Topkaya, S., Desbiez, C., & Ertunc, F., (2019). *Presence of Cucurbit Viruses in Ankara and Antalya Province and Molecular Characterization of Coat Protein Gene of Zucchini Yellow Mosaic Virus Turkish Isolates.*

Tripathi, S., Suzuki, J. Y., Ferreira, S. A., & Gonsalves, D., (2008). *Papaya ringspot virus*-P: Characteristics, pathogenicity, sequence variability and control. *Mol. Plant Pathol., 9*(3), 269–280.

Trkulja, V., Salapura, J. M., Curkovic, B., Stankovic, I., Bulajic, A., Vucurovic, A., & Krstic, B., (2013). First report of *tomato spotted wilt virus* on gloxinia in Bosnia and Herzegovina. *Plant Dis., 97*(3), 429–429.

Tsai, W. S., Abdourhamane, I. K., Knierim, D., Wang, J. T., & Kenyon, L., (2010). First report of *zucchini yellow mosaic virus* associated with leaf crinkle and yellow mosaic diseases of cucurbit plants in Mali. *Plant Dis., 94*(7), 923.

Tsai, W. S., Hu, C. J., Shung, D. P., Lee, L. M., Wang, J. T., & Kenyon, L., (2011). First report of *squash leaf curl Philippines virus* infecting chayote (*Sechium edule*) in Taiwan. *Plant Dis., 95*(9), 1197.

Turina, M., Tavella, L., & Ciuffo, M., (2012). Tospoviruses in the Mediterranean area. *Adv. Virus Res., 84*, 403–437.

Tzanetakis, I. E., & Martin, R. R., (2004). Complete nucleotide sequence of a strawberry isolate of *beet pseudo yellows virus. Virus Genes, 28*(3), 239–246.

Ullman, D. E., Sherwood, J. L., & German, T. L., (1997). Thrips as vectors of plant pathogens. *Thrips as Vectors of Plant Pathogens,* 539–565.

Usher, L., Sivparsad, B., & Gubba, A., (2012). Isolation, identification and molecular characterization of an isolate of *zucchini yellow mosaic virus* occurring in Kwazulu-Natal, South Africa. *South Afr. J. Plant Soil, 29*(2), 65–71.

Van, K. Y., & Van, D. H. J. M., (1959). Virusziekten van de komkommer in Nederland. *Tijdschrift over Plantenziekten, 65*(6), 257–271.

Vani, S., & Varma, A., (1993). Properties of *cucumber green mottle mosaic virus* isolated from water of river Jamuna. *Indian Phytopathol., 46*, 118–122.

Varma, A., & Giri, B. K., (1998). Virus diseases. In: Nayar, N. M., & More, T. A., (eds.), *Cucrbits* (pp. 225–245). Oxford and IBH, Publishing house Pvt. Ltd. New Delhi, India.

Varma, A., & Malathi, V. G., (2003). Emerging Gemini virus problems: A serious threat to crop production. *Ann. Appl. Biol., 142*(2), 145–164.

Varveri, C., Vassilakos, N., & Bem, F., (2002). Characterization and detection of *cucumber green mottle mosaic virus* in Greece. *Phytoparasitica, 30*(5), 493–501.

Vengadesan, G., Anand, R. P., Selvaraj, N., Perl-Treves, R., & Ganapathi, A., (2005). Transfer and expression of *npt II* and *bar* genes in cucumber (*Cucumis satavus* L.). *In Vitro Cellular Devtl. Biol. Plant, 41*(1), 17–21.

Verma, R. K., Mishra, R., & Gaur, R. K., (2016). Genetic variability of the replicase (NIb) gene of *papaya ringspot virus* in Northern India indicates common ancestry with isolates from China and Taiwan. *J. Plant Pathol., 105*–110.

Webber, B. C., (1982). *The Biology of the Ambrosia Beetle Xylosandrus germanus (Blandford) (Coleoptera: Scolytidae) and its Effects on Black Walnut* (p. 0338). PhD Dissertation, Southern Illinois University.

Wintermantel, W. M., (2004). Pumpkin (*Cucurbita maxima* and *C. pepo*), a new host of beet-pseudo yellows virus in California. *Plant Dis., 88*(1), 82.

Wintermantel, W. M., Laura, L. H., Arturo, A. C., & Eric, T. N., (2009). A new expanded host range of *cucurbit yellow stunting disorder virus* includes three agricultural crops. *Plant Dis., 93*(7), 685–690.

Wisler, G. C., Duffus, J. E., Liu, H. Y., & Li, R. H., (1998). Ecology and epidemiology of whitefly-transmitted closteroviruses. *Plant Dis., 82*, 270–280.

Woudt, L. P., De Rover, A. P., De Haan, P. T., & Van, G. M. Q. J. M., (1993). Sequence analysis of the RNA genome of cucumber chlorotic spot virus (CCSV), A whitefly transmitted closterovirus. *Book of Abstract, 9th Int. Cong. of Virol.* (Vol. 326). Glasgow.

Wu, H. J., Meng, L. I., Ni, H. O. N. G., Bin, P. E. N. G., & GU, Q. S., (2020). Molecular and biological characterization of melon-infecting *squash leaf curl China virus* in China. *J. Intr. Agr., 19*(2), 570–577.

Wu, H. W., Yu, T. A., Raja, J. A., Wang, H. C., & Yeh, S. D., (2009). Generation of transgenic oriental melon resistant to *zucchini yellow mosaic virus* by an improved cotyledon-cutting method. *Plant Cell Rep., 28*(7), 1053–1064.

Xiang, H. Y., Shang, Q. X., Han, C. G., Li, D. W., & Yu, J. L., (2008). First report on the occurrence of *cucurbit aphid-borne yellows virus* on nine cucurbitaceous species in China. *Plant Pathol., 57*(2), 390–390.

Xie, Y., & Zhou, X. P., (2003). Molecular characterization of *squash leaf curl Yunnan virus*, a new begomovirus and evidence for recombination. *Arch. Virol., 148*(10), 2047–2054.

Xie, Y., Jiang, T., & Zhou, X., (2006). Agroinoculation shows *tobacco leaf curl Yunnan virus* is a monopartite begomovirus. *Eur. J. Plant Pathol., 115*(4), 369–375.

Xing, K., Zhu, X., Peng, X., & Qin, S., (2015). Chitosan antimicrobial and eliciting properties for pest control in agriculture: *Agro. Sust. Dev., 35*, 569–588.

Yakoubi, S., Desbiez, C., Fakhfakh, H., Wipf-Scheibel, C., Marrakchi, M., & Lecoq, H., (2008). Biological characterization and complete nucleotide sequence of a Tunisian isolate of *Moroccan watermelon mosaic virus. Arch Virol., 153*(1), 117–125.

Yakoubi, S., Desbiez, C., Fakhfakh, H., Wipf-Scheibel, C., Marrakchi, M., & Lecoq, H., (2007). Occurrence of *cucurbit yellow stunting disorder virus* and cucumber vein yellowing virus in Tunisia. *J. Plant Pathol., 417*–420.

Yakoubi, S., Desbiez, C., Fakhfakh, H., Wipf-Scheibel, C., Marrakchi, M., & Lecoq, H., (2008b). Biological characterization and complete nucleotide sequence of a Tunisian isolate of *Moroccan watermelon mosaic virus. Arch Virol., 153*(1), 117–125.

Yakoubi, S., Lecoq, H., & Desbiez, C., (2008a). *Algerian watermelon mosaic virus* (AWMV): A new potyvirus species in the PRSV cluster. *Virus Genes, 37*(1), 103–109.

Yamashita, S., Doi, Y., Yora, K., & Yoshino, M., (1979). Cucumber yellows virus: Its transmission by the greenhouse whitefly *Trialeurodes vaporariorum* Westwood, and the yellowing disease of cucumber and muskmelon caused by the virus. *Ann. Phytopathol. Soc. Jpn., 45*, 484–496.

Yazdani-Khameneh, S., Aboutorabi, S., Shoori, M., Aghazadeh, A., Jahanshahi, P., Golnaraghi, A., & Maleki, M., (2016). Natural occurrence of *tomato leaf curl New Delhi virus* in Iranian cucurbit crops. *The Plant Pathol. J., 32*(3), 201.

Yeh, S. D., & Chang, T. F., (1995). Nucleotide sequence of the N gene of watermelon silver mottle virus, a proposed new member of the genus tospovirus. *Phytopathology, 85*(1), 58–64.

Yeh, S., Sun, I., Ho, H., & Chang, T., (1995). Molecular cloning and nucleotide sequence analysis of the S RNA of *watermelon silver mottle virus. Tospoviruses Thrips of Floral Vegetable Crops, 431,* 244–260.

Yilmaz, M. A., Ozaslan, M., & Ozaslan, D., (1989). *Cucumber vein yellowing virus* in Cucurbitaceae in Turkey. *Plant Dis., 73*(7).

Yoon, J. Y., Choi, G. S., Choi, S. K., Hong, J. S., Choi, J. K., Kim, W., Lee, G. P., & Ryu, K. H., (2008). Molecular and biological diversities of *cucumber green mottle mosaic virus* from cucurbitaceous crops in Korea. *J. Phytopathol., 156*(7, 8), 408–412.

Yoon, J. Y., Min, B. E., Choi, S. H., & Ryu, K. H., (2001). Completion of nucleotide sequence and generation of highly infectious transcripts to cucurbits rom full length cDNA clone of *Kyuri green mottle mosaic virus. Arch. Virol., 146,* 2085–2096.

Yoon, J. Y., Min, B. E., Choi, S. H., & Ryu, K. H., (2002). Genome structure and production of biologically active *in-vitro* transcript of cucurbit infecting *zucchini green mottle mosaic virus. Phytopathology, 92,* 156–163.

Yousif, M. T., Kheyr-Pour, A., Gronenborn, B., Pitrat, M., & Dogimont, C., (2007). Sources of resistance to *watermelon chlorotic stunt virus* in melon. *Plant Breed, 126*(4), 422–427.

Yuki, V. A., Rezende, J. A. M., Kitajima, E. W., Barroso, P. A. V., Kuniyuki, H., Groppo, G. A., & Pavan, M. A., (2000). Occurrence, distribution, and relative incidence of five viruses infecting cucurbits in the state of Sao Paulo, Brazil. *Plant Dis., 84,* 516–520.

Zhang, G., Ma, L., Wang, P., Liu, Y., Zhang, G., & Hu, Z., (2007). Effect of salicylic acid on *papaya ringspot virus* disease. *Chin. J. Trop. Agric., 6,* 20–22.

Zhang, T., Zheng, Q., Yi, X., An, H., Zhao, Y., Ma, S., & Zhou, G., (2018). Establishing RNA virus resistance in plants by harnessing CRISPR immune system. *Plant Biotech. J., 16*(8), 1415–1423.

Zhang, Y. J., Li, G. F., & Li, M. F., (2009). Occurrence of cucumber green mottle mosaic virus on cucurbitaceous plants in China. *Plant Dis., 93*(2), 200.

Zhao, M. F., Chen, J., Zheng, H. Y., Adams, M. J., & Chen, J. P., (2003). Molecular analysis of *zucchini yellow mosaic virus* isolates from Hangzhou, China. *J. Phytopathol., 151*(6), 307–311.

Zheng, L., Rodoni, B. C., Gibbs, M. J., & Gibbs, A. J., (2010). A novel pair of universal primers for the detection of potyviruses. *Plant Pathol., 59,* 211–220.

Zheng, Y. X., Chen, C. C., Yang, C. J., Yeh, S. D., & Jan, F. J., (2008). Identification and characterization of a tospovirus causing chlorotic ringspots on *Phalaenopsis Orchids. Eur. J. Plant Pathol., 120*(2), 199–209.

Zitter, T. A., Hopkins, D. L., & Thomas, C. E., (1996). *Compendium of Cucurbit Diseases.* The American Phytopathogical Society.

Zouba, A. A., Lopez, M. V., & Anger, H., (1998). *Squash yellow leaf curl virus*: A new whitefly-transmitted poty-like virus. *Plant Dis., 82*(5), 475–478.

Index

For Product Safety Concerns and Information please contact our EU
representative GPSR@taylorandfrancis.com
Taylor & Francis Verlag GmbH, Kaufingerstraße 24, 80331 München, Germany